PHYTOCHEMICAL INDUCTION BY HERBIVORES

PHYTOCHEMICAL INDUCTION BY HERBIVORES

Edited by

DOUGLAS W. TALLAMY
Department of Entomology & Applied Ecology
University of Delaware
Newark, Delaware

MICHAEL J. RAUPP
Department of Entomology
University of Maryland
College Park, Maryland

A Wiley-Interscience Publication
JOHN WILEY & SONS, INC.
New York / Chichester / Brisbane / Toronto / Singapore

Library of Congress Cataloging in Publication Data:

Phytochemical induction by herbivores / edited by Douglas W. Tallamy and Michael J. Raupp.

    Includes bibliographical references.
    1. Pesticidal plants. 2. Plant defenses. 3. Herbivores—Ecology.
I. Tallamy, Douglas W. II. Raupp, Michael J.

SB292.A2P49 1991
582'.023—dc20   90-24394
ISBN 0-471-63241-4 (alk. paper)   CIP

Printed in the United States of America

10 9 8 7 6 5 4 3 2 1

CONTRIBUTORS

I. BALDWIN, Department of Biological Sciences, State University of New York, Buffalo, NY 14226

D. BAZELY, York University, Department of Biology, 4700 Keele Street, North York, Ontario, Canada M3J 1P3

J. H. BENEDICT, Texas Agricultural Experiment Station, Texas A&M University, Corpus Christi, TX 78410

J. P. BRYANT, Institute of Arctic Biology, University of Alaska, Fairbanks, AK 99775-0180

J. F. CHANG, Phero Tech, Inc., Vancouver, British Columbia

T. P. CLAUSEN, Department of Chemistry, University of Alaska, Fairbanks, Alaska 99775-0180

J. S. COLEMAN, Biological Research Laboratories, 130 College Place, Syracuse University, Syracuse, NY 13244

K. DANELL, The Swedish University of Agricultural Sciences, Faculty of Forestry, Department of Wildlife Ecology, UMEA, Sweden H 5-90183

P. J. EDWARDS, Department of Biology, University of Southampton, Southampton SO9 3TU, United Kingdom

S. H. FAETH, Department of Zoology, Arizona State University, Tempe, AZ 85287-1501

T. FAGERSTRÖM, Department of Theoretical Ecology, Lund University, Ecology Building, S-223 62 Lund, Sweden

D. C. FISCHER, Office of Agricultural Entomology, University of Illinois and Illinois Natural History Survey, Champaign, IL 61820

R. M. GIBBERD, Department of Biology, University of Southampton, Southampton SO9 3TU, United Kingdom

S. E. HARTLEY, Institute of Terrestrial Ecology, Banchory Research Station, Hill of Brathens, Glassel, Banchory, Kincardineshire AB3 4BY, United Kingdom

E. HAUKIOJA, Department of Biology, University of Turku, SF-20500 Turku 50, Finland

C. JONES, Institute of Ecosystem Studies, The New York Botanical Garden, Mary Flagler Cary Arboretum, Box AB, Millbrook, NY 12545

R. KARBAN, Department of Entomology, University of California, Davis, CA 95616

M. KOGAN, Integrated Plant Protection Center, Oregon State University, Gilmore Annex 100, Corvallis, OR 97331-3904

J. H. LAWTON, Imperial College at Silwood, Centre for Population Biology, Department of Pure and Applied Biology, Ascot, Berks SL5 7PY, United Kingdom

E. S. MCCLOUD, Department of Entomology, University of Illinois, Urbana, IL 61801

J. H. MYERS, The Ecology Group, Departments of Zoology and Plant Science, University of British Columbia, Vancouver, Canada V6T 1Z4

S. NEUVONEN, Department of Forest Protection, The Finnish Forest Research Institute, SF-01301 Vantaa, Finland

P. NIEMELÄ, Department of Forest Protection, The Finnish Forest Research Institute, SF-01301 Vantaa, Finland

D. PROVENZA, Department of Range Science, College of Natural Resources, UMC 52, Utah State University, Logan, UT 84322

K. F. RAFFA, Department of Entomology, University of Wisconsin, Madison, WI 53706

M. J. RAUPP, Department of Entomology, University of Maryland, College Park, MD 20742

P. B. REICHARDT, Department of Chemistry, University of Alaska, Fairbanks, AK 99775-0180

C. S. SADOF, Department of Entomology, Purdue University, West Lafayette, IN 47907

D. W. TALLAMY, Department of Entomology and Applied Ecology, University of Delaware, Newark, DE 19717-1303

J. TOUMI, Department of Ecology, Lund University, Ecology Building, S-223 62 Lund, Sweden

R. A. WERNER, Institute of Northern Forestry, 308 Tanana Drive, Fairbanks, AK 99775-5500

J. L. WOLFSON, Department of Entomology, Purdue University, West Lafayette, IN 47907

S. D. WRATTEN, Department of Biology, University of Southampton, Southampton SO9 3TU, United Kingdom

PREFACE

Phytochemical Induction by Herbivores is the first compilation of reviews dealing with plant responses to challenge from herbivores. The book represents two decades of efforts to test the hypothesis that phytochemical, morphological, and phenological processes triggered by plant tissue damage are defensive adaptations against herbivores.

The studies presented in this volume focus on several controversial aspects of the induction phenomenon. The defensive nature of induced phytochemicals remains inconclusive: Chemical changes provoked by herbivory could be active responses to discourage further herbivory, passive consequences of shifts in the availability of carbon and nutrient resources, or contributions to plant repair processes. The role of allelochemical production costs in driving the evolution of inducible responses is also unresolved and there is little consensus as to the specificity of induction processes. Inducible phytochemicals have been cited as a force driving the foraging patterns and population cycles of herbivores, though empirical support for these claims has been elusive. Even the justification for distinguishing inducible from constitutive compounds is waning; in many earlier studies we simply have failed to recognize the dynamic properties of so-called constitutive allelochemicals.

Readers of this volume will quickly notice that inducible responses are rarely as predictable or directional as theory would have us believe. Extreme variation within and between studies has hampered our understanding of induction. Much of this variation has been traced to a number of design-related environmental factors including light intensity, soil nutrient and water quality, temperature, root binding, and elicitor differences. However, there is also variation inherent in the induced response itself. Seemingly, induction in some, if not all, plants is most accurately characterized as an increase in the variation of allelochemical titer rather than in uniform concentration.

This volume is divided into three parts. Part I comprises discussions treating primarily the physiological, morphological, and phytochemical responses of plants to herbivore attack.

vii

The first chapter challenges the perception that herbivore-induced changes in plants are elicitor-specific adaptations that increase resistance to attack by herbivores. *Coleman* and *Jones* argue that biotic and abiotic factors of the environment alter induced responses of plants in predictable ways. Attributes such as plant genotype, ontogeny, and phenology also shape patterns of resource acquisition and allocation, which, in turn, can influence induced responses. By taking a phytocentric perspective of phytochemical induction, they provide a framework for testing hypotheses concerning the induced responses of plants.

The second chapter continues to emphasize the importance of patterns of resource allocation in understanding phytochemical induction. *Baldwin* provides a well-documented case of alkaloid induction in wild tobacco plants. In addition to detailing the physiological mechanisms of induction, he discusses evidence supporting the notion that the induction of alkaloid defenses may be costly for *Nicotiana*.

Quaking aspen is the model system used by *Clausen, Bryant, Reichardt,* and *Werner* to examine long- and short-term induction of phytochemicals, particularly carbon-based allelochemicals. Long-term reductions in the suitability of aspen leaves may be linked to short-term changes in leaf size and photosynthetic rates following defoliation.

In Chapter 4 *Tuomi, Fagerström,* and *Niemalä* contrast two evolutionary scenarios of plant defense. The first assumes a primary function for defense resulting in competition for limited reserves with other primary processes such as growth and reproduction. The latter assumes a secondary function for defense that may be hierarchical in nature and can thereby account for only a portion of the observed variation in levels of carbon-based secondary compounds. Fixed and flexible models of resource allocation are used to examine these scenarios under differing environmental constraints.

Hartley and *Lawton* examine rapidly induced changes in birch phytochemicals and find complex and highly variable responses. Factors such as the method of injury, be it herbivore-associated or artificial, alter the induced response. Moreover, changes in allelochemicals following injury are relatively small compared to inherent within- and between-plant variability. This raises doubt concerning the adaptive significance of these responses as insect defenses: The selective value of rapidly induced changes in birch phenolic compounds may be in their role as antimicrobial agents.

Drawing on extensive studies of *Betula, Populus,* and *Salix* in boreal systems, *Bryant, Danell, Provenza, Reichardt, Clausen,* and *Werner* examine physiological mechanisms for long-term induction by herbivores. Observed changes in phytochemicals are not active defenses but result instead from reallocation of secondary metabolites due to passive nutrient stress, the reversion to juvenile growth forms, and changes in plant architecture following defoliation and twig removal. The implications of these changes in plant structure and chemistry for population dynamics of mammals are discussed.

Part II reviews our current understanding of the impact of inducible plant responses on herbivore fitness, behavior, population dynamics, and community structure.

Tallamy and *McCloud* revisit previous interpretations of the effects of inducible cucurbitacins on the fitness and feeding behavior of squash and cucumber beetles. They provide insights into the evolution and maintenance of induced responses in cucurbits.

One general outcome of defoliation is a reduction in the suitability of subsequent foliage as a food resource for herbivorous insects. *Raupp* and *Sadof* in Chapter 8 and *Edwards, Wratten,* and *Gibberd* in Chapter 9 examine the effects of induced changes in plant quality on foraging patterns of insects. Raupp and Sadof conclude that phytochemical induction helps explain spatial distributions of leaf beetles utilizing willows.

Edwards, Wratten, and Gibberd detail how wound-induced changes may shape patterns of defoliation at several spatial scales and alter competitive interactions among plants.

One of the best-known examples of rapid phytochemical induction involves the class of compounds known as proteinase inhibitors. *Wolfson* reviews evidence supporting and refuting the role of proteinase inhibitors as effective herbivore defenses. She also details ontogenetic variation in the production of proteinase inhibitors in response to wounding in tomatoes.

Questions have arisen regarding the strength of the selective force exerted by herbivores, particularly insects, on plants. In one well-studied association between bark beetles and their coniferous hosts, herbivores can produce high levels of mortality under certain environmental conditions. In Chapter 11 *Raffa* reviews the substantial body of literature concerning the interactions between bark beetles and conifers. Specific and general elicitors induce morphological and chemical changes in conifers. The strength and magnitude of induced changes may be mitigated by a variety of environmental factors and thereby differentially influence the population dynamics of bark beetles.

Neuvonen and *Haukioja* return to the birch system in Chapter 12 with a summary of their pioneering studies of induced phytochemical responses. They detail evidence for the rapid and delayed effects of induction on the success of herbivores, and emphasize experimental pitfalls and limitations in the design and interpretation of induction studies. The importance of induced responses, particularly delayed resistance, in determining the population dynamics of herbivores is also discussed.

While the first portion of Part II focuses primarily on phytochemical aspects of the induced response, chapters by *Faeth* (Chapter 13) and *Myers* and *Bazely* (Chapter 14) broaden our perspective on induced responses beyond phytochemistry by scrutinizing phenological, morphological, and architectural changes in plants following herbivory. Faeth summarizes and interprets induction studies that find considerable variation in plant chemical, nutritional, phenological, and morphological responses. Through his studies of oak folivores, he emphasizes the contribution of phenological processes such as leaf abscission to the induced responses of oaks. The direct and indirect effects of herbivory on interacting members of the second and higher trophic levels is detailed.

Myers and Bazely bring together a scattered literature on the induction and anti-herbivore function of plant physical defenses. They examine the evidence for predicted defensive roles of physical structures such as thorns, prickles, spines, and hairs, and find support for these predictions largely wanting. Changes in the expression of these structures following herbivory may be linked to a reversion from mature to juvenile growth forms in plants. Physical structures such as thorns and spines are best viewed as complex adaptations to a variety of environmental pressures of which defense from herbivores may play a minor role.

In Part III authors consider the implications of employing inducible defenses to mitigate herbivory in agronomic systems.

In Chapter 15 *Kogan* and *Fischer* review an extensive body of literature concerning the responses of soybeans to attack by pathogens and herbivores. They discuss the classes of phytochemicals induced in soybeans with particular emphasis on the phytoalexins. Elicitors, specificity, and effects of these allomones on pathogens and insects are detailed. The potential of incorporating inducible plant responses as a component of integrated pest management programs is also discussed.

In Chapter 16 *Benedict* and *Chang* examine the complex interactions among bacteria, cotton plants, and herbivorous insects such as the boll weevil. Microbial flora of the cotton phylloplane are capable of inducing changes in the secondary chemistry of cotton. This response is highly variable and mediated by a variety of factors including dose of microbial inoculum, plant age, and plant genotype. Although induced changes have the potential to alter the suitability of cotton for closely associated herbivores, Benedict and Chang conclude that bacterially induced resistance is not likely to be an effective management tactic for boll weevils.

Karban concludes the volume by reviewing some of the practical limitations of theory concerning the impact of induced plant changes on the population dynamics of herbivores. Citing his own studies with mites in agricultural systems he provides insights into how these problems can be addressed. Finally, Karban reviews the potential role of inducible plant resistance as a viable component of pest-management programs.

Newark, Delaware
College Park, Maryland
June 1991

DOUGLAS W. TALLAMY

MICHAEL J. RAUPP

CONTENTS

PART I   PHYTOCHEMICAL RESPONSES TO
         HERBIVORE ATTACK   1

**1 A PHYTOCENTRIC PERSPECTIVE OF PHYTOCHEMICAL
    INDUCTION BY HERBIVORES**   3

James S. Coleman and Clive G. Jones

1   Introduction / 3
2   A Phytocentric Model / 5
    2.1   Plant Genotype / 7
    2.2   Plant Phenology and Ontogeny / 7
    2.3   Abiotic Environment / 8
    2.4   Biotic Environment / 9
    2.5   Interactions of Components / 9
    2.6   Resource Suitability to Herbivores / 10
    2.7   Herbivore Effects on Plant Functions / 11
3   The Phytocentric Model and Phytochemical
    Induction / 11
    3.1   Plant Genotype and Induction / 11
    3.2   Leaf Ontogeny and Induction / 28
    3.3   Plant Phenology and Induction / 31
    3.4   Abiotic Environment and Induction / 33
    3.5   Biotic Environment and Induction / 34

4 Summary / 35
   Acknowledgments / 37
   References / 37

2 DAMAGE-INDUCED ALKALOIDS IN WILD TOBACCO   **47**
Ian T. Baldwin

1 Introduction / 47
2 The Mechanisms of Induction / 48
   2.1 Sites of Alkaloid Synthesis / 48
   2.2 Damage Cues / 51
   2.3 Alkaloidal Responses to Real and Simulated
        Herbivory / 53
   2.4 Environmental and Physiological Constraints / 55
3 Effects on Insect Herbivores / 56
4 Phenotypic Costs of Inducible Alkaloidal Defenses / 57
   4.1 Experimental Measures of Cost / 58
   4.2 Why May Induced Alkaloid Production Be
        Costly? / 60
5 Conclusion / 64
   Acknowledgments / 65
   References / 65

**3 LONG-TERM AND SHORT-TERM INDUCTION IN
    QUAKING ASPEN: RELATED PHENOMENA?**   **71**
*Thomas P. Clausen, Paul B. Reichardt, John P. Bryant, and
Richard A. Werner*

1 Introduction / 71
2 Natural History of the Quaking Aspen–Large Aspen
   Tortrix Interaction in Alaska / 72
3 Secondary Metabolites of Quaking Aspen / 73
4 Short-Term Induction in Quaking Aspen / 74
5 Long-Term Induction in Quaking Aspen / 76
6 Possible Interactions between Long- and Short-Term
   Induction in Quaking Aspen / 79
7 Summary / 81
   Acknowledgments / 81
   References / 81

4 CARBON ALLOCATION, PHENOTYPIC PLASTICITY, AND INDUCED DEFENSES   **85**

Juha Tuomi, Torbjörn Fagerström, and Pekka Niemelä

1 Introduction / 85
2 Carbon Allocation and Defense / 86
    2.1 Defense and Trade-offs / 86
    2.2 Allocation Models / 87
    2.3 Components of Carbon-Based Defenses / 90
3 Phenotypic Flexibility of Carbon-Based Defenses / 90
    3.1 Constitutive Defenses / 90
    3.2 Induced Responses / 92
4 Responding Units: Plants or Modules? / 94
    4.1 Modules and Plant Fitness / 94
    4.2 Responses to Localized Disturbances / 96
5 Triggering Induced Responses / 98
6 Conclusions / 100
    Acknowledgments / 101
    References / 101

5 BIOCHEMICAL ASPECTS AND SIGNIFICANCE OF THE RAPIDLY INDUCED ACCUMULATION OF PHENOLICS IN BIRCH FOLIAGE   **105**

S. E. Hartley and J. H. Lawton

1 Introduction / 105
2 Rapid Phytochemical Induction in Damaged Birch Leaves / 108
    2.1 Initial Experiments / 108
    2.2 Is Insect Attack the Same as Artificial Damage? / 111
    2.3 Problems of Interpretation / 112
    2.4 The Induction of Phenolic Biosynthesis: Phenylalanine Ammonia Lyase / 114
    2.5 Summary of Rapidly Induced Changes in Birch Foliage / 117
3 The Effect of Wound-Induced Changes on Insects / 118
    3.1 Preference Tests with Mature Leaves and Late Instar Caterpillars / 118
    3.2 Tests with Young Leaves and Early Instars / 122
    3.3 *Spodoptera, Apocheima*, and Manipulated Phenolics / 123

3.4   Links between Laboratory and Field / 125

3.5   Field Experiments / 125

3.6   Summary of Effects on Insects / 126

4   Conclusion / 127

References / 129

PART II   IMPACT OF INDUCIBLE
PHYTOCHEMICALS ON HERBIVORE
FITNESS, BEHAVIOR, POPULATION
DYNAMICS, AND COMMUNITIES   133

6   EFFECTS OF MAMMAL BROWSING ON THE
CHEMISTRY OF DECIDUOUS WOODY PLANTS   135

*J. P. Bryant, K. Danell, F. Provenza, P. B. Reichardt,
T. A. Clausen, and R. A. Werner*

1   Introduction / 135

2   Long-Term Chemical Responses of Deciduous Woody
Plants to Herbivory / 136

2.1   Leaves / 136

2.2   Twigs / 139

3   Physiological Mechanisms of Long-Term Responses to
Herbivory / 143

3.1   Leaves / 143

3.2   Twigs / 145

4   Implications for Plant Defense Theory / 147

4.1   Leaf Responses to Herbivory / 147

4.2   Twig Responses to Herbivory / 148

5   Summary / 149

Acknowledgments / 149

References / 149

7   SQUASH BEETLES, CUCUMBER BEETLES, AND
INDUCIBLE CUCURBIT RESPONSES   155

Douglas W. Tallamy and Eric S. McCloud

1   Introduction / 155

2   Cucurbitacin Induction in *Cucurbita* / 156

3   *Cucurbita*–Squash Beetle Interactions / 161

4   *Cucurbita*–Cucumber Beetle Interactions / 171

5   Cucurbitacins and the Assumption of Defense / 175

6   Conclusion / 177

Acknowledgments / 177

References / 178

8 RESPONSES OF LEAF BEETLES TO INJURY-RELATED CHANGES IN THEIR SALICACEOUS HOSTS   **183**

Michael J. Raupp and Clifford S. Sadof

1 Introduction / 183
2 Responses of Willows and Associated Herbivores to Injury / 184
   2.1 Physical and Chemical Responses of Willows to Injury / 184
   2.2 Short-Term Responses of Weeping Willows to Leaf Injury / 185
   2.3 Behavioral Responses of *Plagiodera versicolora* to Injured Willow Leaves / 187
   2.4 Physiological Responses of *Plagiodera versicolora* to Injured Willow Leaves / 189
3 Injury-Related Changes in Leaf Suitability and Spatial Patterns of Herbivory / 190
4 Injury-Related Changes in Leaf Suitability and Temporal Patterns of Herbivory / 193
5 Impact of Injury-Related Changes in Leaves on Higher Trophic Levels / 196
6 Summary / 197
   Acknowledgments / 198
   References / 198

9 THE IMPACT OF INDUCIBLE PHYTOCHEMICALS ON FOOD SELECTION BY INSECT HERBIVORES AND ITS CONSEQUENCES FOR THE DISTRIBUTION OF GRAZING DAMAGE   **205**

P. J. Edwards, S. D. Wratten, and R. M. Gibberd

1 Introduction / 205
2 Effects of Damage on Plant Palatability to Insects / 206
   2.1 Laboratory Bioassays of Leaf Palatability / 206
   2.2 Effects of Damage on Leaf Palatability: Between Leaf Effects / 206
   2.3 Effects of Localized Damage on within-Leaf Grazing Patterns / 208
3 Field Bioassays of between-Leaf Effects of Damage / 212
4 Effects of Leaf Damage on Insect Foraging: An Experiment / 212
   4.1 Movement between Leaves / 214
   4.2 Feeding Damage / 214

5   Discussion / 216
References / 220

**10   THE EFFECTS OF INDUCED PLANT PROTEINASE
INHIBITORS ON HERBIVOROUS INSECTS**   **223**
Jane L. Wolfson

1   Introduction / 223
2   Insect Digestive Proteinases / 225
3   Proteinase Inhibitors / 226
    3.1   Plant Proteinase Inhibitors / 226
    3.2   Impact of Proteinase Inhibitors on Insect Growth
          and Development / 228
4   Induction of Proteinase Inhibitors in Plants / 232
    4.1   Tomatoes: The Model System / 232
    4.2   Effects of Plant Age and Environmental Conditions
          on Leaf Quality of Foliage from Wounded Tomato
          Plants / 233
    4.3   The Impact of Tomato Proteinase Inhibitor on
          Larval Growth / 236
    4.4   Effect of Plant Age on Proteinase Inhibitor
          Induction / 237
5   Conclusion / 238
Acknowledgments / 239
References / 239

**11   INDUCED DEFENSIVE REACTIONS IN CONIFER–BARK
BEETLE SYSTEMS**   **245**
Kenneth F. Raffa

1   Introduction / 246
2   Life History and Biology of Bark Beetles / 246
3   Role of Induced Reactions in Conifer Resistance / 247
4   Induced Conifer Response to Bark Beetle Attack / 248
    4.1   Histological Changes / 248
    4.2   Chemical Changes / 248
    4.3   Biological Effects / 250
5   Elicitation of Induced Responses / 253
    5.1   Biotic and Purified Elicitors / 253
    5.2   Sequential Stages of Induction / 256

6   Beetle Counteradaptations to Induced Defenses / 256

   6.1   Exhaustion of Host Defenses / 257

   6.2   Tolerance and Avoidance of Host Toxins / 257

   6.3   Integration of Colonization Strategies / 260

   6.4   Conflicting Rate Reactions: Accumulation versus Depletion of Allelochemicals / 261

   6.5   Reproductive Costs of Mass Attack / 262

7   Factors Affecting the Extent of Induced Responses / 263

   7.1   Host Stress / 263

   7.2   Effects of Prior Exposure, Inoculum Load, and Site of Attack / 264

   7.3   Seasonal and Age-Related Variation / 264

   7.4   Genetic and Environmental Components of Constitution and Induced Defenses / 264

8   Rose of Host Resistance in the Population Dynamics and Management of Bark Beetles / 266

   8.1   Population Dynamics / 266

   8.2   Management Considerations / 267

       Acknowledgment / 268

       References / 268

12   THE EFFECTS OF INDUCIBLE RESISTANCE IN HOST FOLIAGE ON BIRCH-FEEDING HERBIVORES   277

S. Neuvonen and E. Haukioja

1   Introduction / 277

2   Inducible Resistance in the Birch–*Epirrita* System / 279

   2.1   Effects of Rapid Inducible Resistance on *Epirrita* Performance / 279

   2.2   Effects of Delayed Inducible Resistance of *E. autumnata* / 281

   2.3   Counteradaptations by *Epirrita* to Inducible Resistance / 283

   2.4   The Evolutionary Origin of Delayed Inducible Resistance / 283

3   The Variable Role of Inducible Resistance in Different Birch–Herbivore Systems / 284

4   Conclusion / 287

   References / 288

**13 VARIABLE INDUCED RESPONSES: DIRECT AND
INDIRECT EFFECTS ON OAK FOLIVORES**   **293**

Stanley H. Faeth

1   Introduction / 293
    1.1   Variable Consequences of Induced
          Responses / 294
2   Mitigating Factors in Induced Responses / 300
    2.1   Variation in Chemical Changes / 300
    2.2   Variation in Nutritional Changes / 303
    2.3   Variation in Phenological and Morphological
          Changes / 304
    2.4   Are Phenological Changes Defensive? / 305
3   Other Complicating Factors / 306
    3.1   Life Histories of Participating Insect Species / 306
    3.2   Behavior of Phytophagous Insects / 308
    3.3   Induced Responses and Intraspecific
          Competition / 308
    3.4   Induced Responses and the Third Trophic
          Level / 310
4   Future Directions / 312
    4.1   Know Thy Effects / 312
    4.2   Know Thy Response / 314
5   Conclusion / 314
    Acknowledgments / 315
    References / 315

**14 THORNS, SPINES, PRICKLES, AND HAIRS: ARE THEY
STIMULATED BY HERBIVORY AND DO THEY DETER
HERBIVORES?**   **325**

Judith H. Myers and Dawn Bazely

1   Introduction / 325
2   Types of Physical Defenses / 326
    2.1   Hairs as Plant Defense / 326
    2.2   Thorns, Spines, and Prickles / 327
3   Changes in Plant Physical Characteristics Following
    Herbivory / 331
4   Environmental Stress and Physial Defense
    Structures / 337
5   Discussion / 338
    Acknowledgments / 340
    References / 340

PART III   AGRICULTURAL IMPLICATIONS OF INDUCIBLE PHYTOCHEMICALS   345

15 INDUCIBLE DEFENSES IN SOYBEAN AGAINST HERBIVOROUS INSECTS   **347**

Marcos Kogan and Daniel C. Fischer

1   Introduction / 347

2   Soybean Phytoalexins / 349

   2.1   Biosynthesis of Isoflavonoids and Relationships to Other Plant Phenolics / 350

   2.2   Soybean Resistance to Fungal Pathogens / 352

   2.3   *Phytophthora* Resistance and Phytoalexins Accumulation in Soybean / 353

   2.4   Elicitors of Phyloalexins / 355

3   Effects of Soybean Phytoalexins on Herbivore Behavior and Nutritional Physiology / 357

   3.1   Specific Phytoalexins / 358

   3.2   Phenolic Acids and Their Derivatives / 362

4   Does Insect Feeding Induce Resistance in Soybean? / 365

   4.1   Induction by Previous Herbivory: Greenhouse Plants / 365

   4.2   Induction by Previous Herbivory: Field Plants / 366

   4.3   Postingestive Effects of Induced Resistance / 368

5   Chemical Nature of Allomonal Compounds Induced by Herbivory / 369

6   Enzymatic Inhibition / 370

7   Ecological Effects of Inducible Defenses and Their Potential Use in Integrated Pest Management / 371

   7.1   Potential Applications in Soybean Integrated Pest Management / 372

Acknowledgments / 373

References / 373

16 BACTERIALLY INDUCED CHANGES IN THE COTTON PLANT–BOLL WEEVIL PARADIGM   **379**

J. H. Benedict and J. F. Chang

1   Introduction / 379

2   Background / 381

3 Boll Weevil Responses to Bacterially Treated
   Cotton / 382
   3.1 Field Free-choice Studies of Plant Damage and
        Yield / 382
   3.2 Boll Weevil Preference, Survival, and Reproductive
        Biology / 384
   3.3 Boll Weevil Pheromone Production / 386
4 Plant Semiochemical Responses to Bacterial
   Treatment / 388
   4.1 Airborne Monoterpenes from Flower Buds and
        Leaves / 388
   4.2 Terpene Aldehydes and Tannins in Flower Bud and
        Leaf Tissues / 391
5 Conclusions and Prospects / 394
   Acknowledgments / 395
   References / 396

17 **INDUCIBLE RESISTANCE IN AGRICULTURAL
    SYSTEMS**   **402**
    Richard Karban

1 Introduction / 403
2 Do Plant Changes Affect Herbivore Populations? / 405
   2.1 Scale / 405
   2.2 Compensatory Mortality / 406
   2.3 Timing / 407
   2.4 Extent of Initial Damage / 407
   2.5 Environmental Conditions / 408
3 Induced Resistance of Cotton to Spider Mites / 408
   3.1 Problems of Scale / 410
   3.2 Timing / 411
   3.3 Extent of Initial Damage / 412
   3.4 Environmental Conditions / 413
4 Practical Consequences / 414
5 Conclusion / 415
   References / 416

SPECIES INDEX   **421**

SUBJECT INDEX   **427**

PHYTOCHEMICAL INDUCTION BY HERBIVORES

PART I

PHYTOCHEMICAL RESPONSES TO HERBIVORE ATTACK

CHAPTER 1

A PHYTOCENTRIC PERSPECTIVE OF PHYTOCHEMICAL INDUCTION BY HERBIVORES

JAMES S. COLEMAN[1]
Department of Biological Sciences, Stanford University, Stanford, CA 94305

CLIVE G. JONES
Institute of Ecosystem Studies, The New York Botanical Garden, Mary Flagler Cary Arboretum, Box AB, Millbrook, NY 12545

1 Introduction
2 A phytocentric model
  2.1 Plant genotype
  2.2 Plant phenology and ontogeny
  2.3 Abiotic environment
  2.4 Biotic environment
  2.5 Interactions of components
  2.6 Resource suitability to herbivores
  2.7 Herbivore effects on plant functions
3 The phytocentric model and phytochemical induction
  3.1 Plant genotype and induction
  3.2 Leaf ontogeny and induction
  3.3 Plant phenology and induction
  3.4 Abiotic environment and induction
  3.5 Biotic environment and induction
4 Summary
  References

1 INTRODUCTION

The biochemistry of plant tissues may change following mechanical damage or herbivory (Green and Ryan 1972, Ryan and Green 1974, Bryant 1981,

[1] Current address: Biological Research Laboratories, 130 College Place, Syracuse University, Syracuse, NY 13244. (315) 443–3748.

3

Schultz and Baldwin 1983, Baldwin and Schultz 1985). Rhoades (1979, 1985) and Haukioja (1980) hypothesized that these phytochemical changes increased plant resistance to subsequent herbivore attack and thus evolved as active defenses, because many reports showed that herbivore fitness or feeding rates were reduced on previously damaged plants in comparison to controls (Table 1). Given the large data base, it seems reasonable to accept that damage to plant tissues can result in altered chemical and nutritional qualities of plants that subsequently reduce rates of herbivore feeding and growth, at least in the short term (see Neuvonen and Haukioja, this volume, Chapter 12).

Nevertheless, a number of important issues must be addressed before the role of phytochemical induction in the defense of plants from herbivory can be accurately assessed. First, many studies on phytochemical induction contain large amounts of uncontrolled or unexplained variation within treatments, which has created controversial interpretation of results. For example, Fowler and Lawton (1985) argued that there was insufficient evidence to support a defensive role for phytochemical induction, because many experimental studies of induced resistance were poorly designed, were analyzed with inappropriate statistics, and had excessive amounts of variation within experimental treatments. They also suggested that the relatively small changes in herbivore fitness that were reported in these relatively short-term experiments would not result in significant changes in long-term herbivore population dynamics or subsequent plant damage.

This second point was echoed by Myers and Williams (1987) who suggested that recent studies on induced resistance overemphasized the effect of herbivory on short-term changes in leaf chemistry, and underemphasized how these changes in leaf chemistry would actually affect the biology and long-term population fluctuations of herbivorous insects.

Third, it is unclear whether reductions in foliage quality of some plants following leaf damage are the result of active responses of plants to herbivory or simply the result of passive deterioration of foliage quality that might be caused by tissue injury or removal. For example, Tuomi et al. (1984) suggested that the reduction in foliage quality of birch trees grown on nutrient poor soils to *Epirrita autumnata* larvae following herbivory or mechanical leaf damage was the result of decreases in the amount of nitrogen available to herbivores in leaves, because removal of leaf tissues subsequently reduces plant nitrogen stores.

Finally, the assumption that induced responses of plants are evolved defenses implies that a given response should be initiated by the feeding of a specific herbivore species or group of species and should not be a generalized plant response to many abiotic and biotic stresses. Whether this is actually the case is unclear. For instance, accumulation of phenolic compounds in plant leaves may follow herbivory and mechanical leaf damage (Schultz and Baldwin 1983, Hartley and Lawton, this volume, Chapter 5), pathogen infection (Cruickshank 1980), exposure to high light (Waring et al. 1985, Larsson et al. 1986, Mole et al. 1988, Mole and Waterman 1988) or exposure to nutrient limitation (Gershenzon 1984).

We believe that addressing the issues stated above will require a deeper understanding of the process of phytochemical induction and the factors that control and constrain the ability of plants to exhibit induced responses. Hence, we take a phytocentric perspective of phytochemical induction by herbivores by focusing on characteristics of plants that are likely to affect their ability to respond rapidly to herbivory. Specifically, we focus on how plant genotype, phenology, ontogeny, and interaction with the abiotic and biotic environments can control and constrain the process of phytochemical induction. We assert that this perspective not only helps to resolve some of the issues presented above, but that it also aids in predicting the species and age of plants that are most likely to exhibit phytochemical induction, the environmental conditions that promote or inhibit induced responses, and the characteristics of specific herbivores and their feeding habits that may make them likely or unlikely to be affected by phytochemical changes.

2   A PHYTOCENTRIC MODEL

A phytocentric model of phytochemical induction by herbivores is shown in Figure 1. By definition, phytochemical induction results in changes in the resource quality and suitability of plant tissues for herbivores. The model shows that the biochemical and physical characteristics of plant tissues that determine resource quality are the outcome of plant resource acquisition, allocation, and partitioning. These plant functions are constrained or influenced by many factors including the impact of herbivores. Herbivore fitness is related to plant functions in two ways. First, herbivore fitness is at least partially determined by the resource quality of host plants. Second, the removal of plant tissues by herbivores influences the ability of plants to function and, in turn, influences the resource quality and suitability of host plants for subsequent herbivore feeding.

The purpose of this model is to lay a framework for interpreting herbivore-induced changes in the resource quality and suitability of plant tissues in light of numerous other factors that influence and constrain plant functions. Specifically, this model emphasizes plant processes of resource acquisition, allocation, and partitioning. Carbon-based resources are acquired through photosynthesis, while nutrients (e.g., nitrogen and phosphorus) and water are usually acquired by root uptake. These resources are subsequently allocated into specific biochemical pathways and are translocated to other parts of plants. Eventually, these resources are partitioned to growth of vegetative or reproductive tissues, maintenance of living tissues, repair of damaged tissues, production of chemical or structural defenses against pathogens and herbivores, and senescence of tissues (Mooney 1972). The specific patterns of resource acquisition, allocation, and partitioning into these var-

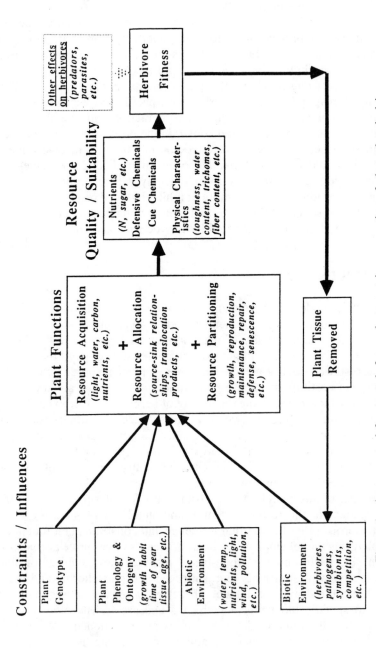

Figure 1 A conceptual framework for a phytocentric perspective of phytochemical induction by herbivores.

ious processes varies greatly within and between plant species (Mooney 1972).

Four categories of factors can constrain the acquisition of resources, and influence the patterns of resource allocation and partitioning within plants:

1. Plant genotype
2. Plant phenology and ontogeny
3. Abiotic environmental factors
4. Biotic environmental factors

We first examine these model components and then show how they can be used to design induction experiments with reduced amounts of unexplained variation in data relative to many previous studies, and to predict situations in which induction is more or less likely to occur.

2.1 Plant Genotype

Plant genotype is a major determinant of how plants acquire and utilize resources. Genotype accounts for large differences in resource acquisition between species (e.g., C_3 versus C_4 plants), and resource utilization for growth and reproduction (i.e., early versus late successional species (Bazzaz 1979)). Genotypic effects also account for large differences between plants of closely related species. For example, wild radish (*Raphanus sativus* × *raphanistrum*) and "cherry belle" radish (*R. sativus* cv. "cherry belle") have similar physiological capacities for assimilating carbon and acquiring nutrients and water (Mooney and Chiariello 1984), but they have very different patterns of resource allocation and partitioning which leads to large differences in their biomass production (Mooney and Chiariello 1984, Mooney et al. 1988). Wild radish attains a much larger biomass than "cherry belle" radish because it allocates proportionally more resources to growth of roots and shoots, while "cherry belle" radish allocates preferentially to the formation of a storage organ (hypocotyl). Subsequently, wild radish plants have a much larger fraction of their biomass dedicated to resource acquisition than "cherry belle" plants so they can acquire more resources and accumulate more biomass (Mooney and Chiariello 1984).

2.2 Plant Phenology and Ontogeny

Resource allocation and partitioning change as plants develop. For example, many seedlings allocate proportionally more of their resources into the growth of shoots and roots than do mature plants, which may shift allocation to favor the development of reproductive organs (e.g., flowers and fruits) or storage tissues (e.g., ray parenchyma in woody tissue) (Kozlowski 1971a,b, Waring and Schlessinger 1986). As leaves develop, they show extensive changes in their photosynthetic and respiration rates, nutrient and chemical composition, and anatomical characteristics as they proceed from metabolite sinks to metabolite sources and eventually senesce (Isebrands and Larson 1973, Larson et al. 1980, Coleman 1986).

Plant species also show very different resource allocation patterns during the season, which can be ascribed to genotypically determined phenological differences. Indeterminate or free-growing trees (e.g., *Populus* spp., *Liriodendron* spp.) can allocate resources to the production of shoots throughout the growing season and continually produce leaves. Many shoots of these plants contain leaves of all developmental stages (Spurr and Barnes 1980). In contrast, determinate or fixed-growing trees (e.g., *Quercus* spp., *Carya* spp.) produce most of their leaves in a single flush (although some species also produce second and third flushes), and shoot extension usually ends early in the growing season (Spurr and Barnes 1980). Leaves of determinate trees synchronously develop, and resources acquired by leaves are normally not used for shoot growth until the following season (Kozlowski 1971b, Spurr and Barnes 1980).

There are other phenological effects in addition to growth habit. For example, trees in temperate forests tend to allocate resources to the growth of shoots early in the season, but to roots later in the season (Kozlowski 1971b, Pallardy and Kozlowski 1979, Larcher 1980). Photosynthetic rates of many plants decline during the growing season (Treharne and Eagles 1970, Zima and Sestak 1979, Nelson et al. 1982), and concentrations of secondary compounds in plant tissues can show extensive seasonal variation (Schultz et al. 1982, Jones 1983, Haukioja et al. 1985a, Lindroth and Batzli 1986). On a shorter temporal scale, plants show diurnal variation in resource acquisition and allocation. For example, cottonwood leaf characteristics, such as the dry weight of a mature leaf, can vary by as much as 30% diurnally (Dickson 1987), and translocation of starch from leaves in the dark period results in extensive reductions of dry matter (Geiger 1986, Dickson 1987). Diurnal fluctuations in secondary compounds are also known to occur in some plants (Adams and Haganon 1977).

2.3 Abiotic Environment

Plants show flexibility in adjusting physiological processes to environmental conditions (Larcher 1980, Chapin et al. 1987, Schliting 1986). This phenotypic plasticity results in adjustments of plant resource acquisition, allocation, and partitioning that can ameliorate environmental limitations on plant growth, such as drought, shading, or nutrient deficiency (Bradshaw 1965, Bloom et al. 1985, Hunt and Nicholls 1986, Robinson 1986, Schliting 1986, Chapin et al. 1987). Changes in allocation appear to be directed toward making all environmental factors equally limiting to plant growth (Bloom et al. 1985, Hunt and Nicholls 1986, Robinson 1986, Szaniawski 1987). For example, plants exposed to root stress (e.g., water or nutrient limitation) allocate proportionally more resources to root growth (Chapin 1980, Rufty et al 1981, Szaniawski 1983, Clarkson 1985). Similarly, exposure of plants to shoot stress (e.g., shading, air pollution) tends to result in allocation of relatively more resources to shoot growth (Bloom et al. 1985, Lechowicz 1987, Mooney et al. 1988).

Environmental stresses limiting plant growth may also cause shifts in the partitioning of plant resources to different processes (Waring and Pitman 1985, Bazzaz et al. 1987, Chapin et al. 1987). For example, shaded willow plants supplied with adequate nutrients produced less carbon based secondary compounds than similar plants under high light and adequate nutrient conditions (Waring et al. 1985, Larsson et al. 1986). Many tree species often allocate fewer resources to reproduction during periods of unfavorable environmental conditions (Kozolowski 1971b, Spurr and Barnes 1980, Bazzaz et al. 1987). Ozone air pollution decreases resource acquisition by photosynthesis in many plants (Reich and Amundson 1985, Lechowicz 1987) and in bean plants results in allocation of proportionally more resources for maintenance and repair of leaf tissues (Amthor and Cumming 1988).

2.4 Biotic Environment

Interactions with herbivores, pathogens, and symbionts also change resource acquisition, allocation, and partitioning of plants. Mychorrhizal or nitrogen-fixing symbioses increase rates of resource acquisition, but plants also allocate proportionally more resources to roots in order to support symbiotic microorganisms (see Chapin 1980, Clarkson 1985, Szaniawski 1987). Infestation of plants with aphids or rust pathogens affects the rates of plant photosynthesis, respiration, and nutrient uptake, and causes shifts in allocation of resources between roots and shoots (Kosuge 1978, Ayres 1982, Wood et al. 1985, Hawkins et al. 1986a,b). Many herbaceous and woody plants allocate proportionally more resources to shoots following defoliation (Caldwell et al. 1979, Bassman and Dickmann 1982, 1985, McNaughton 1983), whereas root consumption often results in increased allocation of resources to root growth (Sutton 1967, Caldwell et al. 1979, Chapin and Slack 1979, Ingham and Detling 1984, Andersen 1987). Crawley (1985) showed that small amounts of herbivory ($<10\%$) on oak trees greatly reduced reproductive output in comparison to undamaged trees. Herbivore and pathogen damage are known to change the secondary chemistry of some plants (Green and Ryan 1972, Ryan and Green 1974, Cruickshank 1980, Schultz and Baldwin 1983, Kuc and Presig 1984), suggesting a shift in plant resource partitioning to the production of chemical defenses (see McLaughlin and Shriner 1980, Hartley and Lawton, this volume, Chapter 5).

2.5 Interactions of Components

Genotypic, phenological, ontogenetic, and environmental influences on plant functions act simultaneously and may interact. For example, Norden et al. (1986) documented genotype × environment interactions in the resource partitioning of peanut genotypes to reproduction. Differences in aphid resistance between goldenrod genotypes may only be expressed when plants are water stressed (Maddox and Cappuccino 1986). Similarly, genotypic resistance of cereals to rust and powdery mildew fungi may only be expressed under specific environmental conditions (Browder 1985). The effect of water stress on growth and resource partitioning of barley is modified by preinfection of plants with powdery mildew, indicating abiotic X biotic environmental interactions (Ayres 1982).

2.6 Resource Suitability to Herbivores

In general, the most suitable tissues for herbivores have high concentrations of nitrogen (Fox and McCauley 1977, Mattson 1980, Scriber and Slansky 1981), adequate water (Feeny 1976, Scriber and Slansky 1981), low leaf toughness (e.g., fiber, lignin, silica) (Feeny 1976, Raupp 1985), low concentrations of secondary compounds (Feeny 1976, Fox and McCauley 1977, Rhoades 1985), and few, if any, anatomical barriers, such as trichomes and sclerids (Myers 1987, Tingey and Laubengayer 1981). The processes of plant resource acquisition, allocation, and partitioning determine these biochemical and anatomical characteristics of plant tissues and, therefore, the suitability of plants to herbivores. Plant genotype is a major determinant of quality and quantity of secondary compounds and anatomical defenses that affect herbivores (Berenbaum 1981, Bell 1984). Leaf age is an important modifier of many plant–insect interactions because the contents of nutrients, secondary compounds, toughness, water, and fiber all change as leaves age (Isebrands and Larson 1973, Jepson 1983, Raupp and Denno 1983, Hartnett and Bazzaz 1984, Coleman 1986, Potter and Kimmerer 1986). Abiotic environmental influences on plants can change the concentration and quality of nitrogen (White 1984, Mattson and Haack 1987), secondary compounds (Rhoades 1983b, Gershenzon 1984, Mattson and Haack 1987), water and fiber contents (Lewis 1984, Mattson and Haack 1987), and the expression of anatomical defenses (Bell 1984, Myers 1987). Changes in the resource suitability of plant tissues to herbivores that occur following herbivore attack are extensively dealt with in other chapters.

While it is possible to make some generalizations about resource quality, specific interactions and adaptations of herbivores to host plants may also be critical determinants of resource suitability. For example, many insects use plant defensive chemicals as host-specific feeding or oviposition cues (Dethier 1980, Jermy 1983), and can detoxify such compounds (Brattsten 1979). Some insects use these compounds as a nutritional substrate (Rosenthal et al. 1977, Bernays and Woodhead 1982). Other insects may be deterred by these compounds and not oviposit or feed on plants that are otherwise perfectly adequate for insect growth and survivorship (Gupta and Thorsteinson 1960, Jermy and Szentsi 1978).

2.7 Herbivore Effects on Plant Functions

The amount, distribution, and rate of herbivory affects plant resource acquisition, allocation, and partitioning. Herbivores that remove plant tissues most certainly decrease the amount of tissue available to plants for resource acquisition, and also disrupt normal patterns of resource allocation. These changes, in turn, influence plant resource suitability for herbivores. Our model does not predict the specific outcome of herbivore damage on subsequent resource suitability of plants. This will depend on specific plant processes and their interactions such as other constraints and influences on plant functions, specific physiological, biochemical and anatomical responses of plants to the impact of additional tissue removal, and specific requirements and characteristics of herbivores.

3 THE PHYTOCENTRIC MODEL AND PHYTOCHEMICAL INDUCTION

3.1 Plant Genotype and Induction

A number of aspects of plant genotype may determine whether induction occurs and whether it is detectable to an investigator. It is important to recognize these factors to reduce variation in the results of induction experiments and to obtain information that will help resolve questions surrounding the specificity of induced responses and the effect of short-term plant responses on herbivore populations. First, the capacity of a plant to exhibit any induced response will be at least partially under genetic control (Ellingboe 1982, Heath 1982), so failure to discriminate inducible versus noninducible genotypes could result in extensive variation in data and make detection of treatment effects difficult. For example, some genotypes of flax exhibit induced resistance to rust pathogens, while others do not (Ellingboe 1982, Heath 1982, Coffey and Wilson 1983). It is easy to see the difficulty that would arise from interpreting an induction experiment where differences between these genotypes were not accounted for.

Second, different genotypes may have different thresholds of damage before exhibiting induced responses, and may show different responses once induced. The same level of herbivore damage would, therefore, produce different degrees of resistance or biochemical responses in different genotypes. For example, pine genotypes taken from different geographic areas exhibited different biochemical responses and degrees of induced resistance after exposure to similar amounts of damage (Leather et al. 1987). Similar results were obtained in studies on induced resistance of birch (Haukioja and Hahimaki 1985, Neuvonen and Haukioja 1984).

Third, the ability of plant genotypes to express induced responses may be differentially affected by the same environmental conditions (e.g., some genotypes may be inducible only under specific water and nutrient regimes).

Fourth, constitutive resistance of plant genotypes to herbivores may mask induced resistance. Genotypes with extremely high levels of constitutive resistance may not show detectable increases in resistance following herbivore damage, even if biochemical induction occurred.

Finally, differences in the degree of constitutive resistance in genetically mixed plant populations might lead to some erroneous conclusions regarding induction. For instance, if a population was randomly divided into control and damaged plants in such a way that the group receiving mechanical damage had a higher mean constitutive resistance, then the apparently greater resistance of damaged plants relative to controls might be attributed to induced responses. However, in this scenario, these differences in resistance would have merely been a function of differences in genotypically determined constitutive resistance of control and damaged plants.

Plant genotype may well account for much of the error variance reported in many phytochemical induction studies. For example, Coleman et al. (1987) partitioned the variance associated with two different plant genotypes, four stages of leaf ontogeny, and ozone exposure (with charcoal-filtered air controls) by ANOVA from experiments on effects of prior ozone exposure on the resistance of eastern cottonwood to leaf rust. We showed that ozone exposure significantly increased resistance of cottonwood to leaf rust with an approximate 50% reduction in pathogen reproduction on ozone-treated plants. We also showed that plant genotype accounted for six times as much variance in data as the ozone treatment. An ANOVA placing plant genotype into the error term resulted in a nonsignificant ($p > .05$) treatment effect.

The genotypic component was not accounted for in 75% of the phytochemical induction studies that we reviewed (Table 1). In general, studies that accounted for plant genotype showed particularly large increases in plant resistance to herbivores following plant damage in comparison to studies that did not account for genotypic effects. For example, Karban and Carey (1984) used specific genotypes of cotton plants in their experiments and showed an approximate 200% reduction in the population growth of mites on previously damaged plants. Edwards et al. (1986) showed an unwavering preference of *Spodoptera littoralis* larvae for undamaged tomato plants of the same genotype in comparison to damaged plants. Fowler and Lawton (1985) reviewed a number of studies and argued that there was little evidence for induced resistance. Most of the studies they reviewed, however, did not state that the genotypic relationships of plants were known. Interestingly, differences in insect growth, development, or reproduction between damaged or control plants never exceeded 40% in these studies, and were generally much less (Haukioja and Niemela 1977, 1979, Wallner and Walton 1979, Edwards and Wratten 1982, Rhoades 1983a, Valentine et al. 1983, Raupp and Denno 1984, Wratten et al. 1984).

Variation between plant genotypes in traits affecting induction may be useful for examining the ecological and evolutionary role of phytochemical induction in plant defense. For example, if induction is a defensive response,

TABLE 1 Summary of Some Published Reports on Phytochemical Induction Indicating the Results Obtained, and Whether the Plant Characteristics of Genotype, Leaf Ontogeny, Phenology, and Interactions with the Abiotic and Biotic Environment Were Controlled for in Experiments

Plants/ Herbivores	Response of Herbivores on Recently Damaged Plants in Comparison to Controls	Plant Genotype	Leaf Ontogeny	Phenology (Age, Shoot Type, etc.)	Abiotic History (Nutrients, Light, etc.)	Biotic History (Pathogens, Symbionts, etc.)	Source
Birch Tree Studies							
Birch (Betula pendula)/Coleophora serratella	When area proximal to leaf mines damaged, larva moved within damaged leaf but not necessarily to a new leaf. Larval development time increased, and was associated with increased phenolics in leaves.	?[1]	?	Small trees Shoot type - ?	?	?	Bergelson et al. (1986)
Birch (Betula papyrifera) Balsam poplar (Populus balsamifera), Aspen (P. tremuloides), Green Alder (Alnus crispa)/Snowshoe Hares	Juvenile shoots of clipped plants had high concentrations of terpenes and phenolic resins which decreased palatability of tissue for hares.	discussed, but not accounted for	NA[2]	yes	?	?	Bryant (1981)

13

TABLE 1 Summary of Some Published Reports on Phytochemical Induction Indicating the Results Obtained, and Whether the Plant Characteristics of Genotype, Leaf Ontogeny, Phenology, and Interactions with the Abiotic and Biotic Environment Were Controlled for in Experiments (*Continued*)

Plants/ Herbivores	Response of Herbivores on Recently Damaged Plants in Comparison to Controls	Plant Genotype	Leaf Ontogeny	Phenology (Age, Shoot Type, etc.)	Abiotic History (Nutrients, Light, etc.)	Biotic History (Pathogens, Symbionts, etc.)	Source
Birch (*B. pendula*, *B. pubescens*)/ Moose browsing & insect community	Increased amounts of herbivory on browsed trees.	?[3] (Natural damage was used for defining treatments)	?	Plant age Shoot type - ?	?	?	Dannell & Huss-Danell (1985)
Birch (*B. pubescens*), Oak (*Quercus robur*), Alder (*Alnus glutinosa*), Poplar (*Populus nigra*)/Snails	Feeding rates inhibited on birch, but no effects were found on other species.	?	?	?	?	?	Edwards and Wratten (1982)
Birch (*B. pendula* & *B. pubescens*)/Herbivore community	No differences in amounts of leaf tissue damaged.	?	?	Saplings Shoot type - ?	?	?	Fowler & Lawton (1985)

Plant (Herbivore)						Reference
Birch (*B. pubescens*)/ *Apochemia pilosaria* & herbivore community	*Apochemia pilosaria* larval growth and development was inhibited. Decreased numbers of many herbivores on damaged plants, but no difference in amounts of leaves damaged.	?	Half expanded leaves were damaged	?	?	Fowler & MacGarvin (1986)
Birch (*B. pubescens*)/ *Eppirita autumnata* (2 strains)	Decreased larval growth. Some differences in the degree of induced resistance between the different birch provenances. Resource quality of leaves was most reduced on damaged and adjacent leaves and least reduced on leaves not proximal to a damaged leaf.	? (2 provenances)	? (bioassayed adjacent leaves)	?	?	Haukioja & Hahimaki (1985)

TABLE 1 Summary of Some Published Reports on Phytochemical Induction Indicating the Results Obtained, and Whether the Plant Characteristics of Genotype, Leaf Ontogeny, Phenology, and Interactions with the Abiotic and Biotic Environment Were Controlled for in Experiments (*Continued*)

Plants/ Herbivores	Response of Herbivores on Recently Damaged Plants in Comparison to Controls	Plant Genotype	Leaf Ontogeny	Phenology (Age, Shoot Type, etc.)	Abiotic History (Nutrients, Light, etc.)	Biotic History (Pathogens, Symbionts, etc.)	Source
Birch (*B. pubescens*)/ *Oporinia autumnata*	Increased larval development time.	?	? (bioassayed leaves above damaged leaf)	Dwarf shoots	?	?	Haukioja & Niemela (1977)
Birch (*B. pubescens*)/ *O. autumnata, Brephos parthenias, Pristophora* sp., *Pteronidea* sp., *Dineura viridirorsata, Eriogaster lanestris*	Increased time for larval development for *O. autumnata, B. parthenias, Pristophora* sp. and *Pteronidea* sp.). Decreased larval growth (*O. autumnata, E. lanestris, Pteronidea* sp.). Reduced development time & increased growth (*D. virididorsata*). Old leaves did not show induced effects.	?	? (Bioassayed adjacent leaves)	Bioassayed throughout the season. Dwarf shoots.	?	?	Haukioja & Niemela (1979)

Birch (*B. pubescens*)/ *E. autumnata*	Reduced larval growth. Greatest reduction on fertilized trees. Insect damage caused a stronger induced response than artificial damage.	?	?	Tree age.	Fertilization	?	Haukioja & Neuvonen (1985)
Birch (*B. pubescens*)/ *E. autumnata*	Reduced larval growth. Stronger induced response when frass was added to damaged trees.	?	?	?	Nutrient poor sites with frass added.	?	Haukioja et al. (1985)
Birch (*B. pubescens*)/ *E. autumnata*	Slightly higher pupal wts. and larval survivorship on browsed plants. Greater difference between leaves of upper and lower crown than between treatments.	?	?	Long shoots	Upper & lower crown.	?	Neuvonen & Danell (1987)

TABLE 1 Summary of Some Published Reports on Phytochemical Induction Indicating the Results Obtained, and Whether the Plant Characteristics of Genotype, Leaf Ontogeny, Phenology, and Interactions with the Abiotic and Biotic Environment Were Controlled for in Experiments (*Continued*)

Plants/ Herbivores	Response of Herbivores on Recently Damaged Plants in Comparison to Controls	Plant Genotype	Leaf Ontogeny	Phenology (Age, Shoot Type, etc.)	Abiotic History (Nutrients, Light, etc.)	Biotic History (Pathogens, Symbionts, etc.)	Source
Birch (*B. pubescens*)/ *E. autumnata*	Larval growth reduced. No compensatory consumption even though the resource quality of leaves declined on damaged plants. Induced responses differed in strength on trees from different sites.	? (two different sites)	?	?	?	?	Neuvonen & Haukioja (1984)
Grey Birch (*B. populifolia*) oak (*Quercus velutina*)/ Gypsy moth (*Lymantria dispar*)	Oak defoliation reduced sugar levels which were correlated with reduced pupal wts. Greater variation in leaf chemistry between years than between treatments. No effects on grey birch.	?	Considered but not accounted for.	?	Soils & light	?	Valentine et al. (1983)

System/Species	Effect						Reference
Grey Birch (B. populifolia) oak (Q. velutina)/Gypsy moth (Lymantria dispar)	Pupal wts reduced on both birch and oak. Increased larval development time and mortality on oak.	?	?	?	Soils and light	?	Wallner & Walton (1979)
Birch (B. pubescens, B. pendula)/Spodoptera litteralis, Orgyia antiqua	Feeding rates reduced on leaves if damage was done in spring. Effects were smaller as the season progressed.	? (one tree of each species)	? (Bio-assayed adjacent leaves)	Bioassayed throughout season. Shoot type - ?	?	?	Wratten et al. (1984)
Birch (B. pubescens)/Birch ahids (Eucheraphis punctipennis)	No effect	?	? (Bio-assayed adjacent leaves)	Seedlings	Temperature & Photoperiod	?	Wratten et al. (1984)

19

TABLE 1 Summary of Some Published Reports on Phytochemical Induction Indicating the Results Obtained, and Whether the Plant Characteristics of Genotype, Leaf Ontogeny, Phenology, and Interactions with the Abiotic and Biotic Environment Were Controlled for in Experiments (*Continued*)

Plants/ Herbivores	Response of Herbivores on Recently Damaged Plants in Comparison to Controls	Plant Genotype	Leaf Ontogeny	Phenology (Age, Shoot Type, etc.)	Abiotic History (Nutrients, Light, etc.)	Biotic History (Pathogens, Symbionts, etc.)	Source
Other Indeterminate and Semi-Indeterminate Trees							
Feltleaf Willow (*Salix alaxensis*)/ Snowshoe hare	Shading increased palatability of both browsed (simulated) and control plants. Fertilization had no effect on induced responses. Clipped plants were less palatable than controls.	yes	NA	yes	Fertilization and light.	?	Bryant (1987)
Lime (*Tilia* spp.)/ Lime aphids (*Eucallipterus tillae*)	Aphid growth and development was inhibited on pre-infested trees	?	?	yes	?	?	Dixon & Barlow (1979) Barlow & Dixon (1980)
Red Alder (*A. rubra*)/ *Malacosoma califonicum pluviale* and *Hyphantria cunea*	Fall webworm grew larger on trees which had experienced moderate defoliation. Larval growth did not differ on regrowth vs. mature leaves.	? (used natural level of defoliation as treatments)	?	?	?	?	Myers & Williams (1984) Williams & Myers (1984)

System	Results					Reference
Red Alder (*A. rubra*)/ *M. californicum pluviale*	No short term effect. Reduced larval growth on foliage from trees in areas of prolonged defoliation.	?	? (Bio-assayed adjacent leaves)	Young trees	?	Myers and Williams (1987)
Willow (*Salix babylonica* and *S. alba*)/ *Plagiodera versicolora*	Beetles had lower fecundity; larvae had longer development times and reduced weights. No difference in adult feeding preference.	?	? (Branches had same number of leaves. Adjacent leaves were bio-assayed)	Mature trees	?	Raupp & Denno (1984)
Red Alder (*A. rubra*)/ *M. californicum* Sitka willow (*Salix stichensis*)/*Hyphantria cunea, M. californicum*	Reduction in larval growth rates, adult fecundity, and larval survivorship on alder. Slight reduction in larval growth rates on willow.	?	Discussed but not accounted for.	Tree size	?	Rhoades (1983b)

TABLE 1 Summary of Some Published Reports on Phytochemical Induction Indicating the Results Obtained, and Whether the Plant Characteristics of Genotype, Leaf Ontogeny, Phenology, and Interactions with the Abiotic and Biotic Environment Were Controlled for in Experiments (*Continued*)

Plants/ Herbivores	Response of Herbivores on Recently Damaged Plants in Comparison to Controls	Plant Genotype	Leaf Ontogeny	Phenology (Age, Shoot Type, etc.)	Abiotic History (Nutrients, Light, etc.)	Biotic History (Pathogens, Symbionts, etc.)	Source
Sycamore (*Acer pseudoplatanus*) Syc. Aphid (*Drepanosiphum platonoides*)	No effect on aphid growth, but aphids were more likely to migrate from damaged plants. Aphids were most affected by leaf age. Aphid infestation resulted in smaller mature leaves.	?	yes	yes	?	?	Wellings & Dixon (1987)
Annual Crop Plants							
Cucurbita moschata/ Acalymma vittata Epilachna tredecimonata	Feeding rates enhanced for *A. vittata* but inhibited for *E. tredecimonata*.	?	?	?	?	?	Carroll & Hoffman (1980)
Tomato/*Spodoptera litteralis*	Feeding preference for controls detected within 8 hrs after damage. Leaf position relative to damaged leaf had no effect on strength of response.	yes	Categorized as young, intermediate, old	Seedling	Temperature, Photoperiod	?	Edwards et al. (1986)

				Seedlings			
Cotton/spider mites (Tetranychus urticae)	Mites preferred control plants.	yes	yes	yes	Temperature, Photoperiod	?	Harrison and Karban (1986)
Cotton/spider mites (T. urticae)	Reduced mite population growth.	yes	yes	yes	Temperature, Photoperiod	?	Karban & Carey (1984)
Cotton/spider mites	Reduced mite population growth in 4 out of 6 trials on mechanically damaged plants.	yes	yes	yes	Temperature, Photoperiod	?	Karban (1985)
Cotton/spider mites	Reduced mite population growth on plants in the field, but to a lesser degree than was found in controlled environmental studies.	yes	cotyledons-damaged; bio-assayed-all leaves.	yes	?	?	Karban (1986)

23

TABLE 1 Summary of Some Published Reports on Phytochemical Induction Indicating the Results Obtained, and Whether the Plant Characteristics of Genotype, Leaf Ontogeny, Phenology, and Interactions with the Abiotic and Biotic Environment Were Controlled for in Experiments (*Continued*)

Plants/ Herbivores	Response of Herbivores on Recently Damaged Plants in Comparison to Controls	Plant Genotype	Leaf Ontogeny	Phenology (Age, Shoot Type, etc.)	Abiotic History (Nutrients, Light, etc.)	Biotic History (Pathogens, Symbionts, etc.)	Source
Cotton/spider mites	Reduced mite population growth on pre-infested plants when the initial mite population grew poorly. No effect when initial mite population grew well. Variation in photoperiod had no effect on induced resistance.	yes	yes	yes	Temperature, Photoperiod	?	Karban (1987)
Sugar beet/beet fly (*Pegomya betae*)	Increased larval mortality.	yes	?	Seedlings	Temperature	?	Rottger & Klingauf (1976)
Zucchini/squash beetle	Beetle trenched leaves apparently to avoid induced defenses, and was unaffected by induced responses.	yes	?	?	?	?	Tallamy (1985)

Determinate Trees

Species	Findings						Reference
Norway spruce (*Picea abies*), Sitka spruce (*Picea stichensis*)/Spruce aphid (*Elatobium abietinum*)	Aphids grew larger on previously infested plants. Magnitude of responses differed with spruce species.	?	?	yes	Temperature, Photoperiod	?	Fisher (1987)
Lodgepole pine (*Pinus contorta*)/Pine beauty moth (*Panolis flammea*)	Larval growth rates and survival decreased on defoliated trees. Oviposition preference for controls. Trees from different provenances exhibited different induced responses. No induced response by old trees tested in the field.	? (2 provenances)	?	Seedlings and mature trees	?	?	Leather et al. (1987)

25

TABLE 1 Summary of Some Published Reports on Phytochemical Induction Indicating the Results Obtained, and Whether the Plant Characteristics of Genotype, Leaf Ontogeny, Phenology, and Interactions with the Abiotic and Biotic Environment Were Controlled for in Experiments (*Continued*)

Plants/ Herbivores	Response of Herbivores on Recently Damaged Plants in Comparison to Controls	Plant Genotype	Leaf Ontogeny	Phenology (Age, Shoot Type, etc.)	Abiotic History (Nutrients, Light, etc.)	Biotic History (Pathogens, Symbionts, etc.)	Source
Oak (*Q. robur*)/ *Phyllonorycter herrisella*	Leaf damage severely reduced miner survivorship and was sometimes related to leaf nutrient status.	?	? (for bioassays)	yes	Lower canopy leaves were used.	?	West (1985)
Other (Grasses & Herbaceous Plants)							
Ragwort (*Senecio jacobaea*)/Cinnebar moth (*Tyria jacobaea*)	Larval growth rate was slightly increased.	?	Young leaves	yes		?	Crawley & Nachapong (1984)
Alfalfa/Prairie vole & Tallgrass prairie/meadow vole	Induced resistance could not explain vole population cycles. Vole feeding did reduce resource quality of plants, but the variation within the season was greater than the variation between treatments.	?	?	Partitioned seasonal	?	?	Lindroth & Batzli (1986)

Nettle (*Urtica diocia*)/*Aglais urticae* | Higher larval growth rates on regrowth foliage. Regrowth foliage quality was very similar to the foliage quality of normal young leaves. | ? | Regrowth vs. mature | ? | ? | ? | Pullin (47)

Summary of effects of plant damage on the subsequent fitness or feeding rates of herbivores.[4]

Number of studies showing reduced fitness or feeding rates of herbivores — 27

Number of studies showing increased fitness or feeding rates of herbivores — 7

Number of studies showing no effects on herbivore fitness or feeding rates — 8

[1] ?, It was unclear, to us, from the published report whether this factor was controlled for.

[2] NA, not applicable.

[3] Information in parentheses is an experimental procedure that relates to the plant characteristic, but does not constitute control for that characteristic in the reported experiments.

[4] Some individual studies reported negative effects for one herbivore–plant interaction, but reported positive or no effects for another herbivore–plant interaction.

then genotypes of a given species that exhibit induced responses should be more fit in the presence of herbivores than plants that do not; assuming that herbivores present formidable selective pressures on these plants, and that all other selective forces are relatively equal. Furthermore, genotypic variance in inducibility of plants may be a useful tool for research questions relating to the cost of plant defenses.

The general lack of consideration given to plant genotype in some studies may have resulted in unsubstantiated conclusions. For example, Myers and Williams (1984), Williams and Myers (1984), and Danell and Huss-Danell (1985) all used natural levels of herbivore damage as the basis for choosing treatment and control plants. Trees that were moderately damaged by herbivory in the field were used as damage-treatment plants, and undamaged trees were used as controls. Myers and Williams (1984) and Williams and Myers (1984) found that moderately damaged trees were sometimes more susceptible to herbivore damage. Danell and Huss-Danell (1985) found that trees moderately browsed by moose were more susceptible to subsequent insect herbivory than unbrowsed trees. These authors suggested that their results indicated a lack of induced defensive responses in treatment plants. However, it is quite possible that their results were simply reflecting the initial differences in the genetic bases of resistance of these plants to herbivory.

3.2   Leaf Ontogeny and Induction

Differences in physiology, biochemistry, and anatomy of leaves of different ages may have important implications for induction studies. Leaves of different ages may respond very differently to damage. The age of leaves selected for assays of induced resistance may affect results, because herbivores often show strong preferences for leaves of certain ages (Phillips 1976, Raupp and Denno 1983, Coleman 1986). Additionally, leaves of different ages may respond differently to stimuli applied to other leaves (Larson 1980, Dickson and Larson 1981, Watson and Casper 1984). Thus, controlling the age of leaves initially damaged and subsequently bioassayed may reduce the substantial amounts of variation within treatments that is often reported in induction studies (see Fowler and Lawton 1985, Bergelson et al. 1986), and aid in interpreting whether induced changes are active or passive plant responses. For example, in our experiments on the effects of ozone on cottonwood rust resistance (Coleman et al. 1987), leaf age accounted for two times the amount of variance in cottonwood resistance to leaf rust than did ozone treatment. Nevertheless, many studies of phytochemical induction by herbivores have not explicitly stated how leaf age was accounted for (but see Edwards et al. 1986, Hartley and Lawton this volume, Chapter 5, Table 1).

Most phytochemical induction studies (Table 1) used annual plants, seedlings, or mature trees with indeterminate (free-growing) growth habits as experimental organisms, so the following discussion will be directed at these plants. Here, expanding leaves act as sinks for metabolites from mature leaves, and these resources are used to complete leaf development and expansion (Larson et al. 1980, Dickson and Larson 1981, Coleman 1986). When expanding leaves are damaged, large changes in their biochemistry and physiology occur, because processes of leaf expansion and development are severely interrupted. For example, Wellings and Dixon (1987) showed that aphids infesting an expanding sycamore leaf compete with the leaf for plant resources exported from other leaves. When expanding sycamore leaves are deprived of resources by aphid feeding, the leaves develop to a smaller size, and subsequently support a smaller population of aphids when they are mature (Wellings and Dixon 1987). Therefore, Wellings and Dixon (1987) suggested that this case of induced resistance was probably a result of passive deterioration of plant resources and was not an active defensive response of sycamore trees.

The development of expanding leaves might also be affected by damage to older leaves. As we previously mentioned, mature leaves are responsible for providing expanding leaves with metabolites necessary for growth (Larson et al. 1980, Dickson and Larson 1981, Coleman 1986). Damage to a mature leaf would affect the capacity of that leaf to acquire resources and subsequently export them to expanding leaves. Thus, damage to mature leaves would be expected to decrease the amount of resources that expanding leaves have available to use during development, and might result in a situation similar to that reported in Wellings and Dixon (1987).

One would not expect that damage to expanding leaves would affect undamaged older leaves, because old leaves have completed development and receive very little, if any, metabolites from expanding leaves (Larson and Dickson 1973, Larson et al. 1980, Dickson and Larson 1981, Coleman 1986). Furthermore, mature leaves would be unlikely to exhibit substantial biochemical changes after damage, because these leaves do not usually show much plasticity to environmental conditions (Larcher 1980, Haukioja et al. 1985b, Coleman 1986). For example, damage to fully expanded birch leaves did not induce resistance in those mature leaves, but damage to expanding leaves induced resistance in those damaged young leaves (Haukioja and Niemela 1979, Wratten et al. 1984, Hartley and Lawton, this volume, Chapter 5).

Data regarding physiological leaf ontogeny can be used to design experimental protocols that will most likely uncover rapidly induced responses of plants to damage. We suspect that the strongest rapidly induced response of plants to damage will occur if (1) damage is applied to young leaves, and bioassays for induced resistance are conducted with those damaged leaves (but see Hartley and Lawton, this volume, Chapter 5, and Neuvonen and Haukioja, this volume, Chapter 12), (2) damage is applied to young leaves, and bioassays are conducted with young leaves that are subsequently produced, because damaged young leaves would eventually supply metabolites to new leaves, (3) damage is applied to older leaves, and bioassays are conducted with expanding leaves. It should be noted that these predictions are based on patterns of resource allocation in plants and may not shed light on whether induced responses of plants are active or passive.

Not all leaves on a plant shoot are equally connected by the plant vascular system (Larson and Dickson 1973, Watson and Casper 1984). This differential connectivity of leaves by the plant plumbing system (the plumbing principle) has important implications for induction studies because it suggests that leaves are differentially affected by changes in the physiology or biochemistry of other leaves. For example, Larson and Dickson (1973), Dickson and Larson (1981) and Larson (1980) mapped the entire vascular system of eastern cottonwood seedlings, and then used this map to predict the destination of radioactively labeled translocation products. They showed that the degree of vascular connectivity of one cottonwood leaf to another could be predicted from the phyllotaxy of a cottonwood plant. The greatest connectivity was between leaves on the same orthostichy (vertically aligned on the stem), with the weakest connectivity between adjacent leaves on the stem (those produced from consecutive nodes) (Larson and Dickson 1973, Larson 1980). They also showed that import of carbon by expanding leaves was compartmentalized within leaves. Different source leaves always supplied relatively restricted regions within specific sink leaves (Larson and Dickson 1973, Larson 1980, Dickson and Larson 1981).

It has been suggested that herbivore damage to leaves might result in a signal being translocated from damaged leaves to other leaves (Haukioja and Niemela 1979, Rhoades 1983a,b, Haukioja and Hahimaki 1985, Hartley and Lawton, this volume, Chapter 5). Consequently, a number of induction experiments have used bioassays of leaves adjacent to damaged leaves to test for induced resistance of plants to herbivory (Haukioja and Niemela 1979, Raupp and Denno 1983, Wratten et al. 1984, Haukioja and Hahimaki 1985, Myers and Williams 1987). It is clear from studies on eastern cottonwood and other plants that adjacent leaves would be least likely to receive a signal through the translocation stream because they are the most weakly connected by plant vasculature (Larson and Dickson 1973, Larson 1980, Larson et al. 1980, Dickson and Larson 1981). In cottonwoods, we would predict that translocated signals would most likely appear in expanding sink leaves on the same orthostichy (i.e., five leaves above the damaged leaf in the case of an eastern cottonwood seedling). The map of cottonwood vasculature would also enable one to predict the exact region within an expanding cottonwood leaf that would receive a signal via the vascular system from a damaged mature leaf (Larson and Dickson 1973, Larson 1980). Thus, given a knowledge of leaf ontogeny and vascular development, it should be possible to design experiments to investigate mechanisms by which induction signals are translocated within plants, and subsequently aid in resolving the issue of active versus passive responses.

Herbivores show different preferences for young and old leaf tissue. For example, Cates (1980) argued that specialist herbivores tend to feed on young leaf tissue because they can probably detoxify defenses in young tissues and effectively utilize the relatively high nitrogen levels in young leaves. In contrast, Cates (1980) argued that generalists are unable to detoxify defenses in young leaves, and therefore prefer older leaves. One experimental procedure in induction studies involves defoliating or clipping plants and subsequently bioassaying newly flushed plant tissues for resistance to herbivores. Pullin (1987) and Crawley and Nachapong (1984) examined the response of specialist herbivores in this type of experiment and found that these herbivores grew as well or better on regrowth plant tissue in comparison to controls. In contrast, studies examining the responses of generalist herbivores on newly flushed regrowth tissue have often found these tissues to be more resistant to herbivory than controls (Haukioja and Niemela 1979, Bryant 1981, 1987). It seems unreasonable to conclude that the resistance of regrowth young tissue to generalist herbivores is evidence for a defensive role of phytochemical induction. Rather, it seems more likely to be a predictable outcome of the inherent unsuitability of young plant tissue for many generalist herbivores (Cates 1980, Coleman 1986). If regrowth plant tissues were more resistant to specialist herbivores, or if regrowth is far more toxic to generalists than the initial plant tissues, then this might be construed as evidence for active production of defenses in newly flushed leaves as a response to defoliation.

3.3 Plant Phenology and Induction

The season of the year, age, and growth habits of plants are all important variables that can affect the expression and detection of induced resistance. Lindroth and Batzli (1986), for example, found a greater degree of variation in the production of phenolics by grassland plants over the growing season than they did between defoliation treatments. Regarding growth habit, Wallner and Walton (1979) and Valentine et al. (1983) showed that two trees with different patterns of leaf production (red and black oak and gray birch) exhibited different changes in their resistance to gypsy moth after defoliation. In this case, oak foliage exhibited greater decreases in the resource suitability of leaves following defoliation than did gray birch. A study exemplifying the importance of plant age showed that pine seedlings exhibited increased resistance to pine beauty moth damage one year after they had been damaged, but mature pine trees did not show this response (Leather et al. 1987).

Plant phenological information my be used to design induction studies with the greatest chance of success in detecting induced responses. Indeterminate trees and tree seedlings of determinate species allocate photosynthetic resources to growth of shoots during a given season, whereas determinate trees allocate photosynthate from a given year to growth of shoots the following year (Spurr and Barnes 1980). Since free-growing trees are continuously allocating resources to production of new leaves, damage to leaves this year might invoke a rapid response that is detectable in newly produced leaves.

On the other hand, growth habits of fixed-growing trees would make detection of rapid induced responses more unlikely, because resources may not be allocated for leaf production until the following growing season (Spurr and Barnes 1980). Fox and Bryant (1984) showed that clipped indeterminate plants (poplar, alder, birch) responded with production of juvenile tissue which was unpalatable to hares, but fixed-growing plants (larch, spruce) did not respond in this manner. However, their study examined the regrowth characteristics of woody plant tissues and may not be readily comparable to studies that examine the reflushing of leaf tissue.

Finally, plants usually allocate resources to production of shoot tissues early in the spring (Kozlowski 1971b, Pallardy and Kozlowski 1979, Larcher 1980), so this would be the most likely time that damage to foliage would induce a rapid response in growing shoots. This is consistent with findings that induced resistance of birch to herbivores was most detectable after damage to leaves in the spring, but damage to leaves later in the season produced less pronounced changes in resistance (Haukioja and Niemela 1979, Wratten et al. 1984, Hartley and Lawton, this volume, Chapter 5).

The lack of attention to plant phenological variables makes interpretation of some induction studies difficult. Seventeen of the studies that we reviewed examined effects of leaf damage to birch trees (*Betula pubescens*, *B. pendula*, *B. populifolia*, *B. papyrifera*) on subsequent resistance of leaves or tissues to herbivory (Table 1). Birch trees present a particularly difficult problem in these kinds of studies because they produce two kinds of shoots. Long shoots resemble free-growing plants because they continue to expand and produce leaves throughout the growing season (Spurr and Barnes 1980). In contrast, short or dwarf shoots resemble determinate tissue because they produce leaves in a single flush in the spring and shoot expansion ends quite early in the growing season (Spurr and Barnes 1980). Long shoots may contain leaves of all ages, whereas short shoots will contain leaves of relatively the same age. Leaves from these different shoots would be expected to show different responses to damage of other leaves on the same or a different shoot because of their different patterns of resource allocation.

Most studies using birch leaves to assay induced responses have not explicitly stated whether leaf damage or leaf bioassays were conducted on leaves from short shoots or long shoots, nor did they explicitly state how leaf age was controlled for (12 reports in all, see Table 1; but also see Hartley and Lawton, this volume, Chapter 5, and Neuvonen and Haukioja, this volume, Chapter 12). Thus, it is difficult to make comparisons between these studies. Nevertheless, we think that knowledge regarding branch phenology of birch trees and the responses of long and short shoots to damage, coupled with the responses of herbivores to damage of these shoots, might be useful for understanding the processes of resource acquisition and allocation in birch, and the significance of induction.

3.4 Abiotic Environment and Induction

Experimental plants or plant tissues growing under different conditions may be physiologically and biochemically distinct, and their responses to damage of leaf or shoot tissues could be different. For example, Waring et al. (1985) and Larsson et al. (1986) showed that identical genotypes of willow trees had very different biochemical and physiological characteristics after growing in environments with different availabilities of light and nutrients. Furthermore, it was suggested that nutrient status and resource availability of plants may largely determine their abilities to exhibit induced responses to herbivores (Bryant et al. 1983, Tuomi et al. 1984, Bryant 1987). Bryant (1987) showed that clipped and shaded willow plants produced regrowth stem tissues with lower concentrations of secondary compounds than did clipped and unshaded trees, presumably because shading reduced carbon allocation to those chemicals.

Failure to consider differences in abiotic environmental conditions may make it difficult to detect induced resistance to herbivory. Neuvonen and Danell (1987) showed that variation in leaf biochemistry and resistance of birch leaves to *E. autumnata* was greater between leaves in the upper crown versus the lower crown than variation between leaves from damage treatment or control groups. Although they did not speculate on the reasons for differences between leaves in different crown positions, the differences are probably related to the variation in light availability or are simply a function of leaf age or shoot type. Inconsistency of data within and between field studies on phytochemical induction by herbivores may well be due to abiotic factors. Most studies do not report the present or past nutrient characteristics of soils, light availability, water availability, or atmospheric pollution levels for plants that are assayed (Table 1). Since plants have extensive plasticity in their physiological and biochemical responses to environmental conditions, it is almost impossible to accurately compare plants that are growing, or have grown, under even slightly different abiotic conditions.

A knowledge of plant responses to abiotic factors under controlled environmental conditions may be useful in interpreting the ecological significance of phytochemical induction. For example, Tuomi et al. (1984) hypothesized that responses of herbivores on birch trees from Scandinavian forests was a consequence of passive deterioration of food quality due to damage. These birches grow in particularly poor soils and defoliation may result in a depletion of available nitrogen for subsequent plant growth. They suggested that fertilization of birch trees would prevent this passive deterioration in foliage quality, and should subsequently reduce the degree of apparent induced resistance of trees to herbivores. Haukioja and Neuvonen (1985) tested this hypothesis by fertilizing birch trees, exposing them to damage, and measuring their subsequent suitability for the development of *E. autumnata* larvae. Larvae actually developed better on unfertilized trees than on fertilized trees. This result was presented by Haukioja and Hahimaki

(1985) as strong evidence against the passive deterioration hypothesis and they further suggested that it indicated a defensive role for induction.

We feel that using knowledge regarding the response of plants to abiotic factors coupled with their response to defoliation may be a useful approach for designing future experiments to examine the defensive role of phytochemical induction. It has been observed numerous times that phenolic levels of leaves are reduced when plants are shaded (Bryant et al. 1983, 1987, Waring et al. 1985, Larsson et al. 1986, Bryant 1987, Mole et al. 1988, Mole and Waterman 1988), and this may be due to a reduction in the amount of carbon available to plants (Bryant et al. 1983, 1987, Mole et al. 1988). If plants are shaded and then defoliated, and still show increases in phenolics relative to unshaded plants, this would be evidence that allocation of carbon to phenolics in damaged plants is a priority plant process and might subsequently be considered a defensive response. Thus, if plant biochemical responses to abiotic factors occur in a manner that is not predicted by their known responses to abiotic environmental conditions, that would provide evidence suggesting other ecological functions for the response, such as defense.

The use of bags to protect foliage or to cage herbivores can also affect plant biochemistry. Leaves that are bagged are necessarily shaded and should have reduced levels of phenolics. Comparison of bagged versus unbagged leaves might only reveal differences relating to environmental effects. Comparison of damaged bagged leaves versus undamaged bagged leaves might actually reveal differences in leaf biochemistry that are due to interactions of biotic (herbivore damage) and abiotic environmental (shade) effects and not due to herbivory alone.

3.5   Biotic Environment and Induction

Although many authors of induction studies reported that they controlled the amount of previous herbivore damage to plants, not a single report stated whether plants were previously free from pathogen infection or had microbial symbioses with mychorrizal fungi or nitrogen-fixing bacteria. Yet, it is essential that we understand the biochemical responses of plants to biotic factors other than insects if the question of specificity of induced responses is to be resolved. It is well known that pathogens and symbionts can induce marked changes in plant biochemistry, nutrient status, and physiology that affect subsequent resistance to herbivores (Chapin 1980, Cruickshank 1980, Larcher 1980, Ajayi and Dewar 1982, Ellingboe 1982, Jones 1984, Kuc and Presig 1984, Karban et al. 1987) and some degree of pathogen infection or plant–microbe symbioses are extremely common in natural plant populations (Larcher 1980, Dinoor and Eshed 1984). Furthermore, herbivores may transmit plant pathogens to plants (Carter 1973, Maramorosch and Harris 1979, Jones 1984), and leaf damage may expose plant tissues to pathogen infection (Dodman 1979). Exposing plants to herbivores or leaf damage may simultaneously expose them to pathogens, and it is difficult to determine whether pathogens or herbivores are responsible for induced responses of plants unless this is explicitly studied. Minimizing differences in the degrees of pathogen infection and symbiotic interactions will certainly reduce variation in data. Additionally, examination of induced responses of plants to many biotic factors will most certainly aid in resolving whether a given induced response is specific to a given set of organisms.

Most plants are utilized by a whole community of consumers. If defoliation of plants induces resistance to one herbivore, but results in susceptibility to another herbivore or pathogen, it becomes problematical to argue that the induced response is defensive. For example, Fowler and MacGarvin (1986) showed that mechanical damage to expanding birch leaves resulted in reduced suitability of foliage for *Apochemia pilosaria*, but damaged plants received as much damage by the herbivore community as control plants. Therefore, induced resistance of birch to *A. pilosaria* did not result in an overall decrease in damage to plants. Furthermore, Hartley and Lawton (1987, this volume, Chapter 5) found that damaged birch leaves were actually more palatable to some birch-feeding caterpillars. Thus, the question of whether induced responses of plants to herbivores evolved as defenses will have to be answered for communities of herbivores and pathogens on plants.

4 SUMMARY

Since phytochemical induction by herbivores is a response of plants to herbivory or tissue damage, we have presented a phytocentric framework that explicitly examines how plant genotype, ontogeny, phenology, and interactions of plants with the abiotic and biotic environment may affect induced phytochemical responses. Specifically, we depict how these factors influence plant resource acquisition, allocation, and partitioning, and how this influences the resource quality and suitability of plant tissues for herbivores. This framework can be used as a guide by researchers in designing induction experiments, and may help researchers reduce variation in data, answer some important unresolved questions regarding the ecological significance of phytochemical induction, and aid in predicting the types and ages of plants and plant tissues that are most likely to show rapid induced responses.

Accounting for plant genotype in phytochemical induction experiments can reduce variation in data by controlling for differences in the inducibility of genotypes, variation in the degree or type of induced response that different genotypes might show, masking of induced resistance by the strong constitutive resistance of some genotypes, and variation in genotypic constitutive resistance between individuals within an experimental population. Furthermore, differences in the inducibility of plant genotypes can also be used to test hypotheses on the role of phytochemical induction in plant defense and the metabolic cost of plant defenses.

Controlling for the age of leaves that are damaged and bioassayed may reduce variation in data by accounting for differences in the inducibility of different aged leaves and the leaf age preferences of different herbivores. Knowledge of leaf ontogeny may also be used to design experiments to examine how translocation of induction signals from damaged to undamaged leaves takes place, and thus provide information on whether induced responses are active defenses or passive resource deterioration. Based on knowledge of leaf ontogeny, we predict that the following damage and bioassay schemes will most likely produce rapid induced responses in plants:

1. Damage to young leaves and bioassays of damaged leaves
2. Damage to young leaves and bioassays of subsequently produced leaves
3. Damage to mature leaves and bioassays of expanding leaves

Accounting for stages of plant development, seasonal effects, growth habits of plants, and the type of shoot (long or short) used in bioassays can reduce variation in data from induction studies, because these factors may exert strong effects on the ability of plants to exhibit induced responses. Based on our knowledge of how plant phenology relates to patterns of plant resource acquisition, allocation, and partitioning, we predict the following:

1. Free-growing plants and seedlings of both fixed and free-growing species will most likely show rapid induced responses to tissue damage.
2. Rapid induced responses of plants are most likely to occur early in the growing season.

Plants that have grown, or are growing, under different abiotic conditions can have very different physiological, biochemical, and anatomical characteristics. Thus, variation in data on induced resistance might be large if damage treatment and control plants have experienced even slightly different abiotic environments. Additionally, plants show predictable physiological responses to certain environmental stresses. Knowledge of these responses allows researchers to test whether phytochemical responses to a given regime of herbivory and environmental stress are short-term physiological adjustments to stress or are likely to be evolved defenses.

Plant pathogens and symbionts can affect phytochemistry. Thus, variation in data can be significantly reduced if these factors are controlled in induction studies. Knowledge regarding the community of consumers that feed on a plant can be used to test whether induced resistance of that plant is a general defense against all consumers, induced resistance is specific to only one consumer and does not affect other consumers, or if induced resistance to one consumer confers susceptibility of the plant to other consumers.

ACKNOWLEDGMENTS

The authors thank Hal Mooney and Fakhri Bazzaz for their patience and support of JSC during the preparation of this manuscript, and Susan Harrison, Steward Pickett, Juli Armstrong, and Barbara Gartner for ideas and criticisms. Concepts presented here were developed during research funded by NSF (BSR-8516679; CGJ, W. H. Smith; 88–17519, CGJ); Mary Flagler Cary Charitable Trust (CGJ, JSC); Electrical Power Research Institute (via Hal Mooney); Andrew W. Mellon Foundation through F. Herbert Bormann (JSC). Contribution to the program Institute of Ecosystem Studies, The New York Botanical Garden.

REFERENCES

Adams, R. P., and A. Haganon. 1977. Diurnal variation in the volatile terpenoids of *Juniperus scopulatum* (Cupressaceae). *Am. J. Bot.* 64:278–285.

Ajayi, O., and M. Dewar. 1982. The effect of barley yellow dwarf virus on honey dew production by cereal aphids, *Sitobion avenae* and *Metapolphium dirhordium*. *Ann. Appl. Biol.* 100:203–212.

Amthor, J. S., and J. R. Cumming. 1988. Low levels of ozone increase bean leaf maintenance respiration. *Can. J. Bot.* 66:724–726.

Andersen, D. A. 1987. Below-ground herbivory in natural communities: a review emphasizing fossorial animals. *Quart. Rev. Biol.* 62:261–286.

Ayres, P. G. 1982. Water stress modifies the influence of powdery mildew on root growth and assimilation import of barley. *Physiol. Plant Pathol.* 21:283–293.

Baldwin, I. T., and J. C. Schultz. 1985. Rapid changes in tree leaf chemistry induced by damage: evidence for communication between plants. *Science* 221:277–278.

Barlow, N. D., and A. F. G. Dixon. 1980. *Simulation of Lime Aphid Population Dynamics*. Wageningen, The Netherlands: Centre for Agricultural Publishing and Documentation.

Bassman, J. H., and D. Dickmann. 1982. Effects of defoliation in the developing leaf zone of young *Populus × Euramericana* plants. II. Translocation of ^{14}C assimilated carbon. *Forest Sci.* 28:599–612.

Bassman, J. H., and D. Dickmann. 1985. Effects of defoliation in the developing leaf zone on young *Populus × Euramericana* plants. III. Distribution of ^{14}C photosynthate after defoliation. *Forest Sci.* 31:358–366.

Bazzaz, F. A. 1979. The physiological ecology of plant succession. *Annu. Rev. Ecol. Syst.* 10:351–372.

Bazzaz, F. A., N. R. Chiariello, P. D. Coley, and L. F. Pitelka. 1987. Allocating resources to reproduction and defense. *BioScience* 37:58–67.

Bell, A. A. 1984. Morphology, chemistry and genetics of *Gossypium* adaptations to pests. *Rec. Adv. Phytochem.* 18:197–230.

Berenbaum, M. 1981. Patterns of furanocoumarin distribution and insect herbivory in the Umbelliferae: plant chemistry and community structure. *Ecology* 62:1254–1266.

Bergelson, J., S. Fowler, and S. Hartley. 1986. The effects of foliage damage on casebearing moth larvae, *Coleophora seratella*, feeding on birch. *Ecol. Entomol.* 11:241–250.

Bernays, E. A., and S. Woodhead. 1982. Plant phenols utilized as nutrients by a phytophagous insect. *Science* 216:201–203.

Bloom, A. J., F. S. Chapin III, and H. A. Mooney. 1985. Resource limitation in plants: an economic analogy. *Annu. Rev. Ecol. Syst.* 16:363–392.

Bradshaw, A. D. 1965. Evolutionary significance of phenotypic plasticity in plants. *Adv. Genet.* 13:115–155.

Brattsten, L. B. 1979. Biochemical defense mechanisms in herbivores against plant allelochemicals. In G. A. Rosenthal and D. H. Janzen (eds.), *Herbivores: Their Interactions with Secondary Plant Metabolites*. New York: Academic, pp. 199–270.

Browder, L. E. 1985. Parasite:host:environment specificity in the cereal rusts. *Annu. Rev. Phytopathol.* 25:210–222.

Bryant, J. P. 1981. Phytochemical deterrence of snowshoe hare browsing by adventitious shoots of four alaskan trees. *Science* 213:889–890.

Bryant, J. P. 1987. Feltleaf willow–snowshoe hare interactions: plant carbon/nutrient balance and floodplain succession. *Ecology* 68:1319–1327.

Bryant, J. P., F. S. Chapin III, and D. R. Klein. 1983. Carbon/nitrogen balance of boreal plants in relation to herbivory. *Oikos* 40:357–368.

Bryant, J. P., F. S. Chapin III, P. B. Reichardt, and T. P. Clausen. 1987. Response of winter chemical defense in Alaska paper birch and green alder to manipulation of plant carbon/nitrogen balance. *Oecologia* 72:510–514.

Caldwell, M. M. J. H. Richards, D. A. Johnson, R. S. Nowak, and R. S. Dzurec. 1979. Coping with herbivory: photosynthetic capacity and resource allocation in two semiarid *Agropyon* bunchgrasses. *Oecologia* 50:14–24.

Carroll, C. R., and C. A. Hoffman. 1980. Chemical feeding deterrent mobilized in response to insect herbivory and counter adaptation by *Epilachna tredecimonata*. *Science* 209:414–415.

Carter, W. 1973. *Insects in Relation to Plant Disease*. New York: Wiley.

Cates, R. G. 1980. Feeding patterns of monophagous, oligophagous and polyphagous insect herbivores: the effects of resource abundance and plant chemistry. *Oecologia* 46:22–31.

Chapin, F. S., III. 1980. The mineral nutrition of wild plants. *Annu. Rev. Ecol. Syst.* 11:233–260.

Chapin, F. S. III, A. J. Bloom, C. B. Field, and R. H. Waring. 1987. Plant responses to multiple environmental factors. *Bio Science* 37:49–57.

Chapin, F. S., III, and M. Slack. 1979. Effects of defoliation upon root growth, phosphate absorption and respiration in nutrient limited tundra graminoids. *Oecologia* 42:76–89.

Clarkson, D. T. 1985. Factors affecting mineral nutrient acquisition by plants. *Annu. Rev. Plant Physiol.* 36:77–115.

Coffey, M. D., and W. E. Wilson. 1983. An ultrastructural study of the late blight fungus *Phytophora infestans* and its interactions with the foliage of two potato cultivars possessing different levels of general (field) resistance. *Can. J. Bot.* 61:2669–2685.

Coleman, J. S. 1986. Leaf development and leaf stress: increased susceptibility associated with sink-source transition. *Tree Physiol.* 2:289–299.

Coleman, J. S., C. G. Jones, and W. H. Smith. 1987. The effect of ozone on cottonwood—leaf rust interactions: independence of abiotic stress, genotype and leaf ontogeny. *Can. J. Bot.* 65:949–953.

Crawley, M. J. 1985. Reduction in oak fecundity by low density insect populations. *Nature* 314:163–164.

Crawley, M. J., and M. Nachapong. 1984. Facultative defenses and specialist herbivores? Cinnebar moth (*Tyria jacobaeae*) on the regrowth foliage of ragwort (*Senecio jacobaea*). *Ecol. Entomol.* 9:389–393.

Cruickshank, I. A. M. 1980. Defenses triggered by the invader. In J. Horsfall and E. B. Cowling (eds.), *Plant Disease,* Vol. 5. New York: Academic, pp. 247–255.

Danell, K., and K. Huss-Danell. 1985. Feeding by insects and hares on birches earlier affected by moose browsing. *Oikos* 44:75–81.

Detheir, V. G. 1980. Evolution of receptor sensitivity to secondary plant substances with special reference to deterrents. *Am. Natur.* 115:45–66.

Dickson, R. E. 1987. Diurnal changes in leaf chemical constituents and ^{14}C partitioning in cottonwood. *Tree Physiol.* 3:157–172.

Dickson, R. E., and P. R. Larson. 1981. ^{14}C fixation, metabolic labeling patterns and translocation profiles during leaf development in *Populus deltoides*. *Planta* 152:461–470.

Dinoor, A., and N. Eshed. 1984. The role of pathogens in natural plant communities. *Annu. Rev. Phytopathol.* 22:443–466.

Dixon, A. F. G., and N. D. Barlow. 1979. Population regulation in the lime aphid. *Zool. J. Linn. Soc.* 67:225–237.

Dodman, R. L. 1979. How the defenses are breached. In J. Horsfall and E. B. Cowling (eds.), *Plant Disease,* Vol. 4. New York: Academic.

Edwards, P. J., and S. D. Wratten. 1982. Wound-induced changes in palatability of birch (*Betula pubescens* Ehrh. ssp. *pubescens*). *Am. Natur.* 119:816–818.

Edwards, P. J., S. D. Wratten, and H. Cox. 1986. Wound-induced changes in the acceptability of tomato to larvae of *Spodoptera litteralis*. A laboratory bioassay. *Ecol. Entomol.* 10:155–158.

Ellingboe, A. H. 1982. Genetic aspects of active defense. In R. K. S. Wood (ed.), *Active Defense Mechanisms in Plants*. New York: Plenum, pp. 179–192.

Feeny, P. P. 1976. Plant apparency and chemical defense. *Rec. Adv. Phytochem.* 10:1–40.

Fisher, M. 1987. The effect of previously infested spruce needles on the growth of the green spruce aphid *Elatobium abietinum* and the effect of the aphid on the amino acid balance of the host plants. *Ann. Appl. Biol.* 111:33–41.

Fowler, S. V., and J. H. Lawton. 1985. Rapidly induced defenses and talking trees: the devil's advocate position. *Am. Natur.* 126:181–195.

Fowler, S. V., and M. MacGarvin. 1986. The effects of leaf damage on the performance of insect herbivores on birch, *Betula pubescens*. *J. Anim. Ecol.* 55:565–573.

Fox, J. F., and J. P. Bryant. 1984. Instability of the snowshoe hare–woody plant interaction. *Oecologia* 63:128–135.

Fox, L. R., and B. J. McCauley. 1977. Insect grazing on *Eucalyptus* in response to variation in leaf tannins and nitrogen. *Oecologia* 29:145–162.

Geiger, D. R. 1986. Processes affecting carbon allocation and partitioning among sinks. In J. Cronshaw, W. J. Lucas, and R. T. Giaquinta (eds.), *Phloem Transport*. New York: Liss, pp. 375–378.

Gershenzon, J. 1984. Changes in the levels of plant secondary metabolites under water stress and nutrient stress. *Rec. Adv. Phytochem.* 18:273–320.

Green, T. R., and C. A. Ryan. 1972. Wound induced proteinase inhibitors in plant leaves. *Science* 175:776–777.

Gupta, P. D., and A. J. Thorsteinson. 1960. Food plant relationships of the diamondback moth (*Plutella maculipennis*). *Entomol. Exp. Applic.* 3:241–250.

Harrison, S., and R. Karban. 1986. Behavioral responses of spider mites (*Tetranychus urticae*) to induced resistance of cotton plants. *Ecol. Entomol.* 11:181–188.

Hartley, S. E., and J. H. Lawton. 1987. The effects of different types of damage on the chemistry of birch foliage and the responses of birch feeding insects. *Oecologia* 74:432–437.

Hartley, S. E., and J. H. Lawton. 1991. The biochemical aspects and significance of the rapidly induced accumulation of phenolics in birch foliage. In D. W. Tallamy and M. J. Raupp (eds.), *Phytochemical Induction by Herbivores*. New York: Wiley, pp. 105–132.

Hartnett, D. C., and F. A. Bazzaz. 1984. Leaf demography and plant–insect interactions: goldenrods and phloem-feeding aphids. *Am. Natur.* 124:137–142.

Haukioja. E. 1980. On the role of plant defenses in the fluctuations of herbivore populations. *Oikos* 35:202–213.

Haukioja, E., and S. Hahimaki. 1985. Rapid wound-induced resistance in white birch (*Betula pubescens*) foliage to the geometrid *Eppirita autumnata*: a comparison of trees within and outside the outbreak range of the moth. *Oecologia* 65:223–228.

Haukioja, E., and S. Neuvonen. 1985. Induced long-term resistance of birch foliage against defoliators: defense or incidental? *Ecology* 66:1303–1308.

Haukioja. E., and P. Niemela. 1977. Retarded growth of geometrid larva after mechanical damage to leaves of its host tree. *Ann. Zool. Fenn.* 14:48–52.

Haukioja, E., and P. Niemela. 1979. Birch leaves as a resource for herbivores: seasonal occurrence of increased resistance in foliage after mechanical damage of adjacent leaves. *Oecologia* 39:151–159.

Haukioja, E., P. Niemela, and S. Siren. 1985a. Foliage phenols and nitrogen in relation to growth, insect damage and ability to recover after defoliation in the mountain birch, *Betula pubescens* spp. *tortuosa*. *Oecologia* 63:214–222.

Haukioja. E., J. Suomela, and S. Neuvonen. 1985b. Long-term inducible resistance in birch foliage: triggering cues and efficacy on a defoliator. *Oecologia* 65:363–369.

Hawkins, C. D. B., M. J. Aston, and M. I. Whitecross. 1986a. Short term effects of infestation by two aphid species on plant growth and shoot respiration of three legumes. *Physiol. Plant* 68:329–334.

Hawkins, C. D. B., M. J. Aston, and M. I. Whitecross. 1986b. Interactions between aphid infestation and plant growth and uptake of nitrogen and phosphorus by three leguminous host plants. *Can. J. Bot.* 64:2362–2367.

Heath, M. C. 1982. Absence of active defense mechanisms in compatible host–pathogen interactions. In R. K. S. Wood (ed.). *Active Defense Mechanisms in Plants*. New York: Plenum, pp. 143–156.

Hunt, R., and A. D. Nicholls. 1986. Stress and coarse control of root-shoot partitioning in herbaceous plants. *Oikos* 47:149–158.

Ingham, R. E., and J. K. Detling. 1984. Plant–herbivore interactions in a North American mixed grass prairie. III. Soil nematode populations and root biomass on *Cynomys ludovicianus* colonies and adjacent uncolonized areas. *Oecologia* 63:307–313.

Isebrands, J. G., and P. R. Larson. 1973. Anatomical changes during leaf ontogeny in *Populus deltoides*. *Am. J. Bot.* 60:199–208.

Jepson, P. C. 1983. A controlled environment study of the effect of leaf physiological age on the movement of apterous *Myzus persicae* on sugar beet plants. *Ann. Appl. Biol.* 103:173–185.

Jermy, T. 1983. Multiplicity of insect anti-feedants in plants. In D. L. Whitehead and W. S. Bowers (eds.), *Natural Products for Innovative Pest Management*. Oxford: Pergammon, pp. 223–236.

Jermy, T., and A. Szentsi. 1978. The role of inhibitory stimuli in the choice of oviposition sties by phytophagous insects. *Entomol. Exp. Applic.* 24:458–471.

Jones, C. G. 1983. Phytochemical variation, colonization, and insect communities: the case of bracken fern (*Pteridium aquilinum*). In R. F. Denno and M. S. McClure (eds.), *Variable Plants and Herbivores in Natural and Managed Systems*. New York: Academic, pp. 513–538.

Jones, C. G. 1984. Microorganisms as mediators of plant resource exploitation by insect herbivores. In P. W. Price, C. N. Slobodchikoff, and W. S. Gaud (eds.), *A New Ecology: Novel Approaches to Interactive Systems*. New York: Wiley, pp. 51–84.

Karban, R. 1985. Resistance against spider mites in cotton induced by mechanical abrasions. *Entomol. Exp. Applic.* 37:137–141.

Karban, R. 1986. Induced resistance against spider mites in cotton: field verification. *Entomol. Exp. Applic.* 42:239–242.

Karban, R. 1987. Environmental conditions affecting the strength of induced resistance against mites in cotton. *Oecologia* 73:414–419.

Karban, R., and J. R. Carey. 1984. Induced resistance of cotton seedlings to mites. *Science* 223:53–54.

Karban, R., R. Adamchak, and W. C. Schnathorst. 1987. Induced resistance and interspecific competition between spider mites and a vascular wilt fungus. *Science* 235:678–680.

Kosuge, T. 1978. The capture and use of energy by diseased plants. In J. Horsfall and E. B. Cowling (eds.), *Plant Disease,* Vol. 3. New York: Academic, pp. 86–116.

Kozlowski, T. T. 1971a. *Growth and Development of Trees,* Vol. 2, *Cambial Growth, Root Growth, Reproductive Growth.* New York: Academic.

Kozlowski, T. T. 1971b. *Growth and Development of Trees,* Vol. 1, *Seed Germination, Ontogeny, and Shoot Growth.* New York: Academic.

Kuc, J., and C. Presig. 1984. Fungal mechanisms of disease resistance in plants. *Mycologia* 76:767–784.

Larcher, W. 1980. *Plant Physiological Ecology.* Berlin: Springer.

Larson, P. R. 1980. Interactions between phyllotaxis, leaf development, and the primary-secondary vascular transition in *Populus deltoides. Ann. Bot.* 46:757–769.

Larson, P. R., and R. E. Dickson. 1973. Distribution of imported ^{14}C in developing leaves of eastern cottonwood according to phyllotaxy. *Planta* 111:95–112.

Larson, P. R., J. G. Isebrands, and R. E. Dickson. 1980. Sink to source transition of *Populus* leaves. *Ber. Deutch Bot. Ges. Bd.* 93:79–87.

Larsson, S., A. Wiren, L. Lundgren, and T. Ericsson. 1986. Effects of light and nutrient stress of leaf phenolic chemistry in *Salix dasyclados* and susceptibility to *Galerucella lineola* (Coleoptera). *Oikos* 47:205–210.

Leather, S. R., A. D. Watt, and G. I. Forrest. 1987. Insect-induced chemical changes in young lodgepole pine (*Pinus contorta*). The effect of previous defoliation on oviposition, growth and survival of the pine beauty moth, *Panolis flammea. Ecol. Entomol.* 12:275–281.

Lechowicz, M. J. 1987. Resource allocation by plants under air pollution stress: implications for plant–pest–pathogen interactions. *Bot. Rev.* 53:281–300.

Lewis, A. C. 1984. Plant quality and grasshopper feeding: effects of sunflower condition on preference and performance in *Melanoplus differentialis. Ecology* 65:836–847.

Lindroth, R. L., and G. O. Batzli. 1986. Inducible plant chemical defenses: a cause of vole population cycles? *J. Anim. Ecol.* 55:431–439.

Maddox, D. G., and N. Cappuccino. 1986. Genetic determination of plant susceptibility to an herbivorous insect depends on environmental context. *Evolution* 40:863–866.

Maramorosch, K., and K. E. Harris (eds.). 1979. *Leafhopper Vectors and Plant Disease Agents.* New York: Academic.

Mattson, W. J. 1980. Herbivory in relation to plant nitrogen content. *Ann. Rev. Ecol. Syst.* 11:119–162.

Mattson, W. J., and R. A. Haack. 1987. The role of drought in the outbreaks of plant eating insects. *BioScience* 37:110–118.

McLaughlin, S. B., and D. S. Shriner. 1980. Allocation of resources to defense and repair. In J. G. Horsfall and E. B. Cowling (eds.), *Plant Disease,* Vol. 5. New York: Academic, pp. 407–431.

McNaughton, S. J. 1983. Compensatory plant growth as a response to herbivory. *Oikos* 40:329–336.

Mole, S., and P. G. Waterman. 1988. Light-induced variation in phenolic levels in foliage of rain-forest plants. II. Potential significance to herbivores. *J. Chem. Ecol.* 14:23–34.

Mole, S., J. A. M. Ross, and P. G. Waterman. 1988. Light-induced variation in phenolic levels in foliage of rain-forest plants. I. Chemical changes. *J. Chem. Ecol.* 14:1–22.

Mooney, H. A. 1972. The carbon balance of plants. *Annu. Rev. Ecol. Syst.* 3:315–346.

Mooney, H. A., and N. R. Chiariello. 1984. The study of plant function: the plant as a balanced system. In R. Dirzo and S. Sarukhan (eds.), *Perspectives on Plant Population Ecology.* Sunderland, MA: Sinauer, pp. 305–321.

Mooney, H. A., M. Kuppers, G. Koch, J. Gorham, C. Chu, and W. E. Winner. 1988. Compensating effects to growth of carbon partitioning changes in response to SO_2-induced photosynthetic reduction in radish. *Oecologia* 75:502–506.

Myers, J. H. 1987. Nutrient availability and the development of mechanical defenses in grazed plants: a new experimental approach to the optimal defense theory. *Oikos* 49:350–351.

Myers, J. H., and K. S. Williams. 1984. Does tent caterpillar attack reduce the food quality of red alder foliage? *Oecologia* 62:74–79.

Myers, J. H., and K. S. Williams. 1987. Lack of short or long term inducible defenses in the red alder–western tent caterpillar system. *Oikos* 48:73–78.

Nelson, N. D., D. Dickmann, and K. Gottschalk. 1982. Autumnal photosynthesis in short rotation intensively cultured *Populus* clones. *Photosynthetica* 16:321–333.

Neuvonen, S., and K. Danell. 1987. Does browsing modify the quality of birch foliage for *Eppirita autumnata* larvae? *Oikos* 49:156–160.

Neuvonen, S., and E. Haukioja. 1984. Low nutrient quality as defense against herbivores: induced responses in birch. *Oecologia* 63:71–74.

Neuvonen, S., and E. Haukioja. 1991. The effects of inducible responses in host foliage on birch feeding herbivores. In D. W. Tallamy and M. J. Raupp (eds.). *Phytochemical Induction by Herbivores.* New York: Wiley, pp. 277–288.

Norden, A. J., D. W. Gorbert, D. A. Knault, and F. G. Martin. 1986. Genotype × environment interactions in peanut multi-line populations. *Crop Sci.* 26:46–48.

Pallardy, S. G., and T. T. Kozlowski. 1979. Early shoot and root growth of *Populus* clones. *Silvae Gent.* 28:153–156.

Philips, W. M. 1976. The effects of leaf age on feeding 'preference' and egg laying in the Chyrsomelid beetle *Haltica lythri. Physiol. Entomol.* 1:223–226.

Potter, D. A., and T. W. Kimmerer. 1986. Seasonal allocation of a defense investment in *Ilex opaca* Aiton and constraints on a specialist leaf-miner. *Oecologia* 69:217–224.

Pullin, A. S. 1987. Changes in leaf quality following clipping and regrowth of *Urtica dioica* and consequences for a specialist herbivore, *Aglais urticae. Oikos* 49:39–45.

Raupp, M. 1985. Effects of leaf toughness on mandibular wear of the leaf beetle, *Plagiodera versicolora. Ecol. Entomol.* 10:73–79.

Raupp, M. J., and R. F. Denno. 1983. Leaf age as a predictor of herbivore distribution and abundance. In R. F. Denno and M. S. McClure (eds.), *Variable Plants and Herbivores in Natural and Managed Systems*. New York: Academic, pp. 91–124.

Raupp, M. J., and R. F. Denno. 1984. The suitability of damaged willow leaves as food for the leaf beetle, *Plagiodera versicolora*. *Ecol. Entomol.* 9:443–448.

Reich, P. B., and R. G. Amundson. 1985. Ambient levels of ozone reduce net photosynthesis in tree and crop species. *Science* 230:566–570.

Rhoades, D. F. 1979. Evolution of plant chemical defenses against herbivores. In G. A. Rosenthal and D. H. Janzen (eds.), *Herbivores: Their Interaction with Secondary Plant Metabolites*. New York: Academic, pp. 3–54.

Rhoades, D. F. 1983a. Responses of alder and willow to attack by tent caterpillars and fall webworms: evidence for pheromonal sensitivity of willows. In P. A. Hedin (ed.), *Plant Resistance to Insects*. Washington, DC: American Chemical Society, pp. 55–68.

Rhoades, D. F. 1983b. Herbivore population dynamics and plant chemistry. In R. F. Denno and M. S. Mcclure (eds.), *Variable Plants and Herbivores in Natural and Managed Systems*. New York: Academic, pp. 155–220.

Rhoades, D. F. 1985. Offensive–defensive interactions between herbivores and plants: their relevance in herbivore population dynamics and ecological theory. *Am. Natur.* 125:205–238.

Robinson, D. 1986. Compensatory changes in the partitioning of dry matter in relation to nitrogen uptake and optimal variations in growth. *Ann. Bot.* 58:841–848.

Rosenthal, G. A., D. H. Janzen, and D. L. Dahlman. 1977. Degradation and detoxification of canavanine by a specialized seed predator. *Science* 196:658–660.

Röttger, V., and F. Klingauf. 1976. Changes in sugar beet leaves caused by the beet fly, *Pegomya betae* (Muscidae:Anthomyidae). *Z. Angew. Ent.* 82:220–227.

Rufty, T. W., S. D. Raper, and W. A. Jackson. 1981. Nitrogen assimilation, root growth and whole-plant responses of soybean to root temperature, and to carbon dioxide and light in the aerial environment. *New Phytol.* 88:606–619.

Ryan, C. A., and T. R. Green. 1974. Proteinase inhibitors in natural plant protection. *Rec. Adv. Phytochem.* 8:123–140.

Schliting, C. D. 1986. The evolution of phenotypic plasticity. *Annu. Rev. Ecol. Syst.* 17:667–693.

Schultz, J. C., and I. T. Baldwin. 1983. Oak leaf quality declines in response to defoliation by gypsy moth larvae. *Science* 217:149–151.

Schultz, J. C., P. J. Nothnagle, and I. T. Baldwin. 1982. Seasonal and individual variation in leaf quality of two northern hardwood tree species. *Am. J. Bot.* 69:752–759.

Scriber, J. M., and F. Slansky. 1981. The nutritional ecology of immature insects. *Annu. Rev. Entomol.* 26:183–211.

Spurr, S. H., and B. V. Barnes. 1980. *Forest Ecology*, 3d ed. New York: Wiley.

Sutton, R. F. 1967. Influence of root pruning on height increment and root development of out-planted species. *Can. J. Bot.* 45:1671–1682.

Szaniawski, R. K. 1983. Adaptation and functional balance between shoot and root activity of sunflower plants grown at different temperatures. *Ann. Bot.* 51:453–459.

Szaniawski, R. K. 1987. Plant stress and homeostasis. *Plant Physiol. Biochem.* 25:63–72.

Tallamy, D. 1985. Squash beetle feeding behavior: an adaptation against induced cucurbit defenses. *Ecology* 66:1574–1579.

Tingey, W. M., and J. E. Laubengayer. 1981. Defense against the green peach aphid and potato leafhoppers by glandular trichomes of *Solanum berthaultii*. *J. Econ. Entomol.* 74:721–725.

Treharne, K. J., and C. F. Eagles. 1970. Effect of temperature on photosynthetic activity of climatic races of *Dactylis glomerata* L. *Photosynthetica* 4:107–117.

Tuomi, J., P. Niemela, E. Haukioja, S. Siren, and S. Neuvonen. 1984. Nutrient stress: an explanation for plant anti-herbivore responses to defoliation. *Oecologia* 61:208–210.

Valentine, H. T., W. E. Wallner, and P. M. Wargo. 1983. Nutritional changes in host foliage during and after defoliation, and the relation to the weight of gypsy moth pupae. *Oecologia* 57:298–302.

Wallner, W. E., and G. S. Walton. 1979. Host defoliation: a possible determinant of gypsy moth population quality. *Ann. Entomol. Soc. Am.* 72:62–67.

Waring, R. H., and G. B. Pitman. 1985. Modifying lodgepole pine stands to change susceptibility to mountain pine beetles. *Ecology* 66:889–897.

Waring, R. H., and W. H. Schlessinger. 1986. *Forest Ecosystems: Concepts and management*. Orlando, FL: Academic.

Waring, R. H., A. J. S. McDonald, S. Larsson, T. Ericsson, A. Wiren, E. Arnidsson, A. Ericsson, and T. Lohammar. 1985. Differences in chemical composition of plants grown at constant relative growth rates with stable mineral nutrition. *Oecologia* 66:157–160.

Watson, M. A., and B. B. Casper. 1984. Morphogenetic constraints on patterns of carbon distribution in plants. *Annu. Rev. Ecol.* 15:233–259.

Wellings, P. W., and A. F. G. Dixon. 1987. Sycamore aphid numbers and population density. III. The role of aphid induced change in plant quality. *J. Anim. Ecol.* 56:161–170.

West, C. 1985. Factors underlying the late seasonal appearance of the lepidopteran leaf-mining guild on oak. *Ecol. Entomol.* 10:111–120.

White, T. C. R. 1984. The abundance of invertebrate herbivores in relation to the availability of nitrogen in stressed food plants. *Oecologia* 63:90–105.

Williams, K. S., and J. H. Myers. 1984. Previous herbivore attack of red alder may improve food quality for fall webworm larvae. *Oecologia* 63:166–170.

Wood, B. W., W. L. Tedders, and T. R. Thompson. 1985. Feeding influence of three pecan aphid species on carbon exchange and phloem integrity of seedling pecan foliage. *J. Am. Soc. Hortic. Sci.* 110:393–396.

Wratten, S. D., P. J. Edwards, and I. Dunn. 1984. Wound-induced changes in the palatability of *Betula pubescens* and *B. pendula*. *Oecologia* 61:372–375.

Zima, J., and Z. Sestak. 1979. Photosynthetic characteristics during ontogenesis of leaves. 4. Carbon fixation pathways, their enzymes and products. *Photosynthetica* 13:83–106.

CHAPTER 2

DAMAGE-INDUCED ALKALOIDS IN WILD TOBACCO

IAN T. BALDWIN
Department of Biological Sciences, State University of New York, Buffalo,
New York 14226

1   Introduction
2   The mechanisms of induction
    2.1   Sites of alkaloid synthesis
    2.2   Damage cues
    2.3   Alkaloidal responses to real and simulated herbivory
    2.4   Environmental and physiological constraints
3   Effects on insect herbivores
4   Phenotypic costs of inducible alkaloidal defenses
    4.1   Experimental measures of cost
    4.2   Why may induced alkaloid production be costly?
5   Conclusion
    References

1   INTRODUCTION

The Solanaceae is a family with a proclivity for producing alkaloids that interfere with the nervous system of many animals. So specific are these compounds that neurobiologists use the alkaloids of *Datura* and *Nicotiana* to classify cholinergic receptors as muscarinic and nicotinic, respectively. Few other plant defense compounds have received as much attention as the pyridine-based alkaloids of *Nicotiana*, probably as a result of human infatuation with the pleasurable, addictive, and deadly (James 1971) qualities of nicotine.

Pyridine-containing alkaloids occur in many members of the genus *Nicotiana* at levels that are toxic to most herbivores (Gordon 1961, Hassal 1969, Schmetz 1971, Matsumura 1976). These alkaloids constituted one of the commonly used insecticides before the development of synthetic pesticides such as DDT. Clearly, alkaloids function as a constitutive plant defense, presenting a formidable barrier to phytophagy. In addition to func-

47

tioning as a qualitative defense (sensu Feeny 1976), alkaloids may be employed as a quantitative defense by the plant; when present in high concentrations, they are deleterious to insects that customarily feed on alkaloid-containing plants (Parr and Thurston 1972). Scientists have long known that damage to the flowering top of a tobacco plant increases its total alkaloid content (Reuter 1957). Decapitation or topping at the onset of flowering is standard practice in the production of cultivated tobacco: it increases the size, weight, and alkaloid content of leaves (Woetz 1955). Surprisingly, the ecological significance of this chemical response and the quantitative use of constitutive alkaloidal defenses have remained unexplored.

Here I examine substantial increases in the alkaloid content of undamaged leaves of wild tobacco plants (*Nicotiana sylvestris*) which have undergone real and simulated herbivory, not to their flowering tops, or apical or lateral buds, but to their fully expanded leaves. Initially I examine the mechanism that effects changes in alkaloid titers. Supported by some of these mechanistic details, the examination will become more ecological and evolutionary in focus, addressing the environmental and physiological constraints on the response, the question of whether herbivores activate the response, the consequences for herbivore growth, and the phenotypic cost (as measured by the effect on seed set) associated with the alkaloidal response.

2   THE MECHANISMS OF INDUCTION

Field-grown plants subjected to a manual defoliation regime designed to mimic the rate and amount of leaf mass removed by one last-instar tobacco hornworm (*Manduca sexta*) per plant undergo dramatic and rapid (less than 8-day) increases in total alkaloids (438% for leaves at stalk position 5, 319% for leaves at stalk position 8, Fig. 1) when compared to leaves of the same age and position and on undamaged plants (Baldwin 1988a). Despite numerous hypotheses to explain why plants have such responses to damage, we understand little about their mechanisms. By understanding these mechanisms, we will better appreciate the patterns of plant defense and, more importantly, the responsiveness of these mechanisms to selection. In this section I examine the sites of alkaloid synthesis, the types of damage cues, the responses of plants to real and simulated herbivory, and the environmental and physiological constraints on this damage-induced response.

2.1   Sites of Alkaloid Synthesis

The well-documented increase in alkaloid content after topping in cultivated tobacco (*N. tabacum*) is largely due to an increase in nicotine synthesis in the roots (Mizusaki et al. 1973). Alkaloids are then carried to the leaves via the xylem (Dawson 1941). However, nicotine synthesis in the leaves (Bose et al. 1956) and decreased rates of leaf nicotine degradation (Yoshida 1962)

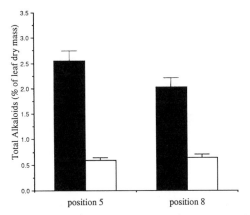

Figure 1   Total alkaloid contents (mean % dry mass ± SE) of leaves from stalk positions 5 and 8 from 20 damaged (dark bars) and undamaged (white bars) paired, full-sib *Nicotiana sylvestris* plants growing in field plots.

may also play a role in the increase in leaf alkaloid titers induced by topping. The extensive literature on cultivated tobacco suggests four hypotheses to explain alkaloid increases in wild tobacco after damage:

1. Lower alkaloid degradation rates in leaves
2. Higher leaf alkaloid synthesis rates
3. Higher transpiration rates allowing for increased transport of nicotine from roots to leaves
4. Increased amounts of alkaloid entering the leaf via the xylem.

The latter could be a result of higher rates of alkaloid synthesis in those roots with xylem traces leading to undamaged leaves, and/or the reloading of alkaloids into these traces. Both of these mechanisms are likely to be energy-intensive. Two of these mechanisms (1 and 3) could conceivably operate at no cost to the plant (see section 4); hence, this putative induced response may be no more than a passive reorganization of a constitutive defense (Baldwin 1989).

Hypothesis 4 appears to best explain how the increase in leaf alkaloids comes about (Baldwin 1989). Consecutive alkaloid samplings of detached leaves indicate that neither the rates of alkaloid degradation nor those of leaf alkaloid synthesis differ between damaged and undamaged plants. Leaf transpiration rates do not differ after damage. However, when xylem fluid is collected with a Scholander pressure chamber and microcapillaries are inserted into the xylem bundles of the petioles and analyzed for alkaloid content, a dramatic difference between damaged and undamaged plants is observed. Xylem fluid entering the leaves of damaged plants has 10 times

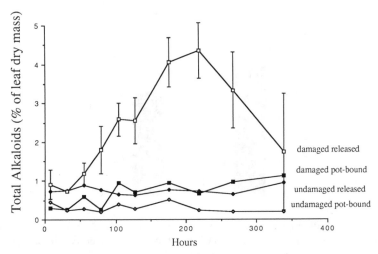

Figure 2   Total alkaloid contents (mean % dry mass ± SE) of leaves from stalk position 5 from 60 damaged and 60 undamaged *Nicotiana sylvestris* plants growing in large pots (released) and in small pots (pot-bound). Plants were randomly sampled, 3 per treatment per time period, over 338 h from the start of leaf damage. The damage regime lasted from 0 to 60 h.

the alkaloid concentration of that collected from undamaged plants. Given that decapitation of cultivated tobacco dramatically increases root alkaloid synthesis (Mizusaki et al. 1973, Saunders and Bush 1979), it is plausible that the alkaloidal response to leaf damage would also include an increase in root alkaloid synthesis.

The hypothesis that leaf damage effects changes in root alkaloid production is also suggested by the observation that pot-bound plants exhibit highly attenuated alkaloidal responses to leaf damage. Greenhouse-grown plants will typically show little change in leaf alkaloid concentrations in response to leaf damage. However, transplanting these plants into pots that contain more soil makes them responsive to damage (Baldwin 1988a). Figure 2 shows the time course of the alkaloidal response to damage as examined in both released and pot-bound greenhouse plants. Plants were grown in small pots, removed, and then either returned to their original pots or transferred to larger pots at the beginning of the experiment. Damage to plants (as described in Baldwin 1989) over a 60-h period removed 38% and 46% of shoot mass (representing a majority of the plant's leaf area) from plants in large and small pots, respectively. Damaged plants in large pots exhibited a four-fold increase in the alkaloid titers of their undamaged leaves as compared with the alkaloid titers of plants in smaller pots which did not respond significantly to damage. The increase began 3 days after the start of damage, reached a maximum at 9 days, and waned to control levels by 14 days (Fig. 2; Baldwin 1989). Although the rate of increase in leaf alkaloid production after damage to greenhouse-grown plants is consistent with the results from field-grown plants, the waning of the response (Fig. 2) may not be. By the end of the experiment, the plants had resided in their larger pots for 20 days since the transplanting and may have again become pot-bound. These results indicate that root stress will strongly influence a plant's alkaloidal response to damage and are consistent with the hypothesis that leaf damage effects changes in root alkaloid production.

2.2 Damage Cues

If root alkaloid synthesis is stimulated by leaf damage, how do the roots know that the leaves have been damaged? I have attempted to distinguish between (1) a positive cue produced by, or correlated with, leaf damage that stimulates root alkaloid synthesis, and (2) a negative one, a continuous cue produced by leaves, the reduction of which by leaf loss initiates the alkaloidal response. One candidate for such a negative cue is the plant hormone, auxin. Topping, which increases the alkaloid content of cultivated tobacco, also removes the apical meristem, an important source of auxin. Young expanding leaves in *N. sylvestris* are known to be sources of diffusible auxin, but older, fully expanded leaves are not (Avery 1935). Moreover, when auxins are applied to excised root cultures (Solt 1957), callus culture (Tabata et al. 1971), or leaves of intact plants (Yasumatsu 1967), nicotine contents of cultivated tobacco are reduced. Exogenously applied auxin can either inhibit or stimulate nicotine synthesis, depending on its concentration (Mizusaki et al. 1973). Consistent with the potential role of auxin as a negative damage cue is the observation that a thin layer of either natural (indole-3-acetic acid) or synthetic (naphth-1-acetic acid) auxin in lanolin base applied to the damaged leaf edge in *N. sylvestris* inhibits the induced alkaloidal response (Fig. 3; Baldwin 1989).

Although exogenously applied auxins clearly have a pronounced effect on alkaloid biosynthesis, a change in endogenous auxin concentration is probably not functioning as a negative cue that activates the response to damage. If this were the case, blocking the transport of this hormone from sites of damage (leaves) to sites of alkaloid synthesis (roots) should mimic leaf damage and cause induction. However, attempts to block the transport of endogenous auxin from the leaves to the roots with an auxin transport inhibitor (TIBA) or steam girdling failed to mimic leaf damage (Baldwin 1989). Moreover, when ethylene, which inhibits auxin transport (Audus 1972), is applied to the foliage of cultivated tobacco (Cutler and Gaines 1971, Kasperbauer and Hamilton 1978) or to *N. sylvestris* (Baldwin, unpublished results), leaf alkaloid concentrations remain unchanged. Finally, results from steam-girdling experiments implicate the presence of a positive cue, produced by or correlated with leaf damage.

Steam girdling kills the phloem, which is the most likely route of transport for an endogenous damage cue; the xylem, the route of alkaloid transport

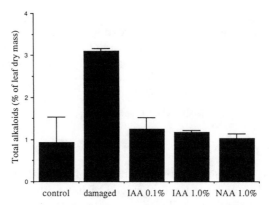

Figure 3   Total alkaloid contents (mean % dry mass ± SD) of leaves from stalk position 4 of 15 plants (3 per treatment) in the following 5 treatments 100 h after the start of leaf damage: control undamaged plants, and 4 damage treatments where 4 leaf positions were cut 5 times over an 80-h period. After each cutting the cut leaf edge was coated with either pure lanolin, or lanolin with napth-1-acetic acid or indole-3-acetic acid (% by mass).

from roots to leaves, remains intact. Transecting the phloem connections between damaged leaves and roots appears to inhibit the induced response. When bolting *N. sylvestris* plants are steam girdled at various positions on the stalk, leaf alkaloid titers are slightly elevated compared to those of un-girdled and undamaged plants. Ungirdled plants suffering only leaf damage and plants damaged below the girdle have high induced alkaloid titers. How-ever, when plants are girdled below the leaf damage, leaf alkaloid titers are no higher than those of girdled plants without leaf damage (Baldwin 1989). These results suggest that phloem-borne positive cues, perhaps similar to those that have been found responsible for the activation of proteinase in-hibitors (see Chapter 1), initiate the processes that cause the alkaloidal re-sponse. Moreover, consistent with the positive cue hypothesis is the ob-servation that the alkaloidal response varies according to the timing and/or the amount of leaf and vein damage, rather than the amount of leaf mass lost (section 2.3).

To examine the species specificity of the damage cue, I grafted jimson-weed (*Datura stramonium*) scions to the rootstocks of *N. sylvestris*. All plants with successful grafts (those in which the jimsonweed scion had grown and started to flower) were transplanted into larger pots. Six plants had approximately 50% of their leaves cut over a 4-day period, and another 6 were left undamaged. Leaf and nectar samples were collected from all plants 9 days after the end of the damage regime. Mean leaf nicotine titers were not significantly higher in damaged plants. Alkaloid titers in the nectar (Fig. 4) of 1-day-old *Datura* flowers were significantly elevated ($p = .011$) from damaged plants (Baldwin, unpublished results). These results suggest that

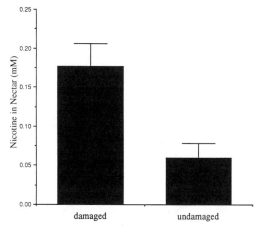

Figure 4 Nicotine concentration (mean mM ± SE) in the nectar of 1-day-old flowers from 6 undamaged and 6 damaged *Datura stramonium* scions grafted to *Nicotiana sylvestris* rootstocks.

the damage cues are constituents of both damaged jimsonweed and tobacco leaves. Similarly, the damage cues that induce proteinase inhibitors have been found in over 20 species of plants (Ryan and Green 1974), and these cues activate not only the induction of proteinase inhibitors but also that of monoterpenes in lodgepole pine (Miller et al. 1986).

2.3 Alkaloidal Responses to Real and Simulated Herbivory

The responses of *Nicotiana* to artificial damage are clear. Yet, to fully understand the potential ecological consequences of these responses, one must know how a plant responds to actual herbivory. To this end, the induced alkaloidal responses in undamaged leaves of plants subjected to herbivory by the larvae of the tobacco hornworm were compared with different simulations of this herbivory, all of which removed the same amount of leaf mass (Baldwin 1988c). Third-instar larvae were placed on plants and allowed to consume any leaf except a previously designated alkaloid sample leaf for approximately 7 h a day for 5 days. Every 30–45 min during the feeding period, each plant was examined for new damage, and the area of new damage was traced. With these tracings, a comparable amount of leaf area was cut every 30–45 min from two other sets of plants. On one set of plants the spatial array of feeding damage was mimicked (the simulated herbivory treatment) and on another set of plants, the tracings were used simply to remove identical leaf mass, without regard to the spatial position of the feeding damage (the pseudosimulated herbivory treatment). Thus, leaf damage mimicked either larval feeding or the timing and amount of leaf damage inflicted

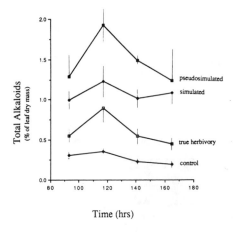

Time (hrs)

Figure 5 Total leaf alkaloid contents (mean % leaf dry mass ± SE) from undamaged stalk position 5 leaves of 20 control, true-damaged by caterpillars, simulated caterpillar-damaged, and pseudosimulated caterpillar-damaged *Nicotiana sylvestris* plants, were determined from 5 plants per sampling time over 165 h from the start of leaf damage. Figure is redrawn from Baldwin (1988c).

by larval feeding. Plants were then analyzed for alkaloid content at various times after the damage regime had ended (Fig. 5; Baldwin 1988c).

Plants on which larvae had fed, in contrast to undamaged plants, responded with a significant increase in alkaloid titers, but this response was significantly lower than that observed in plants damaged to mimic the larval feeding. Moreover, the pseudosimulated herbivory treatment produced the greatest alkaloidal response. Clearly, artificial damage was not able to mimic larval feeding damage. That plants respond less to caterpillar feeding than to mechanical damage is at least partly a result of the location of larval feeding sites.

There are two explanations for the differences in plant responses. First, the larvae could chemically interfere with the ability of the plants to recognize and respond to damage. Second, mechanical simulation may inadequately simulate the physical properties of herbivory. Larval feeding was not mimicked bite by bite, so clearly temporal and spatial differences existed between the real and the simulated herbivory. Cutting leaf tissues with scissors shears cells. This may result in the production of damage cues altogether different from those resulting from caterpillar mandibles.

One physical characteristic of larval feeding not exactly replicated in the simulation was the tendency of larvae to consume tissue around veins in a given area before consuming the veins. An extreme variation of this vein-avoidance behavior was attempted (Baldwin 1988c). Alkaloidal responses from plants with tissues removed between secondary veins were compared with those from plants damaged without regard to veins. Plants subjected to an interveinal damage regime had significantly lower alkaloid titers than did plants damaged without regard to veins (Baldwin 1988c). These experiments underscore the difficulty of simulating herbivory: A plant's response to damage is determined by how one removes leaf area, not merely by the amount of leaf area removed.

2.4   Environmental and Physiological Constraints

A number of environmental factors influence a plant's alkaloidal response to damage. Clearly, soil nitrogen availability is important (Tso 1972), for not only does nicotine contain nitrogen (17%), but the enzymes responsible for nicotine's synthesis do as well. Similarly, soil characteristics that hinder root growth will be influential. Less obvious is the role of water. Conditions of low evapotranspiration (high humidity and low temperature) will greatly slow the water loss from plants and consequently the transport of alkaloids to the leaves. Furthermore, the activities of three enzymes thought to be important in induced nicotine biosynthesis—ornithine decarboxylase, putrescine N-methyltransferase, and N-methylputrescine oxidase—which increase within 24 h of the topping of cultivated tobacco, are inhibited in vitro by nicotine (Mizusaki et al. 1973). Nicotine concentrations in the xylem fluid of damaged plants are in the range of 0.2–0.5 mM (Baldwin 1989), a concentration at which enzyme activity could easily be repressed by at least 50% (Mizusaki et al. 1973, Floss et al. 1974). These considerations suggest that conditions of low evapotranspiration not only could slow but may even truncate the induced response. These predictions were borne out in an experiment (Baldwin 1989) in which transpiration rates were slowed by placing leaves of induced plants in plastic bags. Under these conditions, alkaloid concentrations were reduced to about one-half of those without bags. It is interesting to note that most members of the genus *Nicotiana* are native to xeric habitats (Kostoff 1941), where conditions of high evapotranspiration prevail.

In addition to these environmental considerations, plant architecture may influence the response. Since the ability of a plant to respond to damage is influenced by auxin concentrations, and since reproductive buds and young leaves are rich sources of diffusible auxins (Avery 1935), the number of reproductive stems a plant contains (a trait that varies greatly among the different species in the genus *Nicotiana*) may influence a plant's ability to respond defensively. Furthermore, because the sites of damage and nicotine synthesis are so widely separated, the size attained by plants that use this form of defense may have an upper limit. Nicotine synthesis in the *Nicotiana* species studied here (Tso 1972) is largely restricted to the roots, perhaps because of the inhibitory effects of light (Ohta and Yatazawa 1978). In tree tobacco (*N. glauca*), one of the tallest members of the genus, 85% of the total alkaloid content is anabasine (Saitoh et al. 1985). This alkaloid is produced in both roots and stems (Dawson 1944). Perhaps by utilizing an alkaloid synthesized in the stem, this plant has escaped the height constraints imposed by a root-associated induced response. Finally, plumbing constraints may play an additional role in within-plant alkaloidal responses. Damage cues may travel within the phloem while the alkaloids move in the xylem. How these vascular traces connect between and among roots and leaves will affect the alkaloid distribution within a plant.

3   EFFECTS ON INSECT HERBIVORES

In this section I examine the effects of the induced alkaloidal response on insect herbivores. The choice of two pests of cultivated tobacco, *Manduca sexta* and *Trichoplusia ni*, as the herbivores makes this examination a poor test of the defensive function of this plant response but a good test of the response on adapted herbivores. As tobacco pests, these insects have been consuming uniformly induced foliage since the development of decapitation as standard practice in the cultivation of tobacco. These herbivores are well known for their resistance to tobacco alkaloids, which is, in part (Morris 1983, 1984), because they can excrete nicotine (Self et al. 1964) and nornicotine (Guthrie and Hodgson 1968) efficiently. Thus, a comparison of the performance of these insects on induced and uninduced foliage will allow us to determine whether their alkaloid-resistant physiologies have obviated behavioral or physiological responses to induced foliage.

In an attempt to examine the effects of induced alkaloid titers on tobacco hornworm larvae, I exploited the observation that pot-bound plants are not responsive to damage (section 2.1) and reared freshly hatched larvae on the foliage of damaged and undamaged pot-bound and released plants (Baldwin 1988b). Plants were grown in pots sunk in a field plot filled with soil from the field plot. Before their leaves were clipped, all plants were removed from their pots; half were returned to their sunken pots, and the other half were placed free in the soil. Larvae reared (in the laboratory) on the high-alkaloid foliage of damaged plants released from their pots gained less weight and ate less (57.2% and 45.7% of controls) than did larvae fed low-alkaloid foliage from undamaged released plants. Similarly, larvae performed equally well on the foliage of damaged and undamaged pot-bound plants that had low alkaloid titers. Damaged released plants had higher chlorophyll contents, but this did not negate the negative effects of high alkaloid titers on larval growth (Baldwin 1988b).

To examine the assumption that alkaloid titers were responsible for the larvae's decreased growth, I removed leaves from undamaged field-grown plants and stem-fed them different concentrations of nicotine. The intent was to mimic the process that normally occurs on the plant and produce leaves with different alkaloid titers. Larvae reared on these artificially induced leaves gained only 38.5% of the weight gained by larvae reared on low-alkaloid foliage. The regression coefficients of larval weight gain against leaf alkaloid content from the pot-bound experiment and the stem-feeding experiment were significantly negative but did not differ from each other, which supports the conclusion that leaf alkaloids are likely to account for the poor larval growth on damaged plants with induced alkaloid titers (Baldwin 1988b). However, given that nicotine is metabolically labile in tobacco leaves (section 4.2), degradation products of nicotine may also be involved.

While other studies (Schoonhoven and Meerman 1978) have found reduced efficiencies of digestion (ECDs) with last-instar hornworms reared on

TABLE 1   Leaf Area (mean ± SD) Consumed by Five 4th-Instar *Trichoplusia nì* Larvae Given a Paired Choice Between Low- and High-Alkaloid Content Leaf Portions

Leaf Alkaloid Content (% dry mass leaf)	Replicates	Area Consumed (cm^2)	p
0.246	14	28.8 ± 19.0	.025
1.110	14	13.1 ± 8.9	
0.301	14	26.3 ± 17.4	.019
1.661	14	7.8 ± 12.5	
0.352	14	23.0 ± 21.2	.017
2.130	14	5.5 ± 5.3	

nicotine-containing artificial diets compared with those of larvae fed alka-loid-free diets, the reduced performance of caterpillars raised on the high-alkaloid foliage of my experiments can be attributed to reduced consumption. The conversions of leaf mass to caterpillar mass, as determined by the slopes (mg/cm^2) of a regression of caterpillar mass against leaf area consumed from damaged and undamaged released plants, were not significantly different (Baldwin 1988b).

The importance of behavioral avoidance is more easily demonstrated with the highly mobile larvae of *Tricoplusia ni*. When these larvae were given a choice between leaves with differing alkaloid contents, they consumed more low-alkaloid foliage (Table 1; Baldwin, unpublished results). Similarly, spi-der mites given a choice between damaged and undamaged cotton plants move to the undamaged host (Harrison and Karban 1986), and leaf beetles avoid willow shoots that have been previously damaged (Raupp and Sadof, this volume, Chapter 8). Clearly it would be preferable to examine the pro-tective role of this induced response with natural populations of plants and their associated herbivores (only some of which are also pests of cultivated tobacco); however, these results do demonstrate that alkaloid-tolerant in-sects are affected by the plant response.

4   PHENOTYPIC COSTS OF INDUCIBLE ALKALOIDAL DEFENSES

Manufacturing chemical defenses can tie up limiting resources (for ex-ample, carbon or nitrogen) that otherwise could be used for growth and reproduction. Plants with large investments in chemical defense are thus assumed to incur a phenotypic cost. This assumption, though difficult to test, has received some support (Chew and Rodman 1979, Krischik and Denno 1983, Coley 1986). If chemical defenses are costly, it would be more economical to employ them only when they are needed; thus, the expense of producing chemical defenses in plants is thought to be an important factor selecting for the evolution of inducible defenses (Haukioja and Hakala 1976, Givnish 1986, Harvell 1986). The phenotypic costs of chemical defense can be broken down into a direct resource-based component, the costs resulting from the synthesis, storage, and autotoxicity of the secondary metabolites, and an indirect component, the opportunity forgone when an investment of resources is made in defense rather than in growth and/or reproduction (Givnish 1986, Gulman and Mooney 1986).

The literature on cultivated tobacco contains a number of indications that constitutive alkaloid production may incur genotypic and phenotypic costs. First, the phenotypic costs of alkaloid production may be high because turnover rates of nicotine are high. Robinson (1974) has calculated that as much as 17% of the net carbon fixed in cultivated tobacco passes through the nicotine biosynthetic pathway. In addition, Jones and Kenyon (1961) found a significant negative correlation between leaf yield and nicotine content in varieties of cultivated tobacco. Recurrent selection regimes carried out on cultivated tobacco for high alkaloid content result in decreased yields (Matzinger et al. 1972), a response that substantiates a negative genetic correlation between these traits.

In this section, I first attempt to examine the phenotypic costs associated with this induced response using techniques that uncouple leaf damage from alkaloid production. *Nicotiana sylvestris* is an annual plant; hence, the effects of alkaloid production on its reproductive output should provide a relevant evolutionary measure of the cost of producing these alkaloids. Moreover, the decrements in reproductive output associated with damage-induced alkaloid production potentially represent an integration of both the indirect and direct components of the cost of chemical defense. I then identify the basis for the apparent high reproductive costs associated with the induced response.

4.1   Experimental Measures of Cost

The ability of auxins to completely inhibit the induced alkaloidal response (Fig. 3) was exploited to create groups of damaged plants with suppressed leaf alkaloid titers (Baldwin et al., 1990). Plants with flower buds subjected to identical leaf damage regimes and same-sized undamaged plants were treated with and without auxins, and reared to maturity. Auxin applications to undamaged plants had no significant effect on alkaloid titers or on any measure of reproductive output. Damaged plants not receiving hormone treatments had induced alkaloid titers approximately 1% leaf dry weight higher than did either undamaged plants or damaged plants treated with auxins (about a 100% increase). At the end of the growing season, plants with induced alkaloid titers had both significantly lower (21%) seed mass per capsule and lower (32%) lifetime seed mass than did damaged plants with low uninduced alkaloid titers that had been treated with auxin. Neither seed size nor seed germinability differed among the treatments (Baldwin et al., 1990).

The second approach to uncoupling alkaloid production from leaf damage exploited the observation that both larval feeding and leaf removal which minimized the damage to leaf vasculature produce alkaloidal responses that are significantly lower than those produced by simple leaf clipping, even when the amount of leaf mass removed is identical (section 2.3). In this experiment, the effects of alkaloid induction on plant reproductive output during the two growth stages of *N. sylvestris*, the rosette stage and the flowering stage, were examined separately.

Plants that had developed a flower bud and commenced stem elongation but had yet to produce flowers were subjected to interveinal damage (section 2.3) and to serial leaf clipping whereby leaf tissue was removed without regard to leaf vasculature. Alkaloid titers 16 days after the end of the damage regime were 1.9 times higher and 3.9 times higher than those of undamaged plants for the interveinal and serial damage treatments, respectively. Plants with high induced alkaloid titers (the serially damaged plants) had significantly lower (45%) lifetime seed mass and fewer (38%) capsules than did undamaged plants (Baldwin et al., 1990).

The results of both the auxin and the vein-avoidance experiments are consistent with the hypothesis that this induced alkaloidal response is associated with reduced reproductive output. In the auxin experiment, a 1% change in leaf alkaloid concentration from a low, uninduced titer of 0.75% leaf dry mass was associated with a decrease of 4.76 g (approximately 150 000 seeds) in lifetime seed production. In the vein-avoidance experiment, induction of a 1% change in alkaloid titer above an already induced titer of 1.07% leaf dry mass was associated with a decrease of approximately 3.7 g in seeds (120 000 seeds). However, these measures of phenotypic cost may be confounded by other responses to leaf damage that were not uncoupled from the alkaloidal response to damage. Moreover, each of the two experiments may have technique-specific difficulties which may confound the interpretation of their results: Other auxin-related responses not associated with the alkaloidal response may have increased the reproductive output of auxin-treated damaged plants, and the vein-avoidance experiment may overestimate the reproductive decrement associated with alkaloid production because damage to veins could affect phloem transport and, consequently, growth during the 5-day damage regime. Nevertheless, the results from these two experiments are remarkably similar.

In a separate experiment, a significant negative correlation ($r = -.504$) was found between lifetime seed production and alkaloid content in undamaged plants (Baldwin et al., 1990). This negative correlation suggests that a change of a constitutive leaf alkaloid titer of 1% leaf dry mass is associated with a decrease of about 500 000 seeds in lifetime reproductive output. The higher estimates of cost from the undamaged plants may result from the different durations that the elevated alkaloid titer is sustained and may underscore the cost-effectiveness of employing a chemical defense intermittently.

On the other hand, plants are able to compensate for rosette-stage damage. Rosette-stage plants subjected to herbivory from the larvae of tobacco hornworms had alkaloid titers after damage 1.5 times greater than those of control plants, whereas plants subjected to mechanical clipping had alkaloid titers 2.4 times greater than those of control plants. Interestingly, at maturity the reproductive output of plants from these damage treatments did not differ from that of undamaged plants. Clearly, rosette-stage plants are able to compensate for the costs of leaf damage and those associated with alkaloid production.

A number of factors may account for the ability of plants to compensate for rosette-stage induction and defoliation, yet not for damage incurred during the flowering stage of growth. First, the absolute amount of shoot mass removed was four times as great for the flowering-stage plants, although the proportion of shoot mass removed was comparable. Rosette-stage plants are likely to have a higher photosynthetic rate per gram of shoot, because they lack the large stem of the flowering-stage plants. Moreover, rosette-stage plants have more time to compensate for leaf loss. Finally, the compensatory responses of flowering plants to defoliation may be hormonally constrained by their developing reproductive buds (Akehurst 1981).

4.2   Why May Induced Alkaloid Production Be Costly?

Working under the assumption that induced alkaloidal responses are, in fact, responsible for some portion of the large observed decrements in reproductive output, I examine in this section possibilities for the observed cost. I explore the patterns of biomass allocation in damaged and undamaged plants and follow this with speculations on potential physiological consequences of induced alkaloid production that could influence the direct component of the cost of this response.

The plants whose alkaloid titers are depicted in Figure 1 were subjected to a simulated herbivory regime during the flowering stage of their growth. At maturity, these defoliated plants, which had an approximately fourfold increase in leaf alkaloid titers, produced 78.8% fewer seeds, 63.7% less root mass, and 73.5% less shoot mass compared to the paired control plants. We know from the two experiments above (section 4.1) that a measurable proportion of the decrease in vegetative and reproductive biomass is associated with increases in alkaloid titers. Decrements are also likely associated with other processes known to increase in damaged tissues, such as the costs of formation of other secondary products, the costs of morphological barriers at the sites of damage, (McLaughlin and Shriner 1980), and the lost photosynthetic income from the lost leaf area.

One place to start an examination of the reproductive costs of plant responses to damage is to ask whether damaged plants allocate biomass to

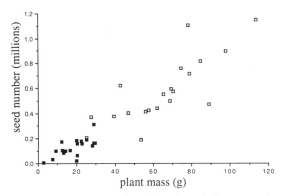

Figure 6   Seed number per plant as a function of total plant mass (root + shoot mass) of damaged (■) and undamaged (□) plants whose alkaloid titers are depicted in Figure 1. Slopes are not significantly different.

vegetative and reproductive structures differently than do undamaged, un-induced plants. A regression of total seed number against total (shoot and root) plant mass indicates that they do not (Fig. 6; Baldwin, unpublished results). The y intercepts for both damaged and undamaged plants were not different from zero, and the slopes did not differ from each other. These results are consistent with observations from many plant species which show reproductive effort to be a function of plant size (Samson and Werk 1986). The costs associated with the responses to damage can be seen simply as a decrement in plant size, which, in turn, corresponds to decreased repro-ductive output (Fig. 6). Moreover, the season-end shoot mass of plants from the vein-avoidance experiment (section 4.1), which had high alkaloid titers, was significantly lower than that of initially same-sized plants with low al-kaloid titers and less vein damage (Baldwin et al., 1990).

The resource requirements of reproduction resemble those of growth (Bazzaz et al. 1987), and thus the plants appear to be allocating resources between growth/reproduction and defense, rather than among growth, re-production, and defense. Defense costs may curtail the total biomass ac-quired by a plant; they do not influence the efficiency with which a vegetative biomass creates reproductive biomass. If this is the case, what then is the reason for the decreases in plant size?

One possibility is that at induced concentrations, these alkaloids are au-totoxic. To explore this possibility, I examined the influence of alkaloid titers normally experienced by the leaf during a response to damage on a leaf's ability to gain carbon. Tobacco leaves from undamaged plants were stem-fed aqueous solutions whose alkaloid concentrations mimicked those found in the xylem fluid of damaged and undamaged plants (Baldwin et al., 1990). The maximum photosynthetic rates of these leaves were then determined. No correlation was observed between leaf alkaloid concentrations and pho-

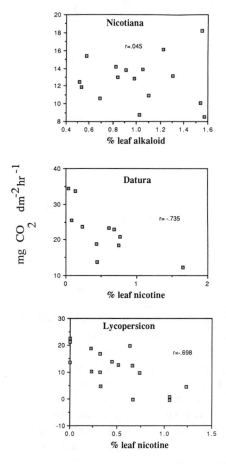

Figure 7 Maximum rates of photosynthesis are not correlated ($r = .045$) with artificially induced leaf alkaloid contents in detached *Nicotiana sylvestris* leaves stem-fed alkaloid solutions, but are negatively correlated ($p < .05$) with leaf alkaloid contents in *Datura stramonium* ($r = -.735$) and *Lycopersicon esculentum* ($r = -.698$) leaves.

tosynthetic rates in *N. sylvestris* foliage. However, in similar experiments (Fig. 7; Baldwin, unpublished data), negative correlations were found with two other solanaceous species: *Lycopersicon esculentum* (tomato) and *Datura stramonium* (jimsonweed). In addition to glycoalkaloids in tomato and tropane alkaloids in jimsonweed, these plants are reported to contain nicotine, albeit in trace amounts (see Leete 1977). These results suggest that physiological adaptations to high leaf alkaloid titers preceded the evolution of nicotine-based inducible responses.

Moreover, plants that can tolerate high concentrations of these alkaloids in their leaf tissues may accrue physiological benefits in addition to those associated with defense. Nicotine is a potent quencher of singlet oxygen (Larson and Marley 1984), which could prove physiologically advantageous if defoliation resulted in rapid increases in light levels for the remaining leaves. Such increases might damage chloroplasts if excited triplet-state chlorophyll reacted with oxygen to form the highly reactive singlet oxygen

(Clayton 1980). If nicotine is playing a photo-protective role in tobacco, such benefits are not realized by plants not adapted to high nicotine titers, for non-*Nicotiana* scions grafted to tobacco roots suffer "typical nicotine damage" (Mothes 1960) if grown in the sun but do not in the shade.

A second approach to identifying a direct component of the costs of alkaloid production is to calculate the amount of nicotine transported from the roots to the leaves of a tobacco plant during an induced response. From the concentration of nicotine in the xylem fluid and measurements of the amount of xylem sap transpired per area of leaf, the amount of alkaloid transported into a leaf during an induced response was estimated to represent a substantial fraction of the total leaf nitrogen content (Baldwin 1989).

These calculations corroborate observations from numerous tracer studies with cultivated tobacco, namely that nicotine has a half-life of about 20 h (Leete and Bell 1959). When labeled nicotine is degraded in a tobacco plant, it is metabolized to primary metabolites such as amino acids, pigments, sugars, and organic acids (Tso and Jeffrey 1959, 1961). The leaves of damaged plants undoubtedly have higher demands for nitrogen in order to support the increase in photosynthetic rates that follows defoliation (Kolodny-Hirsch et al. 1986) and topping (Darkanbaev et al. 1962, Yoshinori et al. 1967). In addition, damaged tobacco may have proteinase inhibitors (Wong et al. 1976) induced after leaf damage, which would increase the nitrogen requirements of its leaves. Nicotine, in addition to its role in defense, is an integral part of a plant's primary metabolism, serving as a well-defended vehicle for nitrogen transport from roots to leaves.

The role of the alkaloidal response in nitrogen transport suggests a hypothesis to account for the apparent costliness of this response for flowering-stage plants. During the early rosette stage, *N. sylvestris* produces small filamentous roots that enlarge and thicken as the floral bud develops. By the time bolting occurs, the roots can be quite massive. If greenhouse-grown plants are completely defoliated just prior to flowering, no new leaves will grow. Yet the plants will still be able to produce mature seeds. Root mass, however, is substantially reduced after seed maturation in defoliated plants (Baldwin, unpublished results), indicating that root storage plays an important role in a plant's compensatory response to defoliation. Moreover, when single leaves are fed labeled carbon dioxide during the flowering stage of growth in cultivated tobacco, the label can be detected in the leaves, stems, and roots, but not in the mature seeds or flowers (Shiroya et al. 1961), indicating that fruit and seeds are not the principal sinks for assimilates fixed after flowering.

In addition to playing a major role in the storage of assimilates, roots are also the principal sites of alkaloid synthesis (Tso 1972). I hypothesize that it is the transport of stored reserves (nitrogen and carbon) in the form of nicotine to leaves, rather than to maturing fruits, and the resource demands of processes (e.g., enzyme induction) required to produce the alkaloidal response that underlie the observed high cost of this induced response. The costs of an alkaloidal response incurred during the rosette stage of growth are, in time, repaid by elevated photosynthetic rates and refoliation. An annual plant developmentally committed to reproduction may not be capable of replenishing the root reserves used in defense.

5 CONCLUSION

Given that so much is known about the biochemistry, physiology, and ecology of these alkaloids, it might not be premature to imagine a few hypothetical scenarios concerning the evolution of these alkaloids and their inducibility. The important consideration is whether these alkaloids and their inducibility are a result of their defensive properties or whether they are merely incidental outcomes of plant primary metabolism. The specificity of the interaction of nicotine with cholinergic receptors, and the resulting havoc that this interaction wreaks in the nervous systems of most animals, strongly suggest that the structure of this alkaloid evolved in response to predation by organisms that use acetylcholine as a neurotransmitter. However, the widespread occurrence of nicotine among plant families suggests otherwise. The ability to produce nicotine is widely distributed phylogenetically; trace amounts of the alkaloid have been found in 24 genera in 12 families (Leete 1977). Only in the genus *Nicotiana* has this biosynthetic potential been consistently developed to defensive concentrations. As we have seen, the ability to use nicotine defensively requires not only the ability to synthesize this alkaloid in large amounts but also the ability to tolerate its presence in the leaves. These considerations suggest that it was the ability to produce and tolerate high concentrations of this alkaloid, rather than the alkaloid's specific structure, that evolved in response to predation.

Presumably, the production of the alkaloid became inducible. This responsiveness to damage could have been selected by ecological considerations; higher quantities may have been needed to deter animals that had evolved resistance to the alkaloids. Inducibility may, therefore, have evolved as a measure to curtail the high costs associated with a nitrogen-based defense. Alternatively, nicotine production may always have been inducible, since it is derived from a biosynthetic pathway, the polyamine pathway, which in plants is currently responsive to stress (Galston 1983) and historically may always have been inducible. The evidence supporting the role of a phloem-borne positive cue activating the alkaloidal response is consistent with the defensive scenario. Some of the details of the alkaloidal responses to different types of leaf damage also support this hypothesis. The large alkaloidal responses to gradual damage, compared to more attenuated responses to sudden leaf damage (Baldwin 1989), would allow the plant to differentiate between damage resulting from biotic agents, which inflict damage over a sustained period of time, and abiotic agents, such as hailstorms, which inflict potentially significant but momentary damage and to which an alkaloidal response would be inappropriate. Other details of the alkaloidal responses do not fit the defensive hypothesis as neatly: namely, the observation that larval feeding elicits smaller alkaloidal responses than do mechanical simulations of such feeding damage. Because plants use endogenously derived damage cues, their defensive responses may be evolutionarily constrained. In contrast to plants, induced defenses in marine byrozoans are activated by cues specific to the predator (Harvell 1986); hence, these induced responses can be specifically tuned to an attacker's presence.

Regardless of its evolutionary origins, the alkaloidal response is currently not purely a defensive response, though historically it may have been; it is now an integral part of a plant's primary metabolism. The quantities of alkaloid that are imported to leaves from the roots to effect a transient increase in leaf alkaloid titer underscore this point. Thus, the defensive response must be considered part of a transport system that moves nitrogen from roots to leaves. In this context, nicotine may be a vehicle for nitrogen transport, shuttling valuable nitrogen along a transport route, the xylem, which is vulnerable to attack by a guild of sap-sucking insects. According to this scenario, increased leaf alkaloid titers may be a by-product of the slow recycling times of the nitrogen vehicles, which happen to serve a defensive role while supplying the nitrogen needed to effect a more general response to leaf damage: the reversal of senescence. These physiological constraints on the induced response, in addition to the potential environmental and architectural constraints mentioned earlier, may profoundly influence the evolution of the defensive role of this induced response.

ACKNOWLEDGMENTS

This work was supported by NSF Grant BSR-8714888 (to I. Baldwin and T. Eisner), by a grant from the Cornell Biotechnology Program, which is sponsored by the New York State Science and Technology Foundation and a consortium of industries, and by NIH Grant AI 02908 (to T. Eisner). A. Spitz, C. Sims, S. Kean, C. Sanchirico, and O. Noblitt are thanked for their technical assistance, B. Chabot for helpful discussion on measurements of photosynthetic rates and use of his IRGA, and D. Tallamy, M. Raupp, E. Wheeler, C. LaMunyon, and three anonymous reviewers for improving earlier drafts of this paper.

REFERENCES

Akehurst, B. C. 1981. *Tobacco*. New York: Longman.
Audus, L. J. 1972. *Plant Growth Substances*. New York: Barnes & Noble.

Avery, G. S. 1935. Differential distribution of a phytohormone in the developing leaf of *Nicotiana*, and its relation to polarized growth. *Torrey Bot. Club Bull.* 62:313–330.

Baldwin, I. T. 1988a. Damaged-induced alkaloids in tobacco: pot-bound plants are not inducible. *J. Chem. Ecol.* 4:1113–1120.

Baldwin, I. T. 1988b. Short-term damage-induced alkaloids protect plants. *Oecologia* 75:367–370.

Baldwin, I. T. 1988c. The alkaloidal responses of wild tobacco to real and simulated herbivory. *Oecologia* 77:378–381.

Baldwin, I. T. 1989. The mechanism of damage-induced alkaloids in wild tobacco. *J. Chem. Ecol.* 151:1661–1680.

Baldwin, I. T., C. L. Sims, and S. E. Kean. 1990. Reproductive consequences associated with inducible alkaloidal responses in wild tobacco. *Ecology* 71:252–262.

Bazzaz, F. A., N. R. Chiariello, P. D. Coley, and L. F. Pitelka. 1987. Allocating resources to reproduction and defense. *BioScience* 37:58–67.

Bose, B. C., H. N. De, I. H. Dalal, and S. Mohammad. 1956. Biogenesis of alkaloids in tobacco plants. *J. Indian Med. Res.* 44:81–89.

Chew, F. S., and J. E. Rodman. 1979. Plant resources for chemical defense. In G. A. Rosenthal and D. H. Janzen (eds.), *Herbivores: Their Interactions with Secondary Plant Compounds*. New York: Academic, pp. 271–307.

Clayton, R. K. 1980. *Photosynthesis, Physical Mechanisms and Chemical Patterns*. Cambridge, England: Cambridge University Press.

Coley, P. D. 1986. Costs and benefits of defense by tannins in a neotropical tree. *Oecologia* 70:238–241.

Cutler, H. G., and T. P. Gaines. 1971. Some preliminary observations on greenhouse-grown tobacco treated with 2-chloro-ethylphosphonic acid at varying pH's. *Tobacco Sci.* 15:100–102.

Darkanbaev, T. R., Z. L. Lukpanov, and Z. H. Kalekenov. 1962. Tobacco pruning. *Sov. Plant Physiol.* 9:60–68.

Dawson, R. F. 1941. The localization of the nicotine synthetic mechanism in the tobacco plant. *Science* 94:396–397.

Dawson, R. F. 1944. Accumulation of anabasine in reciprocal grafts of *Nicotiana glauca* and tomato. *Am. J. Bot.* 31:351–355.

Feeny, P. P. 1976. Plant apparency and chemical defenses. *Recent Advances in Phytochemistry* 10:1–40.

Floss, H. G., J. E. Robbers, and P. F. Heinstein. 1974. Regulatory control mechanisms in alkaloid biosynthesis. *Rec. Adv. Phytochem.* 7:141–178.

Galston, A. W. 1983. Polyamines as modulators of plant development. *BioScience* 33:382–388.

Givnish, T. J. 1986. Economics of biotic interactions. In T. J. Givnish (ed.), *On the Economy of Plant Form and Function*. Cambridge, England: Cambridge University Press, pp. 667–679.

Gordon, H. T. 1961. Nutritional factors in insect resistance to chemicals. *Annu. Rev. Entomol.* 6:27–54.

Gulmon, S. L., and H. A. Mooney. 1986. Cost of defense on plant productivity. In

T. J. Givnish (ed.), *On the Economy of Plant Form and Function*. Cambridge, England: Cambridge University Press, pp. 681–698.

Guthrie, F. E., and E. Hodgson. 1968. Adaptations of insects to nornicotine. *Ann. Entomol. Soc. Am.* 61:545–547.

Harrison S., and R. Karban. 1986. Behavioral responses of spider mites (*Tetranychus urticae*) to induced resistance of cotton plants. *Ecol. Entomol.* 11:181–188.

Harvell, C. D. 1986. The ecology and evolution of inducible defenses in a marine bryozoan: cues, costs and consequences. *Am. Natur.* 128:810–823.

Hassal, C. A. 1969. *World Crop Protection,* Vol. 2. Cleveland, OH: Chemical Rubber Co.

Haukioja, E., and T. Hakala. 1976. Herbivore cycles and periodic outbreaks. Formulation of a general hypothesis. *Reports from the Kevo Subarctic Research Station* 12:1–9.

James, P. D. 1971. *Shroud for a Nightingale*. New York: Scribner.

Jones, G. L., and J. M. Kenyon. 1961. Measured crop performance, tobacco. Research Report, North Carolina Agricultural Experiment Station, 29.

Kasperbauer, M. J., and J. C. Hamilton. 1978. Ethylene regulation of tobacco seedling size, floral induction and subsequent growth and development. *Agron. J.* 70:363–366.

Kolodny-Hirsch, D. M., J. A. Saunders, and F. P. Harrison. 1986. Effects of simulated tobacco hornworm (Lepidoptera: Sphingidae) defoliation on growth dynamics and physiology of tobacco as evidence of plant tolerance to leaf consumption. *Environ. Entomol.* 15:1137–1144.

Kostoff, D. 1941. *Citogenetics of the Genus Nicotiana. Karyosystematics, Genetics, Cytology, Cytogenetics and Phylesis of Tobacco*. Sofia, Bulgaria: State's Printing House.

Krischik, V. A., and R. F. Denno. 1983. Individual, population, and geographic patterns in plant defense. In R. F. Denno and M. S. McClure (eds.), *Variable Plants and Herbivores in Natural and Managed Systems*. New York: Academic, pp. 463–512.

Larson, R. A., and K. A. Marley. 1984. Quenching of singlet oxygen by alkaloids and related nitrogen heterocycles. *Phytochemistry* 23:2351–2354.

Leete, E. 1977. Biosynthesis and metabolism of the tobacco alkaloids. *Proc. Am. Chem. Soc. Symp.* 173:365–388.

Leete, E., and V. M. Bell. 1959. The biogenesis of *Nicotiana* alkaloids: the metabolism of nicotine in *N. tabacum. J. Am. Chem. Soc.* 81:4358–4359.

Matsumura, F. 1976. *Toxicity of Insecticides*. New York: Plenum.

Matzinger, D. F., E. A. Wernsman, and C. C. Cockerham. 1972. Recurrent family selection and correlated response in *Nicotiana tabacum* L.I. 'Dixie Bright 244' × 'Coker 139.' *Crop Sci.* 12:40–43.

McLaughlin, S. B., and D. S. Shriner. 1980. Allocation of resources to defense and repair. In *Plant Disease,* Vol. 5. New York: Academic, pp. 407–431.

Miller, R. H., A. A. Berryman, and C. A. Ryan. 1986. Biotic elicitors of defense reactions in lodgepole pine. *Phytochemistry* 25:611–612.

Mizusaki, S., Y. Tanabe, M. Roguchi, and E. Tamaki. 1973. Changes in the activities of ornithine decarboxylase, putrescine *N*-methyltransferase and *N*-methylputrescine oxidase in tobacco roots in relation to nicotine biosynthesis. *Plant Cell Physiol.* 14:103–110.

Morris, C. E. 1983. Uptake and metabolism of nicotine by the central nervous system of a nicotine-resistant insect, the tobacco hornworm (*Manduca sexta*). *J. Insect Physiol.* 29:807–818.

Morris C. E. 1984. Electrophysiological effects of cholinergic agents in the central nervous system of a nicotine-resistant insect, the tobacco hornworm (*Manduca sexta*). *J. Exp. Zool.* 229:361–374.

Mothes, K. 1960. Alkaloids in the plants. In R. H. F. Manske (ed.), *The Alkaloids, Chemistry and Physiology,* Vol 6. New York: Academic, pp. 1–29.

Ohta, S., and M. Yatazawa. 1978. Effect of light on nicotine production in tobacco tissue culture. *Agric. Biol. Chem.* 42:873–877.

Parr, J. C., and R. Thurston. 1972. Toxicity of nicotine in synthetic diets to larvae of the tobacco hornworm. *Ann. Entomol. Soc. Am.* 65:1185–1188.

Raupp, M. J. 1991. Leaf beetle responses to damage induced changes in their salicaceous hosts. In D. W. Tallamy and M. J. Raupp (eds.), *Phytochemical Induction by Herbivores*. New York: Wiley, pp. 183–204.

Reuter, V. G. 1957. Zustandsbedingte Produktionsanderungen spezifischer Stickstoff-verbindunen in der Tabakpflanze. *Die Kulturpflanze* 5:137–185.

Robinson, T. 1974. Metabolism and function of alkaloids in plants. *Science* 184:430–435.

Ryan, C. A., and T. R. Green. 1974. Proteinase inhibitors in natural plant protection. *Rec. Adv. Phytochem.* 8:123–140.

Saitoh, F., M. Mona, and N. Kawashima. 1985. The alkaloid contents of sixty *Nicotiana* species. *Phytochemistry* 24:477–480.

Samson, D. A., and K. S. Werk. 1986. Size-dependent effects in the analysis of reproductive effort in plants. *Am. Natur.* 127:667–680.

Saunders, J. W., and L. P. Bush. 1979. Nicotine biosynthetic enzyme activities in *Nicotiana tobacum* genotypes with different alkaloid levels. *Plant Physiol.* 64:236–240.

Schmetz, I. 1971. Nicotine and other tobacco alkaloids. In M. Jacobson and D. G. Crosby (eds.), *Naturally Occurring Insecticides*. New York: Dekker, pp. 99–136.

Schoonhoven L. M., and J. Meerman. 1978. Metabolic cost of changes in diet and neutralization of allelochemics. *Entomol. Exp. Applic.* 24:489–493.

Self, L. S., F. E. Guthrie, and E. Hodgson. 1964. Adaptation of tobacco hornworms to the ingestion of nicotine. *J. Insect Physiol.* 10:907–914.

Shiroya, M., G. R. Lister, C. D. Nelson, and G. Krotkov. 1961. Translocation of ^{14}C in tobacco at different stages of development following assimilation of $^{14}CO_2$ by a single leaf. *Can. J. Bot.* 39:855–864.

Solt, M. L. 1957. Nicotine production and growth of excised tobacco root cultures. *Plant Physiol.* 32:488–484.

Tabata, M., Y. Yamamoto, N. Hiraoka, Y. Marumoto, and M. Konoshima. 1971. Regulation of nicotine production in tobacco tissue culture by plant growth regulators. *Phytochemistry* 10:723–729.

Tso, T. C. 1972. *Physiology and Biochemistry of Tobacco Plants*. Stroudsbury, PA: Dowden, Hutchinson & Ross.

Tso, T. C., and R. N. Jeffrey. 1959. Biochemical studies on tobacco alkaloids. I. The fate of labeled tobacco alkaloids supplied to *Nicotiana* plants. *Arch. Biochem. Biophys.* 80:46–56.

Tso, T. C., and R. N. Jeffrey. 1961. Biochemical studies on tobacco alkaloids. IV. The dynamic state of nicotine supplied to *N. rustica. Arch. Biochem. Biophys.* 92:253–256.

Woetz, W. G. 1955. Some effects of topping and suckering flue-cured tobacco. North Carolina Agricultural Experiment Station Technical Bulletin 106.

Wong, P. P., T. Kuo, C. A. Ryan, and C. I. Kado. 1976. Differential accumulation of proteinase inhibitor I in normal and crown gall tissues of tobacco, tomato, and potato. *Plant Physiol.* 57:214–217.

Yasumatsu, N. 1967. Studies on the chemical regulation of alkaloid biosynthesis in tobacco plants. II. Inhibition of alkaloid biosynthesis by exogenous auxins. *Agric. Biol. Chem.* 31:1441–1447.

Yoshida, D. 1962. Degradation and translocation of ^{15}N-labeled nicotine-injected into intact tobacco leaves. *Plant Cell Physiol.* 3:391–395.

Yoshinori, W., S. Wanatabe, and S. Kuroda. 1967. Changes in photosynthetic activities and chlorophyll contents of growing tobacco leaves. *Bot. Mag. Tokyo* 80:123–129.

CHAPTER 3

LONG-TERM AND SHORT-TERM INDUCTION IN QUAKING ASPEN: RELATED PHENOMENA?

THOMAS P. CLAUSEN and PAUL B. REICHARDT
Department of Chemistry, University of Alaska, Fairbanks, AK 99775-0520

JOHN P. BRYANT
Institute of Arctic Biology, University of Alaska, Fairbanks, AK 99775-0180

RICHARD A. WERNER
Institute of Northern Forestry, Fairbanks, AK 99775-5500

1   Introduction
2   Natural history of the quaking aspen–large aspen tortrix interaction in alaska
3   Secondary metabolites of quaking aspen
4   Short-term induction in quaking aspen
5   Long-term induction in quaking aspen
6   Possible interactions between long- and short-term induction in quaking aspen
7   Summary
    References

1   INTRODUCTION

Herbivory can alter the food quality of plants. Although food quality occasionally may increase (Provenza and Malechek 1984, Danell and Huss-Danell 1985, Danell et al. 1985), herbivore-induced changes are usually associated with reduced food quality (Haukioja and Niemelä 1977, Rhoades 1979, Schultz and Baldwin 1982, Bryant et al. 1983). Thus, changes induced by herbivory are often regarded as defenses (but see Fowler and Lawton 1985, Myers 1988). In this paper, we define induction as changes caused by herbivory that reduce the quality of plants as food. This broad definition allows induction to be measured in two ways. First, bioassays using attacked and control plants can directly measure the reduced food quality of attacked plants. Second, measurements of plant chemistry such as increased levels of defensive compounds or decreased levels of nutritional components may be sufficient to infer lower food quality of attacked plants. Our definition, however, does not require that these changes be of sufficient magnitude to significantly affect the feeding behaviors or populations of herbivores.

Long-term induction (LTI; sensu Haukioja 1980) involves changes in a plant that usually occur one or more years following herbivory. In addition, LTI is seldom observed in cases of minor herbivory; it requires severe herbivory to elicit significant long-term plant responses. These properties of LTI may be sufficient to explain population cycles of some herbivores if the reduction in food quality is of sufficient magnitude (Neuvonen and Haukioja 1991, Haukioja et al. 1983, Fowler and Lawton 1985).

In contrast, short-term induction (STI) involves changes in the plant either immediately following herbivory or up to several days later. While this time frame is too short to cause herbivore population cycles (Neuvonen and Haukioja, this volume, Chapter 12), it may have adverse consequences for the herbivore that caused the response, or influence food preferences and feeding behaviors of herbivores encountering the attacked resource shortly thereafter. For example, Raupp and Sadof (this volume, Chapter 8) report that willow leaf beetles eating damaged weeping willow foliage have higher mortality and lay fewer eggs than beetles raised on undamaged foliage.

Here we review what is known about both STI and LTI in quaking aspen (*Populus tremuloides*) foliage that results from simulated attack by insect herbivores. We then demonstrate that the known chemical changes associated with LTI in aspen contribute to but are not on their own sufficient to explain the observed reduction in food quality of these leaves to the semi-specialist, the large aspen tortrix (*Choristoneura conflictana*). Finally, we propose that LTI in quaking aspen is largely the result of an enhancement of its STI responses.

2 NATURAL HISTORY OF THE QUAKING ASPEN–LARGE ASPEN TORTRIX INTERACTION IN ALASKA

The natural histories of quaking aspen, the large aspen tortrix, and their interaction in Alaska have been described in detail elsewhere (Beckwith 1968 and references therein). Populations of the large aspen tortrix outbreak in interior Alaska at intervals of approximately 10–15 years. During peak outbreaks, quaking aspen is severely defoliated (80–100%) within 3–4 weeks of bud break for 2–4 successive years by 2nd- to 5th-instar larvae. Before feeding upon aspen, larvae spend several hours webbing themselves within two or more adjacent leaves (often all leaves on a single short shoot). Over the next 2–4 days the larvae feed on these leaves. Thus, they feed upon damaged leaves for a sufficient period of time to experience the effects of STI.

3   SECONDARY METABOLITES OF QUAKING ASPEN

In addition to condensed tannins, quaking aspen foliage contains as its only major secondary metabolites four phenolic glycosides: salicin, salicortin, tremuloiden, and tremulacin (Fig. 1; Pearl and Darling 1971, Bryant et al. 1987, Lindroth et al. 1987a, Clausen et al. 1989b). An additional compound, 6-hydroxycyclohex-2-ene-1-one (6-HCH) is present in lesser amounts (Clausen et al. 1989b).

The levels of phenol glycosides in the current season's growth of quaking aspen twigs have been determined using an HPLC method reported by Lindroth et al. (1987a,b). Twigs were obtained from aspens that were part of

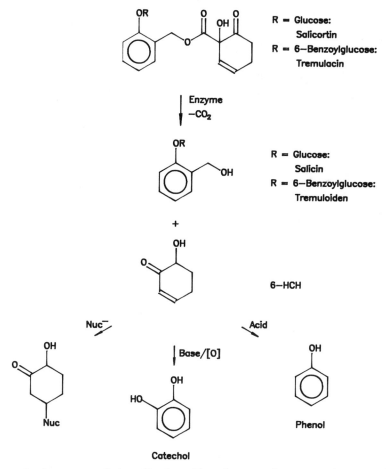

Figure 1   Structures of phenolic glycosides of aspen, the enzymatic conversion of salicortin (or tremulacin) to salicin (or tremuloiden) and 6-HCH, and some possible outcomes of 6-HCH in a herbivore's gut.

an earlier experiment (Bryant et al. 1987). All trees in this study were 3–4 m tall and were manipulated in a $2 \times 2 \times 2$ factorial design the previous summer (nitrogen fertilization, 50 g N/m^2 applied as NH$_4$NO$_3$, phosphorus fertilization, 25 g P/m^2 applied as P$_2$O$_5$; 100% defoliation during mid-growing season). Ten trees were sampled at the crown from each of the eight treatments. These samples were pooled for phytochemical analysis. Both salicortin (10.5 $\pm$ 0.4% dry wt) and tremulacin (6.0 $\pm$ 0.3 dry wt) were present, but no salicin or tremuloiden was present in any sample. Older twigs also contain only salicortin and tremulacin in appreciable although reduced amounts (Clausen et al. 1989).

4   SHORT-TERM INDUCTION IN QUAKING ASPEN

STI in aspen involves two distinct and complementary sets of foliar chemical changes (Fig. 2; Clausen et al., 1989b). In the first change, the levels of salicortin and tremulacin increase up to twofold within 24 h after leaf damage, whereas the levels of salicin and tremuloiden remain constant. These increases in glycoside levels are consistent with the phenolic increases reported by Mattson and Palmer (1987) and Mattson et al. (1987). It is important to note that the observed increases are the result of simulated herbivory and have not been optimized with respect to type and degree of damage, induction time, tree age, and time of damage. For instance, we have chosen an arbitrary 24-h induction period on mature aspen trees that had not undergone any recent severe infestations. STI may be much larger when elicited in the field by insects on previously defoliated trees and when optimal induction times are chosen.

While the above characteristic of STI occurs in the damaged leaf that remains on the tree, another aspect of STI occurs in the consumed portion of the leaf (Fig. 2). Here, both salicortin and tremulacin are rapidly converted (within seconds to minutes) to salicin and tremuloiden, respectively, with 6-HCH as a by-product (Clausen et al. 1989; Fig. 1).

Figure 3 proposes a mechanism for the formation of 6-HCH and its degradation products from salicortin and tremulacin. The mechanism has only a few steps, each of which has biochemical precedent. In this mechanism, ester hydrolysis of salicortin (or tremulacin) results in the formation of salicin (or tremuloiden) and the intermediate carboxylic acid proposed by Lindroth and Pajutee (1987) and Lindroth et al. (1988). Such β-ketoacids tend to be unstable (Hsu et al. 1985) and readily decarboxylate via a 6-membered activated complex to produce an enol which rapidly converts to the corresponding ketone (in this case, 6-HCH). In the case of *Populus balsamifera* (Mattes et al. 1987), this last step must be enzymatically mediated because the 6-HCH isolated from *P. balsamifera* is optically active (the optical activity of 6-HCH from aspen has not been determined). Basic conditions would catalyze the equilibration of 6-HCH and its enol, which, with mild

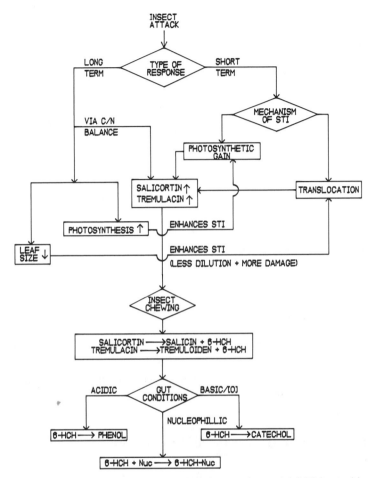

Figure 2   Model showing STI and its possible interactions with LTI in quaking aspen foliage.

oxidants (O_2 or perhaps NAD) would convert to catechol, a highly favored product based on thermodynamics. Under mild acidic conditions, 6-HCH dehydrates to form phenol (Mattes et al. 1987). Finally, 6-HCH can capture nucleophiles via a Michael addition. These potentially harmful products of 6-HCH may occur within the gut of a herbivore and are summarized in Figure 1.

The proposal that 6-HCH and its precursors, salicortin and tremulacin, are harmful to the large aspen tortrix was confirmed by bioassays using artificial diets. When offered in these diets at levels potentially formed in induced leaves, salicortin, tremulacin, and 6-HCH resulted in reduced pupal weights (Clausen et al. 1989b). The deleterious effects of salicortin and tremulacin were confirmed when bioassayed with tiger swallowtail on painted

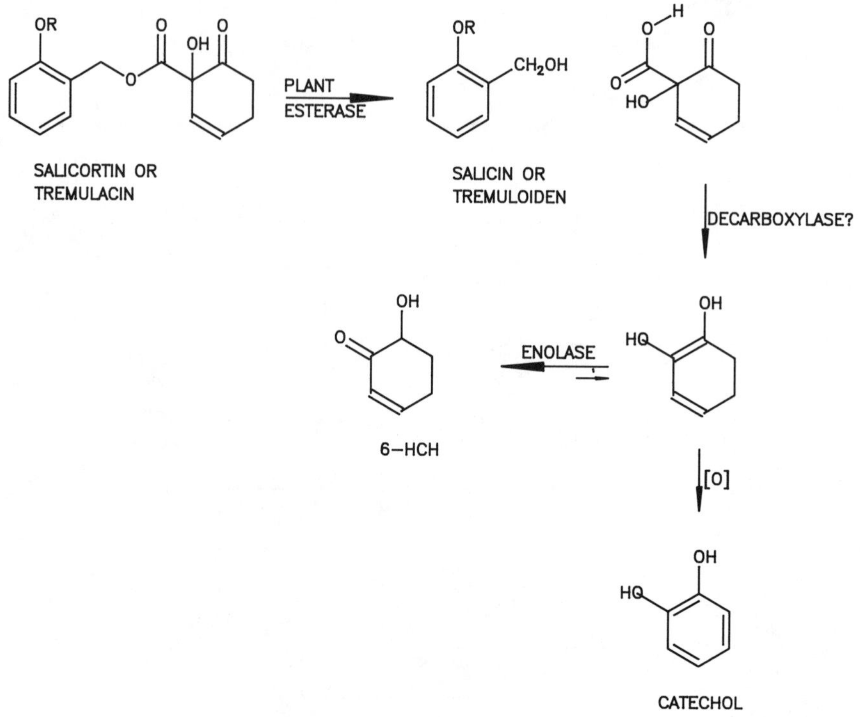

Figure 3   A hypothesized chemical mechanism showing the formation of catechol from salicortin and/or tremulacin.

leaves (Lindroth et al. 1988) and southern armyworm (Lindroth and Peterson 1988).

The occurrence of STI in quaking aspen is also supported by field observations. Scriber (personal communication) observed that swallowtail larvae often move from incompletely eaten leaves of quaking aspen to new undamaged leaves during the course of feeding. Presumably, the swallowtail elicits STI of sufficient magnitude to force it to eventually abandon the leaf that it is consuming. The tortrix, however, webs itself within a group of leaves (sometimes an entire cluster) prior to feeding (Beckwith 1968) and, hence, must deal with any STI responses that it may have elicited.

5   LONG-TERM INDUCTION IN QUAKING ASPEN

We have hypothesized that intensity and time of defoliation were the primary factors causing long-term induction in aspen and subsequent effects on the survival and development of the large aspen tortrix. Long-term induction in aspen was studied by artificial defoliation of 10- to 15-year-old aspen trees using six treatments of 50 and 100% defoliation plus untreated check trees. The following treatments were used to simulate insect defoliation using 10 trees per treatment: early and late defoliation for 1, 2, or 3 years in succession, and untreated checks for a total of 140 trees. Leaves were torn laterally above the leaf petiole junction with about 1 cm of leaf remaining attached to the apical end of the petiole. On trees with 50% defoliation, alternate leaves were removed.

Ten 2nd-instar tortrix were caged in nylon screen sleeve cages on the branches of treated and untreated control trees. The following parameters were recorded: 4th-instar weight, percent pupation, and number of eggs produced by emerging females.

Severe (50–100%) defoliation of aspen produced reflush foliage in early July of the year defoliation occurred, which was of lower food quality to the tortrix. The severity of defoliation was inversely related to food quality (Fig. 4). In addition, repeated defoliations resulted in a significant decline in subsequent food quality (t test, $p < .05$). For instance, tortrix larvae, when reared on aspen that had three successive years of 100% defoliation, had 4th-instar weights reduced by $75 \pm 8.5\%$, pupation success reduced by $60 \pm 5.6\%$, and fecundity reduced by $55 \pm 10.5\%$; larvae reared on aspen defoliated 100% for only one year had 4th-instar weights reduced by $18 \pm 10.6\%$, pupation success reduced by $20 \pm 11\%$, and fecundity reduced by $45 \pm 12\%$.

The chemical basis of LTI in aspen, however, remains elusive. For instance, in a $2 \times 2 \times 2$ factorial experiment with 100% prior defoliation, N fertilization, and P fertilization (Bryant, unpublished results), defoliation of phosphorus-fertilized trees resulted in leaves of lower food quality to the large aspen tortrix larvae relative to leaves of undefoliated phosphorus-fertilized trees. Specifically, pupal weights of larvae reared on these trees were reduced significantly from 144 mg (phosphorus fertilized, undefoliated) to 123 mg (phosphorus fertilized, defoliated) (t test, $p < .05$). The defoliated trees, however, were higher (t test, $p < .001$) in foliar nitrogen and lower (t test, $p < .01$) in condensed tannins, thus ruling out the possibility that tannins or leaf nitrogen were the cause of reduced food quality of the LTI trees.

Of the four glycosides present in aspen, only tremulacin significantly rose in concentration one year following defoliation in the above experiment (t test, $p < .01$; Bryant, unpublished results); levels increased from 0.35% (dry wt; P fertilized, undefoliated) to 2.2% (dry wt; P fertilized, defoliated). If tremulacin at 2% levels causes significant reductions in pupal weights of the large aspen tortrix, it is likely that tremulacin is a major component in LTI in quaking aspen. When 3rd-instar large aspen tortrix larvae were fed excised leaves that had tremulacin painted on at 1.5%, following the procedure of Lindroth et al. (1988), no decrease in pupal weights was observed (Bryant, unpublished results) when compared to larvae fed control (unpainted) leaves. When concentrations were increased to 6%, however, tremulacin did cause

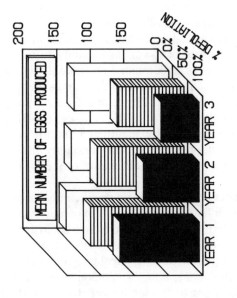

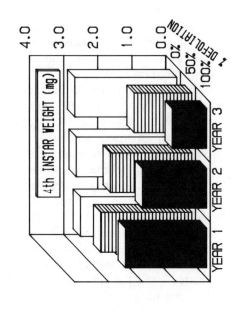

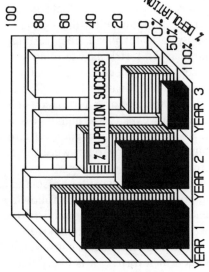

a significant (t test, $p < .01$) decrease in pupal mass of larvae obtained from a 4-year outbreak population, reducing weights from 102.4 ± 4.8mg (control leaves) to 91.4 ± 4.3mg (tremulacin-painted leaves). When larvae were obtained from a first-year outbreak zone, pupal weights were reduced even further from 101.5 ± 4.6mg (control leaves) to 75.5 ± 3.7mg (tremulacin-painted leaves). Hence, tremulacin may be an important factor in quaking aspen LTI if its levels are further enhanced by short-term changes due to herbivory.

6 POSSIBLE INTERACTIONS BETWEEN LONG- AND SHORT-TERM INDUCTION IN QUAKING ASPEN

While we have not been able to reliably predict leaf quality based on leaf chemistry (phenol glycosides, condensed tannins, or nitrogen content), leaf size is an excellent predictor of leaf quality (Fig. 5). Specifically, pupal weights of the large aspen tortrix are inversely related to the size of the leaves on which larvae have fed. Other researchers (Baltensweiler 1985, Werner 1979, Haukioja 1980) have found that LTI is inversely correlated with subsequently reduced leaf size. Bryant et al. (this volume, Chapter 6) argue that the reduction in leaf size following severe defoliation is the result of shifts in the carbon/nutrient status of the plant caused by loss of nutrients (especially nitrogen) during early season defoliation. Hence, the observed reduction in leaf size of LTI aspen trees may not be the result of an active response by the plant, but rather, a natural consequence due to boundary conditions on the growth rate imposed upon the plant by nutrient availability.

We hypothesize that while reduced leaf size is a consequence of LTI by the arguments presented by Bryant et al. (this volume, Chapter 6), it is also a cause of the reduced food value in LTI quaking aspen foliage (Fig. 2). We base this hypothesis on the assumption that elevated levels of salicortin and tremulacin in STI leaves arise from a translocation mechanism (but see below). This is consistent with the observation that in STI, foliar levels of salicortin and tremulacin increase, while levels of salicin and tremuloiden do not: Twigs contain salicortin and tremulacin in high concentrations but no salicin or tremuloiden (Clausen et al. 1989a,b). If the rate of translocation into a damaged leaf is not significantly diminished with reduced leaf size, smaller leaves will experience greater rises in salicortin and tremulacin concentrations than will larger leaves because of less dilution of translocated glycosides. In addition, if STI increases with increasing leaf damage, and neither salicortin nor tremulacin act as significant feeding deterrents, smaller

←——————————————————————————————

Figure 4 Effect of prior defoliation of quaking aspen on the fecundity, pupation success, and larval weight of the large aspen tortrix reared on reflushed foliage following 1, 2, and 3 consecutive years of defoliation.

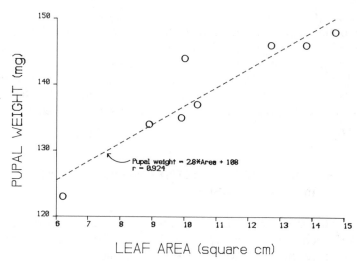

Figure 5   The relationship between aspen leaf size and food quality as measured by pupal weights of the large aspen tortrix larvae raised on trees.

leaves, because they would be more severely damaged over a fixed time period of attack, would respond to STI more intensely than their larger counterparts. Hence, LTI in quaking aspen may be due in part to an enhancement by STI.

A similar interaction between STI and LTI can be hypothesized if photosynthetic gain is the mechanism of the enhancement of salicortin and tremulacin levels in STI leaves (Fig. 2). Bryant (unpublished results) found that prior defoliation of quaking aspen results in increased photosynthetic rates. Other researchers (Waring et al. 1969, Gifford and Marshall 1973, Ollerenshaw and Incoll 1979, Mooney and Chiariello 1984) report similar trends. Hence, if recent leaf damage triggers a response by aspen to produce salicortin and tremulacin from photosynthate, then LTI aspens, because of their known higher photosynthetic rates, will have higher levels of these compounds shortly following damage.

We therefore propose that a plausible mechanism for the reduced quality of quaking aspen foliage following LTI may simply be a consequence of enhanced STI (due to either increased photosynthesis or less dilution during translocation, depending upon the mechanism of STI). In addition, because chewing results in damage to a larger percentage of leaf area in smaller leaves, STI may be enhanced in foliage on previously defoliated trees regardless of the mechanism for enhanced glycoside levels. Hence, the absence of a compelling chemical basis for LTI in quaking aspen may be because STI has been not taken into account.

The above hypothesis regarding the interaction of LTI and STI in quaking aspen has not been tested. We believe, however, that this is a plausible interaction. The assumptions leading to its formulation are reasonable and it would explain the inefficacy of past research to define the chemical basis of LTI because chemical analyses were performed on whole undamaged leaves and, thus, would not have exhibited any short-term increases in their glycoside levels. If this explanation proves to be correct, it may be applicable to other plants that undergo elevations in their defensive secondary metabolites shortly after foliar damage.

7 SUMMARY

Severe simulated insect herbivory of quaking aspen foliage results in replacement foliage having a lower food quality for large aspen tortrix larvae. The chemical basis for the long-term reduction in food quality, however, is not easily defined. Specifically, the levels of quaking aspen's major secondary metabolites, condensed tannins and four phenol glycosides, or nutrients, such as nitrogen, do not suggest a reason for the reduced food quality of previously defoliated aspens. On a time scale of hours, however, quaking aspen responds to simulated insect attack by increasing the levels of its defensive compounds, salicortin and tremulacin. We propose that the basis of the long-term reduction in food quality of previously defoliated aspens may simply be an enhancement of aspen's short-term responses to current herbivory. These assumptions are that either reduced leaf size or enhanced photosynthetic rates, which are known to be associated with long-term induced foliage, will enhance the ability of resultant leaves to respond to short-term changes induced by insect herbivory.

ACKNOWLEDGMENTS

This work was supported by the National Science Foundation (BSR-8416461 and BSR-8500160).

REFERENCES

Baltensweiler, W. 1985. On the mechanisms of the outbreaks of the larch bud moth (*Zeiraphera diniana* Gn; Lepidoptera: Tortricidae) and its impact on the subalpine larch-cembran pine ecosystem. In H. Turner and W. Tranquillini (eds.), *Establishment and Tending of Subalpine Forest: Research and Management*. Proc. 3rd Int. Workshop IUFRO Project Group P1.07–00 Ecology of subalpine zones, Sept. 3–5, 1984, CH-3981 Riederalp, Switzerland.

Beckwith, R. C. 1968. The large aspen tortrix, *Choristoneura conflictana* (Wlkr.), in interior Alaska. *USDA Forest Ser. Res. Note PNW-81*.

Bryant, J. P., F. S. Chapin III, and D. R. Klein. 1983. Carbon/nutrient balance of boreal plants in relation to vertebrate herbivory. *Oikos* 40:357–368.

Bryant, J. P., T. P. Clausen, P. B. Reichardt, M. C. McCarthy, and R. A. Werner. 1987. Effect of nitrogen fertilization upon the secondary chemistry and nutritional value of quaking aspen (*Populus tremuloides* Michx) leaves for the large aspen tortrix (*Choristoneura conflictana* (Walker)). *Oecologia (Berlin)* 73: 513–517.

Bryant, J., K. Danell, F. Provenza, P. Reichardt, T. Clausen, and R. A. Werner. 1991. Effects of mammal browsing upon the chemistry of deciduous woody plants. In D. W. Tallamy and M. J. Raupp (eds.). *Phytochemical Induction by Herbivores*. New York: Wiley, pp. 135–154.

Clausen, T. P., P. B. Reichardt, T. Evans, and J. P. Bryant. 1989a. A simple method for the isolation of salicortin, tremulacin and tremuloiden from quaking aspen (*Populus tremuloides*). *J. Nat. Prod.* 52:207–209.

Clausen, T. P., P. B. Reichardt, J. P. Bryant, R. A. Werner, K. Post, and K. Frisby. 1989b. A chemical model for short-term induction in quaking aspen (*Populus tremuloides*) foliage against herbivores. *J. Chem. Ecol.* 15:2335–2346.

Danell, K., and K. Huss-Danell. 1985. Feeding by insects on birches affected by moose browsing. *Oikos* 44:75–81.

Danell, K., K. Huss-Danell, and R. Bergstrom. 1985. Interactions between browsing moose and two species of birch in Sweden. *Ecology* 66:1867–1878.

Fowler, S. V., and V. H. Lawton. 1985. Rapidly induced defenses and talking trees: The devil's advocate position. *Am. Natur.* 126:181–195.

Gifford, R. M., and C. Marshall. 1973. Photosynthesis and assimilate distribution in *Lolium multiflorum* Lam. following differential tiller defoliation. *Aust. J. Agric. Res.* 24:297–307.

Haukioja, E. 1980. On the role of plant defenses in the fluctuations of herbivore populations. *Oikos* 35:202–213.

Haukioja, E., and P. Niemelä. 1977. Retarded growth of geometrid larva after mechanical damage to leaves of its host tree. *Ann. Zool. Fennici* 14:48–52.

Haukioja, E., K. Kapiainen, P. Niemelä, and J. Toumi. 1983. Plant availability hypothesis and other explanations of herbivore cycles: complementary or exclusive alternatives. *Oikos* 40:419–432.

Hsu, F., G. Nonaka, and N. Shioka. 1985. Acetylated flavanols and proanthocyanidins from Salix sieboldiana. *Phytochemistry* 24:2089–2092.

Lindroth, R. L., and M. S. Pajutee. 1987. Chemical analysis of phenolic glycosides: Art, facts and artifacts. *Oecologia (Berlin)* 74:144–148.

Lindroth, R. L., and S. S. Peterson. 1988. Effects of plant phenols on performance of southern armyworm larvae. *Oecologia (Berlin)* 75:185–189.

Lindroth, R. L., M. T. S. Hsia, and J. M. Scriber. 1987a. Characterization of phenolic glycosides from quaking aspen. *Biochem. Syst. Ecol.* 15:677–680.

Lindroth, R. L., M. T. S. Hsia, and J. M. Scriber. 1987b. Seasonal patterns in the phytochemistry of three *Populus* species. *Biochem. Syst. Ecol.* 15:681–686.

Lindroth, R. L., J. M. Scriber, and M. T. S. Hsia. 1988. Chemical ecology of the tiger swallowtail: mediation of host use by phenolic glycosides. *Ecology* 69:814–822.

Mattes, B. R., T. P. Clausen, and P. B. Reichardt. 1987. Volatile constituents of balsam poplar: the phenol glycoside connection. *Phytochemistry* 26:1361–1366.

Mattson, W. J., and S. R. Palmer. 1987. Changes in levels of foliar minerals and phenolics in trembling aspen (*Populus tremuloides*) in response to artificial defoliation. In W. J. Mattson, J. Levieux, and C. Bernard-Dagen (eds.), *Mechanisms of Woody Plant Defenses against Insects: Search for Pattern*. New York: Springer, pp. 155–168.

Mattson, W. J., R. K. Lawrence, R. A. Haack, D. A. Herms, and P. J. Charles. 1987. Defensive strategies of woody plants against different insect feeding guilds in relation to plant ecological strategies and intimacy of association with insects. In W. J. Mattson, J. Levieux, and C. Bernard-Dagen (eds.), *Mechanisms of Woody Plant Defense against Insects: Search for Pattern*. New York: Springer, pp. 3–38.

Mooney, H. A., and N. R. Chiariello. 1984. The study of plant function: The plant as a balanced system. In R. Dirzo and J. Sarufan (eds.), *Perspectives in Plant Population Ecology*. Sunderland, MA: Sinauer Assoc., pp. 305–323.

Myers, J. S. 1988. The induced defense hypothesis: does it apply to the population dynamics of insects? In K. Spencer (ed.), *Chemical Mediation of Coevolution*. New York: Academic, pp. 345–366.

Neuvonen, S., and E. Haukioja. 1991. The effects of inducible responses in host foliage on birch feeding herbivores. In D. W. Tallamy and M. J. Raupp (eds.), *Phytochemical Induction by Herbivores*. New York: Wiley, pp. 277–291.

Ollerenshaw, J. H., and L. D. Incoll. 1979. Leaf photosynthesis in pure stands of two grasses (*Lolium perenne* and *Lolium multiflorum*) subjected to contrasting intensities of defoliation. *Ann. Appl. Biol.* 92:133–142.

Pearl, I. A., and S. F. Darling. 1971. The structures of salicortin and tremulacin. *Phytochemistry* 10:3161–3166.

Provenza, F. D., and J. C. Malechek. 1984. Diet selection by domestic goats in relation to blackbrush twig chemistry. *J. Appl. Ecol.* 21:831–841.

Raupp, M. J., and C. S. Sadof. 1991. Leaf beetles responses to damaged induced changes in their salicaceous host. In D. W. Tallamy and M. J. Raupp (eds.), *Phytochemical Induction by Herbivores*. New York: Wiley, pp. 183–204.

Rhoades, D. F. 1979. Evolution of plant chemical defense against herbivores. In G. A. Rosenthal Janzen (eds.), *Herbivores: Their Interactions with Secondary Plant Metabolites*. New York: Academic, pp. 4–48.

Schultz, J. C., and I. T. Baldwin. 1982. Oak leaf quality declines in response to defoliation by gypsy moth larvae. *Science* 217:149–151.

Waring, P.F., M. N. Khalifa, and K. J. Treharne. 1969. Enhanced photosynthetic rate by leaf removal. *Nature* 220:453–458.

Werner, R. A. 1979. Influence of host foliage on development, survival, fecundity and oviposition of the spear-marked black moth, *Rheumaptera hastata* (Lepidoptera: Geometridae). *Can. Entomol.* 111:317–322.

CHAPTER 4

CARBON ALLOCATION, PHENOTYPIC PLASTICITY, AND INDUCED DEFENSES

JUHA TUOMI and TORBJÖRN FAGERSTRÖM
Department of Ecology, Lund University, Ecology Building, S-223 62 Lund, Sweden

PEKKA NIEMELÄ
Department of Forest Protection, The Finnish Forest Research Institute, SF-01301 Vantaa, Finland

1 Introduction
2 Carbon allocation and defense
   2.1 Defense and trade-offs
   2.2 Allocation models
   2.3 Components of carbon-based defenses
3 Phenotypic flexibility of carbon-based defenses
   3.1 Constitutive defenses
   3.2 Induced responses
4 Responding units: plants or modules?
   4.1 Modules and plant fitness
   4.2 Responses to localized disturbances
5 Triggering induced responses
6 Conclusion
   References

1 INTRODUCTION

The idea that plants have active antiherbivore defenses has dominated recent plant–herbivore studies. Many of these studies have interpreted potentially defensive characteristics of host plants as counteradaptations to herbivory. The basic tenet of defense theory (Feeny 1970, 1976, Janzen 1974, Orians et al. 1974, Rhoades 1979) is that plants have evolved to achieve optimal partitioning of limited resources between defense and alternative demands (e.g., growth and reproduction) for carbon and mineral nutrients. The optimal constitutive level of defense maximizes plant fitness: that is, the balance between profits (reduced material losses to herbivores) and metabolic costs. Thus, the relative allocation of limited resources between defense and biomass production is considered an evolutionary response of host plants to herbivory.

The incidence of induced defenses has also been analyzed within this evolutionary context. There should be selection for a capacity to respond defensively when high levels of constitutive defenses are not favored due to fluctuating herbivore densities. If defenses are costly, low constitutive defenses are favored during low herbivore densities. An increase in defenses is expected under a high risk of herbivory when the profits of the increased defense level can pay for the costs. Accordingly, the level of induced defenses should gradually decay in the plant population with a declining risk of herbivory. This logic is mainly developed for long-term induced responses (Haukioja and Hakala 1975, Rhoades 1979), but it may also be applied to short-term induced defenses (sensu Haukioja 1982). In both cases, to act as active antiherbivore defenses, such induced responses should (1) be triggered by cues that can be used by the host plant as an indicator for an increased risk of herbivory, and (2) result from an active allocation of limited resources into defenses.

Below we study what will happen to this logic if we relax the assumption of an active or free allocation between defense and biomass production. We study this question in relation to carbon-based allelochemicals, and propose, following Del Moral (1972), that carbon-based secondary substances may have evolved as metabolites that have later acquired additional functions, such as defense against pathogens and herbivores. Consequently, these additional functions may not be sufficient to completely account for the overall variation of carbon-based secondary metabolites. Instead, the total concentrations of carbon-based allelochemicals can depend largely on internal and external constraints limiting plant primary metabolism.

2   CARBON ALLOCATION AND DEFENSE

2.1   Defense and Trade-offs

Chemical defense levels of plants depend on the concentrations of specific secondary compounds and their effectiveness against specific herbivores. Concentrations, in turn, vary as a function of (1) turnover and recirculation rates of secondary compounds and (2) rates at which available resources are invested into secondary metabolism. Fagerström (1989) thus expressed the concentration c of chemical defense substances in an exponentially growing plant as

$$C = \frac{\mu}{\alpha + \lambda\,(1 - \eta)} \tag{1}$$

where μ denotes the current investment rate in defense (invested re-sources·biomass^{-1}·time^{-1}), λ is the current turnover rates of defensive chemicals (time^{-1}), η is the fraction of the turned over defense that is re-circulated into secondary metabolism, and α is the relative growth rate of the biomass of the plant. Hence, the investment rate in defense required to maintain a given concentration of defensive chemicals or

$$\mu = C[\alpha + \lambda (1 - \eta)] \tag{2}$$

is given by that concentration multiplied by the sum of two specific rate constants: the relative growth rate of biomass, α, and the net turnover rate of the defense, $\lambda(1 - \eta)$. This describes a physiological strain required to maintain a given concentration of defensive substances. Such a strain is generally assumed to imply trade-offs between defense and growth. That is, an increase in defense investment should reduce growth, and this reduction thus denotes the maintenance costs of defense.

The relation between defense allocation and associated costs is compli-cated by two factors. First, the resource requirements for primary and sec-ondary compounds contain neither nitrogen nor phosphorus, which are es-sential to growth and protein synthesis (Mooney et al. 1983). To overcome this, we define ρ as a parameter that quantifies the degree of overlap between resource requirements of secondary metabolism and growth. The values of this parameter, introduced and discussed by Fagerström (1989), can vary from 0 (defense drains no resource from growth) to some upper fixed value that corresponds to a complete trade-off between defense and growth. Sec-ond, the growth parameter α in equations 1 and 2 is itself a function of concentration due to the existence of a trade-off between growth and de-fense. When these factors are taken into account, it can be shown (Fager-ström 1989) that the cost in terms of lost growth rate of maintaining a given concentration c is given by

$$K(c) = \rho \, \mu = \frac{\rho c \, [\alpha(0) + \lambda (1 - \eta)]}{1 + \rho c} \tag{3}$$

where $\alpha(0)$ is the relative growth rate of plant biomass with zero concen-tration of defense compounds. This is an increasing convex function of c and, for any fixed concentration, ρ increases the associated costs as ex-pressed by K (for details see Fagerström 1989).

2.2 Allocation Models

Investment rates are essential to the maintenance of given concentrations of allelochemicals. The absolute investment rate to secondary metabolism may be further subdivided into (1) the total amount of resources available to secondary metabolite production and alternative demands for investment, and (2) the proportional allocation of these resources into secondary metabolism. We consider carbon allocation to involve the distribution of the net daily carbon gain C_t between two functional pools: carbon used for secondary metabolism C_d and carbon used for growth C_g. Thus, the maintenance costs of carbon-based defense can be redefined as

$$K(c) = \rho C_d = \rho p C_t \qquad (4)$$

where p is the fraction of plant carbon gain invested in defense. In this expression, the strength ρ of the trade-off between the production of carbon-based allelochemicals and growth indicates how actively plants are allocating available carbon into carbon-based defense. Completely active defense allocation and entirely passive (i.e., $\rho = 0$) antiherbivore resistance represent the extreme end-points of this trade-off continuum. We specify these extreme possibilities in terms of two allocation models which describe variation in C_d in relation to plant carbon and mineral nutrient availability.

In the *fixed allocation model*, we assume a constant proportionality between growth and defense investments regardless of plant nutrient reserves N so that

$$C_d = P_1 C_t(N) \qquad (5)$$

Then the consequent amount of carbon available to growth will be

$$C_g = (1 - P_1) C_t(N) \qquad (6)$$

where P_1 denotes the proportion of available carbon invested into the production of carbon-based allelochemicals instead of growth. Thus, there is in this formula a complete overlap of resource requirements between growth and defense so that, for a constant C_t, an increase in defense level inevitably causes a corresponding decrease in growth and vice versa.

Because constitutive defense is assumed to cause costs on growth, a necessary condition for an increase in P_1 through evolutionary time is that the profits of increased defense exceed the costs. This would be the case in plant populations that are subjected to intense herbivory, and in conditions where plants have low inherent capabilities of compensating for the resources removed by herbivores (Rhoades 1979). Therefore, plants with high levels of constitutive defenses are expected to dominate in nutrient-deficient growing sites where plants cannot readily acquire resources to replace the losses caused by leaf removal (Janzen 1974, McKey et al. 1978). Constitutive defense levels are also expected to be high in slowly growing evergreen trees with resource reserves in leaves, whereas deciduous trees with large belowground resource reserves have better capacities for compensatory growth (Bryant et al. 1983, Coley et al. 1985).

A complete overlap of resource requirements is an idealized case where defense and growth are equally limited by the same resources. Since carbon-based allelochemicals contain no nitrogen or phosphorus required for promoting growth, Bryant et al. (1983) proposed that the accumulation of carbon-based allelochemicals is likely to be favored under conditions where growth is limited more by mineral nutrients than by carbon. Under such conditions, the costs of carbon-based defense would be relatively low since the plant cannot use all the available carbon for growth due to nutrient deficiency. Margna (1977) has presented an equivalent reasoning concerning phenylpropanoid accumulation. He suggested that plant phenols tend to accumulate in plant cells under conditions where protein synthesis is suppressed, whereas conditions favorable to protein synthesis tend to lead to the reverse consequences. Indeed, plant phenols have been shown in several studies to accumulate under nutrient deficiency and to decline after nitrogen application (Del Moral 1972, Gershenzon 1984, Tuomi et al. 1984).

Insofar as secondary metabolite production is a function of resource (Bryant et al. 1983) and substrate (Margna 1977) availability, we can assume that the relative allocation of carbon into defenses is itself a phenotypically flexible trait rather than an evolutionarily fixed characteristic. Tuomi et al. (1988a) proposed such a *flexible allocation model* where carbon is allocated to growth whenever there are sufficient mineral nutrients to construct new cells. The carbon pool available to allelochemicals would thus involve the excess carbon C_e accumulated above growth investments so that

$$C_e = C_t(N) - C_g(N) \tag{7}$$

and C_d could vary as a function of C_e. If we further assume that $C_d = C_e$, the overall proportional allocation of net carbon gain to defense, P_g, or

$$P_2 = C_e/C_t(N) \tag{8}$$

would thus be a phenotypically flexible trait that varies as a function of internal and external conditions modifying C_e relative to C_t. Contrary to the assumptions of our first model, this second model implies no reduction in growth for supporting carbon-based defense, since defense investments vary possibly as a by-product of the excess carbon accumulated above the level that can be used for growth (i.e., $\rho = 0$).

This is also an idealized case since the plant could invest no resources into defense ($C_d = 0$) in such conditions where all carbon can be used to promote growth. This idealization can still be useful though, since it specifies a situation where plant carbon/nutrient balance (Bryant et al. 1983) can phenotypically modify the accumulation of carbon-based secondary compounds. Below we neglect turnover rates of secondary substances, and use absolute defense investments C_d as an index of defense level.

2.3 Components of Carbon-Based Defenses

Although we contrast the fixed and flexible allocation models, both models may, in fact, specify adequate components of carbon-based defenses. Thus, a model with both P_1 and P_2 could presumably better describe actual allocation of carbon between defense and growth. In such a model, active defense allocation could determine some background level of defense investments (i.e., P_1) that imply costs on growth. This background level of defense could also be modified by selection depending on the costs and benefits of defense investments. In addition to such evolutionary modifications, the actual defense level is also subjected to phenotypic variation (i.e., P_2) caused by internal and external conditions constraining plant primary metabolism. Plant carbon/nutrient balance could thus be a factor that accounts for a part of the phenotypic variation in constitutive levels of carbon-based allelochemicals (Bryant et al. 1983), and which also may lead to induced responses of plants to the removal of foliar resources (Tuomi et al. 1984, Tuomi et al. 1988a, Bryant et al. 1988).

Furthermore, our two models specify quantitative changes in defensive investments. The overall investment to defense may not, however, always be the most crucial factor in maintaining plant resistance against specific herbivores. For example, a given defense investment C_d may further be allocated among the production of various specific allelochemicals. This qualitative component of defense allocation may be particularly dependent on the selective importance of specific herbivores over the evolutionary history of the plant population (Tuomi et al. 1988a). Selection may thus eliminate chemical pathways leading to secondary substances that do not prevent herbivore attacks, while other pathways leading to defensively effective substances are favored. Plants may also be selected for an ability to respond to leaf damage through production of these defensively effective allelochemicals (Rhoades 1979, Haukioja and Neuvonen 1985).

3 PHENOTYPIC FLEXIBILITY OF CARBON-BASED DEFENSES

3.1 Constitutive Defenses

Due to the constant proportionality between growth and defense, our first model allows only limited phenotypic flexibility in the concentrations of carbon-based allelochemicals. The main source of nongenetic variation would be the fluctuations in total carbon reserves. Carbon deficiency thus provides a constraint that may efficiently limit the accumulation of carbon-based secondary compounds (Mooney and Gulmon 1982, Gulmon and Mooney 1986, Niemelä et al. 1984b). If, in the present formulation, we assume that the net daily carbon gain C_t increases asymptotically with plant nutrient reserves N (cf. Gulmon and Mooney 1986), then the fixed allocation model would result in a corresponding increase in defense level C_d as well (Fig.

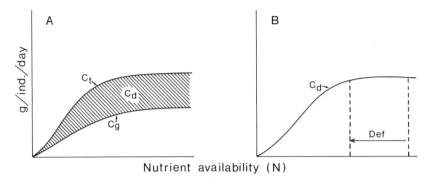

Figure 1   A fixed allocation model (*A*) in which the plant maintains a constant proportional allocation into defense independent of mineral nutrient reserves (N), and (*B*) which implies no change or reduction in defense investments as a response to defoliation-caused nutrient deficiency. C_t = plant net carbon gain, C_2 = carbon diverted to growth, and C_d = carbon allocated into defense.

1*A*). Although the proportional allocation to the production of secondary compounds remains constant, C_d increases asymptotically with *N*. Consequently, nutrient application should increase the absolute concentrations of allelochemicals in plant tissues at nutrient-limited growing sites. This formulation is consistent with the defense theory, expecting lower herbivore resistance under stress conditions (Haukioja and Hakala 1975, Rhoades 1979).

The response of plants obeying the flexible allocation model likely differs from this pattern since then the accumulation of carbon-based allelochemicals depends on the relative sensitivity of photosynthesis and growth to plant nutrient reserves (Tuomi et al. 1988a). If we assume the net carbon gain C_t to be less sensitive to nutrient deficiency than growth C_g (Chapin 1980), the maximum accumulation of surplus carbon C_e above the requirements of growth would take place at moderate nutrient availability. According to the present assumptions of the flexible allocation model, the defense level as indicated by C_d should follow this very same pattern (Fig. 2*A*). Thus, C_d should be low under extreme nutrient deficiency with both suppressed photosynthesis and growth, and also under extremely high nutrient availability where the most part of C_t is diverted to growth. Consequently, this flexible allocation model would lead to a declining level of carbon-based defenses as a response to nutrient application at low and moderate nutrient availabilities. This expectation presupposes that the fertilization treatment actually affects growth more than photosynthesis. At low nutrient availabilities the reverse effects may result, especially if small amounts of fertilizers are applied to the soil (cf. Fig. 2*A*).

The contrast between these two models has two main applications. First, carbon-based secondary compounds may differ with respect to their re-

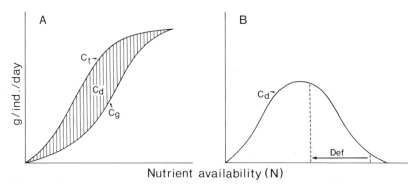

Figure 2   A flexible allocation model (*A*) in which the proportional defense allocation varies as a function of plant nutrient reserves (N), and (*B*) which can result in an increased defense level as defoliation reduces plant mineral nutrient reserves available to growth. C_t = plant net carbon gain, C_2 = carbon diverted to growth, and C_d = carbon allocated into defense. (From Tuomi et al. 1988a.)

sponses to nutrient availability. Some compounds may show a fixed allocation pattern, whereas others may depend more on the constraints limiting plant primary metabolism. This may also have evolutionary importance if these response patterns parallel respectively with substances originally selected to protect the plant (primary allelochemicals) and, on the other hand, metabolites that have acquired defense as an additional function (secondary allelochemicals). Del Moral (1972), for instance, suggested that defense could be an additional function for plant phenols which generally tend to follow a flexible allocation pattern (Gershenzon 1984).

Second, plants may differ in their defense allocation. The fixed allocation pattern would result in a lower maximum growth rate at high nutrient availability than the flexible allocation tactics (Figs. 1*A*, 2*A*). It may also be possible, as predicted by defense theory, that slowly growing evergreen woody plants are selected for a relatively high proportional allocation to defenses; these species could thus show fixed allocation patterns. On the other hand, rapidly growing deciduous species may show more flexible defense allocation, and also higher capacities for compensatory growth at high nutrient availabilities (Bryant et al. 1983, Coley et al. 1985).

3.2   Induced Responses

Constitutive and induced defenses are mainly distinguished in relation to the presence of herbivores (Levin 1971, Haukioja and Hakala 1975, Rhoades 1979). Constitutive defense levels are not affected by herbivores, whereas induced defense refers to the increase in plant resistance as a consequence of herbivore presence. We use induced responses in a wider meaning, namely, to involve all changes in plant resistance as a response to herbivore damage and presence. The resistance of host plants may thus increase or decrease after herbivore attacks, depending on how the removal of foliar resources influences chemical constituents of plan resistance (Niemelä et al. 1984a).

The fixed allocation model does not allow carbon-based defense to increase passively in response to the removal of foliar resources. As herbivores remove foliar carbon pools and reduce photosynthetic leaf area, the amount of carbon available for secondary metabolism is reduced. Since the net carbon gain is a positive function of plant mineral nutrient reserves, the loss of nutrients should cause the same effect, especially at low and moderate nutrient availabilities (Fig. 1*B*). At high nutrient availability there may, however, be no significant response, provided plant nutrient reserves do not decline to a suboptimal level as a consequence of defoliation. Thus, in this model, an active increase (1) in the rate of photosynthesis or (2) in the proportional allocation to defenses is needed to induce higher carbon-based resistance. These changes should also cover the negative effects that the loss of foliar resources and photosynthetic leaf area have on secondary metabolism.

The flexible allocation model allows a variety of induced responses from an increased to a reduced carbon-based defense (Tuomi et al. 1988a). The type of induced responses mainly depends on whether defoliation influences plant carbon more than nutrient reserves, and whether plants are growing on nutrient-rich or poor soils (Fig. 2*B*). Carbon deficiency tends to affect defense in the same way in both models. Thus, the flexible allocation model also implies a reduced carbon-based defense under conditions where defoliation influences carbon reserves more than nutrient reserves. When defoliation influences mineral nutrients more strongly than carbon availability, an accumulation or carbon above the requirements of growth is expected. Consequently, the accumulation of carbon-based secondary metabolites would most likely occur at low and moderate nutrient availabilities (Tuomi et al. 1984).

Complete defoliations of *Betula pubescens* ssp. *tortuosa* and *Salix lapponum*, for instance, are followed in the subsequent growing season by increased levels of foliage phenols and reduced foliar nitrogen content (Fig. 3). These experiments (Tuomi et al. 1984, Niemelä et al., unpublished results) were done in the northernmost Finnish Lapland characterized by nutrient-poor soil (Hinneri 1974). In both species the responses were gradually relaxed during successive years after defoliation. In *B. pubescens* ssp. *tortuosa* foliar nitrogen and phenols returned to the control level within four years, while in *S. lapponum* the relaxation occurred within two years. In both species foliage nitrogen levels tended to relax slightly earlier than foliage phenols (Fig. 3). (For further examples and discussion, see Bryant et al. this volume, Chapter 6).

These examples are consistent with the nutrient stress hypothesis, which predicts an accumulation of carbon-based secondary compounds as a re-

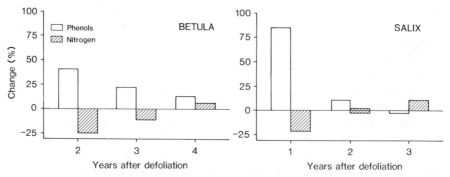

Figure 3   Changes in total foliage phenols (white bars) and nitrogen (shaded bars) as a response to previous complete defoliations of individual trees of *Betula pubescens* spp. *tortuosa* (data from Tuomi et al. 1984) and individual bushes of *Salix lapponum* (data from Niemelä et al., unpublished results) in Finnish Lapland. Negative values indicate a decline and positive values an increase in concentrations relative to controls.

sponse to defoliation-induced deficiency of mineral nutrients (Tuomi et al. 1984). This illustrates a mechanism whereby an improvement in plant resistance is a by-product of the suppression of the alternative chemical pathways supporting protein synthesis (Margna 1977) and growth (Bryant et al. 1983). In terms of resource allocation, such responses do not represent active induced defenses; due to nutrient limitation, the plant cannot allocate carbon or other substrates to alternative pathways. In other words, responses to defoliation-caused resource limitations are expressions of phenotypic flexibility, rather than consequences of active resource allocation.

4   RESPONDING UNITS: PLANTS OR MODULES?

4.1   Modules and Plant Fitness

The defense theory has not clearly specified the units that actually make the decisions of resource allocation to defenses. The implicit proposition of the theory is that defense allocation should maximize the reproductive success of plants, or rather their inclusive fitness (Rhoades 1979). Usually, we perceive entire plants as the evolutionary decision makers, defending most efficiently the plant parts that are most valuable in the currency of fitness and most susceptible to herbivore attacks (Rhoades 1979).

However, plant parts or modules (Harper 1977) may also themselves be partially autonomous functional units (Tuomi et al. 1983, Vuorisalo and Tuomi 1986, Tuomi and Vuorisalo 1989). Even genetic variation in plant resistance may exist within individual plants (Whitham and Slobodchikoff 1981), suggesting that plant parts may be relatively autonomous in their defense allocation (Gill 1986). Watson and Casper (1984) defined the plant units capable of relatively autonomous allocation decisions as integrated physiological units or IPUs. They suggested that the vascular organization of dicotyledons could give rise to sectorial distribution of carbon within individual growing plants (e.g., Oparka and Davies 1985, Watson 1986). Thus, these sectorial root–shoot subsystems could have at least partially autonomous carbon budgets, leading to variation in carbon allocation within plants.

Such a partial autonomy of plant parts can, at least in theory, lead to variation in the constitutive levels of carbon-based allelochemicals. In the fixed allocation model this is possible if plant parts make independent decisions of defense allocation and thereby contribute to the fitness of the entire plant. Whitham and Slobodchikoff (1981) suggested that such mosaicism in plant defenses could especially improve the fitness of long-lived plants relative to insect herbivores with short generation times (see also Antolin and Strobeck 1985). An analogous approach may be applied to induced defenses: If modules make independent allocation decisions, should induced responses be restricted to the attacked modules or should nonattacked modules increase their defense levels as well? The overall fitness of the plant, then, depends on the distribution of expected further herbivory as well as the metabolic costs of induced defenses among semiautonomous modules.

In Table 1 we have assumed that herbivory reduces the fitness of undefended modules by the factor L relative to defended modules, which, in turn, are subjected to higher metabolic costs K due to induced defenses. These fitness components are then specified for both attacked (L_a and K_a) and nonattacked (L_{na} and K_{na}) modules. Table 1 shows the consequent fitness values of a plant with two modules, and Table 2 demonstrates three optimal defense tactics of the plant with different balances between material losses to herbivores and costs of induced defenses.

If induced defenses are costly for both modules relative to the risk of further herbivory, neither the attacked nor nonattacked module should increase its defense level. If, on the other hand, the expected risk of herbivory

TABLE 1   Plant Fitness Expressed in Terms of the Reproductive Success of a Previously Attacked (a) and a Nonattacked (na) Module, which Independently Make Decisions Concerning Induced Defenses[a]

Attacked Module	Nonattacked Module	
	D	ND
D	$2P_0 - (K_a + K_{na})$	$2P_0 - (K_a + L_{na})$
ND	$2P_0 - (L_a + K_{na})$	$2P_0 - (L_a + L_{na})$

[a] D, a module increases proportional allocation into defense; ND, no increase in defense level; P_0, module fitness in the absence of herbivores, K, reduction in the fitness of a module due to increased defense level; L, reduction in the fitness of an undefended module due to herbivory.

TABLE 2   Expected Induced Responses of a Plant with Two Modules that Independently Make Defensive Allocation Decisions[a]

Assumptions	Fitness Components	Induced Responses
1. Low risk of further herbivory equally distributed among modules	$L_a < K_a$ $L_{na} < K_{na}$	Neither attacked nor nonattacked modules respond defensively.
2. High risk of further herbivory equally distributed among modules	$L_a > K_a$ $L_{na} > K_{na}$	Both attacked and nonattacked modules respond defensively.
3. Low risk of further herbivory unequally distributed among modules	$L_a > K_a$ $L_{na} < K_{na}$	Attacked module responds defensively, nonattacked module does not.

[a] Symbols as in Table 1. See text for further discussion.

is higher for the two modules than the costs of induced defenses, then both modules should increase their defense levels as a response to the increased risk of herbivory.

These two situations are applicable to conditions where herbivore attacks increase the risk of further herbivory among both previously attacked and nonattacked modules. Then the increased risk of herbivory should lead to more or less uniform inducible responses throughout the entire plant. These responses should be most probable in long-term induced defenses because herbivore damages are unlikely to be restricted to the same modules during successive years. Short-term induced defenses can also show the same pattern if herbivores can move from one module to another or if the colonization of one module increases the probability of further colonization among the other, previously nonattacked modules.

The above heuristic reasoning is largely consistent with earlier suggestions that the induced responses should spread from attacked to nonattacked parts of the host plants (e.g., Ryan and Green 1974). This is not, however, a necessary pattern in short-term inducible defenses. If the herbivore damage is highly restricted (e.g., leaf miners), the risk of further herbivory may remain low among nonattacked modules so that only the attacked modules should respond defensively to localized feeding (Table 2). In this case, the induced responses should spread over the entire plant only when all or most modules are attacked by herbivores. Localized short-term inducible defenses should also force herbivore damage to be distributed more widely over the foliage (Edwards and Wratten 1983) so that it can be more easily compensated for by the plant relative to highly concentrated herbivore feeding (Watson and Casper 1985).

4.2   Responses to Localized Disturbances

Phenotypic mosaicism of plant defenses can arise as a consequence of independent actions by physically connected modules. In the flexible allocation model this source of variation can be caused by varying resource limitation within individual plants, provided that plant parts are sufficiently autonomous to maintain their own carbon/nutrient balances.

Margna (1977) proposed that substrate availability limits phenylpropanoid accumulation at the cellular level. Phillips and Henshaw (1977) have shown experimentally that, indeed, nitrogen deficiency promotes phenolic production and suppresses protein synthesis in cell cultures of *Acer pseudoplatanus*. Consequently, the relative availability of carbon and mineral nutrients can constrain the secondary metabolism of plant cells. This can also lead to phenotypic mosaicism within the same plant, even within single leaves. Niemelä et al. (1984b), for instance, observed in variegated leaves of *A. pseudoplatanus* that green leaf parts containing cells with normal plastids tended to contain more phenols than white parts with abnormal chloroplasts or parts with both green and white cells.

Plant cells that are separated by cell walls from the external environment can likely maintain their own carbon–nutrient balances. This is not so obvious in the case of plant modules that are more continuously connected with each other (Vuorisalo and Tuomi 1986, Tuomi and Vuorisalo 1989). It is not likely that the modular structure of plants results in fixed functional subsystems, although vascular organization and branching systems can create discontinuity in resource distribution within shoots (Watson and Casper 1984, Watson 1986). In such hierarchically organized branching systems, each level may attain its own carbon/nutrient balance and respond to the perturbations influencing this balance. The units responding to external stresses would thus be a function of the area influenced by the stimuli themselves. If herbivores remove the leaves from a single shoot, a branch, or an entire plant, the responding unit would be the one whose resource balance is perturbed by the damage. Consequently, localized herbivore damages should cause the same changes in plant carbon/nutrient balance at the level of damaged subunits as earlier observed or predicted at the level of entire plants.

Tuomi et al. (1988b) tested this possibility with mountain birch, which was earlier shown to respond to complete defoliations of birch trees. When particular branches were completely defoliated, it was observed during the next summer that total phenols were higher and total nitrogen lower in mature leaves growing on defoliated than undefoliated branches of the same trees (Fig. 4). Branches responded in a similar way as did entire birch trees after complete defoliations (Fig. 3). These localized responses also tended to be stronger on branches that had been defoliated shortly after leaf flush in spring than on branches defoliated in the end of the growing season. Also, leaf size was reduced the next summer on defoliated branches, more so on early defoliated branches (Tuomi et al. 1989). This indicates that such induced responses are associated with localized resource deficiencies suppressing leaf growth and favoring the accumulation of phenolic compounds.

The flexible allocation model allows, at least in principle, relatively localized responses to defoliation. We consider perturbations in plant carbon–

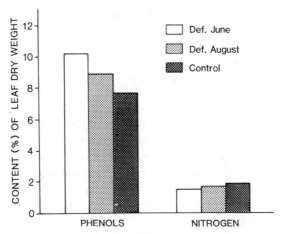

Figure 4   Average total phenols and nitrogen in leaf samples taken from undefoliated branches (black bars) of mountain birch (*Betula pubescens* ssp. *tortuosa*) and from adjacent defoliated branches of same trees. Treated branches were defoliated early (in June, white bars) or late (in August, shaded bars) during the previous growing season (data from Tuomi et al. 1988b).

nutrient balance as a potential mechanism for localized long-term inducible responses, which are more difficult to explain in terms of active allocation models. On the other hand, short-term as well as long-term inducible responses that spread from attacked to nonattacked parts of the plant can less likely be attributed to changes in the carbon/nutrient balance.

5   TRIGGERING INDUCED RESPONSES

A key assumption of the defense theory is that the triggering as well as the relaxation of induced defenses should depend on the risk of herbivory relative to the metabolic costs of an increased defense level (Haukioja and Hakala 1975, Rhoades 1979). Thus, any factor that predictably correlates with the degree of herbivory can function as a potential triggering cue of induced defenses. Cues can be (1) mechanical-associated with (a) the damage of plant tissue (wounding) and/or (b) the removal of plant material, and (2) chemical-associated with the presence of herbivores (e.g., Haukioja and Neuvonen 1985).

Chemical cues potentially allow the most specific interactions if the host plant can recognize the chemical cues of separate herbivore species. The recognition of specific chemical cues, in turn, enables attacks of different herbivore species to induce qualitatively and quantitatively different changes in the secondary metabolism of the plant. Such specific changes are less likely to occur in response to mechanical cues. Mechanical damage can also induce accumulation of secondary compounds that protect the tissues surrounding the injury against pathogen infections.

Consequently, specific chemical cues are likely to be needed to induce a particular resistance response against specific herbivores and pathogens. Wounding, for instance, has been shown to increase the activity of phenylalanine ammonialyase (PAL), the first enzyme of phenylpropanoid metabolism in higher plants (see Hartley and Lawton, this volume, Chapter 5). In some instances, the increase in PAL activity has been associated with the production of ethylene stimulated by wounding. Various cues may differently manipulate phenolic synthesis. Pascholati et al. (1986), for instance, showed that wounding does not result in the accumulation of anthocyanins even though PAL activity is increased. This differs from the results obtained by Heim et al. (1983) who demonstrated an accumulation of anthocyanins in response to fungus infection, whereas PAL activity was not changed. Pascholati et al. (1986) thus concluded that "fungal infection alters phenolic synthesis in one direction whereas wounding alters it in another." Zucker (1983) has also suggested that hydrolyzable tannins could serve as antiherbivore compounds, while condensed tannins would primarily provide resistance against pathogens (cf. Faeth 1986). Consequently, chemical cues specific to herbivores and pathogens may differently alter the production of phenolic compounds and other secondary compounds involved in plant resistance.

Specific chemical interactions between plants and herbivores provide a possibility to separate specific defensive responses from nonspecific chemical changes caused by altered carbon/nutrient balances of the host plant (Tuomi et al. 1988a). Such specific responses should be associated with triggering mechanisms that (1) stimulate secondary metabolism and/or (2) alter chemical pathways of secondary metabolism toward defensively effective compounds. In comparison, the altered carbon/nutrient balance is associated with changes in relative abundances of plant carbon and mineral nutrients that can be favorable or unfavorable to the accumulation of secondary compounds. These nonspecific changes should also equally affect defensively ineffective and effective compounds. This leads to the possibility that specific defensive responses are associated with the enzymatic regulation of secondary metabolism, whereas the plant carbon/nutrient balance mainly affects the amount of precursors and substrates available to the synthesis of secondary metabolites. Both aspects can be essential to induced responses, and their importance can vary depending on the kind of triggering cues and the degree of foliage damage.

The carbon/nutrient balance explanations of induced responses are most unlikely in conditions where the responses are triggered by chemical cues or very slight leaf damage, that is, in wound-induced short-term responses (e.g., Ryan and Green 1974) and especially in responses to chemical cues stimulated by the damage of neighboring trees (for discussion see Baldwin and Schultz 1983, Rhoades 1983, Fowler and Lawton 1985). Alterations in plant carbon/nutrient balance cannot, however, a priori be excluded in short-distance chemical interactions between trees (cf. Haukioja et al. 1985). For example, defoliated trees can compensate for removed foliar nutrients by increased mineral nutrient uptake, which can adversely influence the nutrient availability in the soil for undefoliated trees growing in their immediate vicinity. Furthermore, the degree of damage should always be related to the responding units. Damage that is too slight to affect the carbon–nutrient balance of the entire plant, may still cause local changes in resource availability due to removed foliar resources, or to altered water balance and source–sink relations.

This diversity of possibilities emphasizes a need for a theory of inducible responses that specifies the following:

1. Relations between enzymatic regulation and substrate availability in secondary metabolism.
2. Various physiological changes and responding units associated with herbivore damage.
3. The aspects that may have been modified by herbivory over evolutionary time, as well as those that primarily are expressions of phenotypic plasticity induced by variation in resource availability.

6 CONCLUSION

We contrast two evolutionary scenarios of plant chemical defense. Defense can be the primary function of secondary compounds that are actively supported by the plant in terms of resources drained from alternative demands for carbon and mineral nutrients. In contrast, defense can be an additional or secondary function that can only partially account for the overall variation in secondary metabolism. The latter view is consistent with hypotheses predicting the accumulation of carbon-based secondary compounds as a by-product of suppressed protein synthesis and growth. We formulate fixed and flexible models of carbon allocation that correspond to these evolutionary schemata.

The fixed allocation model leads to an increase in the level of defense with increased availability of mineral nutrients, whereas the flexible allocation model implies declining defense levels at high nutrient availability. The former generates active induced defenses if the plant increases photosynthesis or the proportional allocation into secondary metabolism. However, the latter produces a passive accumulation of carbon-based allelochemicals under conditions where defoliation influences plant mineral nutrients more strongly than carbon reserves.

The responding units in both models can be entire plants or their semi-autonomous physiological subunits. In long-term responses, we propose that an active induced defense implies an increase of defense level among both attacked and nonattacked modules, whereas an altered plant carbon–nutrient balance should lead to localized responses restricted to attacked modules.

Finally, induced responses mediated by the carbon/nutrient balance should be nonspecific, influencing both defensively effective and ineffective compounds, whereas active induced defenses can rely on specific chemical interactions and enzymatic regulation of secondary metabolism.

ACKNOWLEDGMENTS

We are especially grateful to J. P. Bryant, F. S. Chapin III, W. J. Mattson, and T. Vuorisalo for stimulating discussions, which have greatly influenced our present views of defense allocation.

REFERENCES

Antolin, M. F., and C. Strobeck. 1985. The population genetics of somatic mutations in plants. *Am. Natur.* 126:52–62.

Baldwin, I. T., and J. C. Schultz. 1983. Rapid changes in tree leaf chemistry induced by damage: evidence for communication between plants. *Science* 221:277–279.

Bryant, J. P., F. S. Chapin III, and D. R. Klein. 1983. Carbon/nutrient balance of boreal plants in relation to vertebrate herbivory. *Oikos* 40:357–368.

Bryant, J., J. Tuomi, and P. Niemelä. 1988. Environmental constraint of constitutive and long-term inducible defenses in woody plants. In K. Spencer (ed.), *Chemical Mediation of Coevolution*. New York: Academic, pp. 367–390.

Chapin, F. S., III. 1980. The mineral nutrition of wild plants. *Annu. Rev. Ecol. Syst.* 11:233–260.

Coley, P. D., J. P. Bryant, and F. S. Chapin III. 1985. Resource availability and plant anti-herbivore defense. *Science* 230:895–899.

Del Moral, R. 1972. On the variability of chlorogenic acid concentrations. *Oecologia (Berlin)* 9:289–300.

Edwards, P. J., and S. D. Wratten. 1983. Wound induced defenses in plants, and their consequences for patterns of insect grazing. *Oecologia (Berlin)* 59:88–93.

Faeth, S. H. 1986. Indirect interactions between temporally separated herbivores mediated by the host plant. *Ecology* 67:479–494.

Fagerström, T. 1989. Anti-herbivory chemical defense in plants: a note on the concept of cost. *Am. Natur.* 133:281–287.

Feeny, P. P. 1970. Seasonal changes in oak leaf tannins and nutrients as a cause of spring feeding by winter moth caterpillars. *Ecology* 51:565–581.

Feeny, P. P. 1976. Plant apparency and chemical defense. In J. W. Wallace and R. L. Mansell (eds.), Biochemical Interactions between Plants and Insects. *Rec. Adv. Phytochem.* 10:1–40.

Fowler, S. V., and J. H. Lawton. 1985. Rapidly induced defenses and talking trees: the devil's advocate position. *Am. Natur.* 126:181–195.

Gershenzon, J. 1984. Changes in the levels of plant secondary metabolites under water and nutrient stress. *Rec. Adv. Phytochem.* 18:273–320.

Gill, D. E. 1986. Individual plants as genetic mosaics: ecological organisms versus evolutionary individuals. In M. J. Crawley (ed.), *Plant Ecology.* Oxford: Blackwell, pp. 321–343.

Gulmon, S., and H. A. Mooney. 1986. Costs of defense and their effects on plant productivity. In T. J. Givnish (ed.), *On the Economy of Plant Form and Function.* Cambridge: Cambridge University Press, pp. 681–698.

Harper, J. L. 1977. *Population Biology of Plants.* London: Academic.

Hartley, S. S., and J. H. Lawton. 1991. The biochemical basis and significance of rapidly induced changes in birch. In D. W. Tallamy and M. J. Raupp (eds.), *Phytochemical Induction by Herbivores.* New York: Wiley, pp. 105–132.

Haukioja, E. 1982. Inducible defenses of white birch to a geometrid defoliator, *Epirrita autumnata. Proc. 5th Int. Symp. Insect–Plant Relationships.* Pudoc: Wageningen, pp. 199–203.

Haukioja, E., and T. Hakala. 1975. Herbivore cycles and periodic outbreaks. Formulation of a general hypothesis. *Rep. Kevo Subarctic Res. Sta.* 12:1–19.

Haukioja, E., and S. Neuvonen. 1985. Induced long-term resistance of birch foliage against defoliators: test of defensive and non-defensive hypotheses. *Ecology* 66:1303–1308.

Haukioja, E., J. Suomela, and S. Neuvonen. 1985. Long-term inducible resistance in birch foliage: Triggering cues and efficacy on a defoliator. *Oecologia (Berlin)* 65:363–369.

Heim, D., R. L. Nicholson, S. F. Pascholati, A. E. Hagerman, and W. Billett. 1983. Etiolated maize mesocotyls: a tool for investigating disease interactions. *Phytopathology* 73:424–428.

Hinneri, S. 1974. Podzolic processes and bioelement pools in subarctic forest soils at the Kevo Station Finnish Lapland. *Rep. Kevo Subarctic Res. Sta.* 11:26–34.

Janzen, D. H. 1974. Tropical black water rivers, animals and mast fruiting by the Dipterocarpaceae. *Biotropica* 6:69–103.

Levin, D. A. 1971. Plant phenolics: an ecological perspective. *Am. Natur.* 105:157–181.

Margna, U. 1977. Control at the level of substrate supply: an alternative in the regulation of phenylpropanoid accumulation in plant cells. *Phytochemistry* 16:419–426.

McKey, D., P. G. Waterman, C. N. Mbi, J. S. Gartlan, and T. T. Struhsaker. 1978. Phenolic content of vegetation in two African rain forests: ecological implications. *Science* 202:61–63.

Mooney, H. A., and S. L. Gulmon. 1982. Constraints on leaf structure and function in reference to herbivory. *BioScience* 32:198–206.

Mooney, H. A., S. L. Gulmon, and N. D. Johnson. 1983. Physiological constraints on plant chemical defenses. In P. A. Hedin (ed.), *Plant Resistance to Insects.* American Chemical Society Symposium Series 208. Washington, D.C.: American Chemical Society, pp. 21–36.

Niemelä, P., J. Tuomi, R. Mannila, and P. Ojala. 1984a. The effect of previous damage on the quality of Scots pine as food for Diprionid sawflies. *Z. Angew. Entomol.* 98:33–43.

Niemelä, P., J. Tuomi, and S. Sirén. 1984b. Selective herbivory on mosaic leaves of variegated *Acer pseudoplatanus*. *Experimentia* 40:1433–1434.

Oparka, K. J., and H. V. Davies. 1985. Translocation of assimilates within and between potato stems. *Ann. Bot.* 56:45–54.

Orians, G. H., R. G. Cates, D. F. Rhoades, and J. C. Schultz. 1974. Producer-consumer interactions. *Proc. 1st Ecol. Congr.* (Hague 1974), pp. 213–217.

Pascholati, S., R. Nicholson, and L. Butler. 1986. Phenylalanine ammonialyase activity and anthocyanin accumulation in wounded maize (*Zea mays*) mesocotyls. *Phytopathol. Z.* 115–165–172.

Phillips, R., and G. G. Henshaw. 1977. The regulation of synthesis of phenolics in stationary phase cell culture of *Acer pseudoplatanus* L. *J. Exp. Bot.* 28:785–794.

Rhoades, D. F. 1979. Evolution of plant chemical defense against herbivores. In G. A. Rosenthal and D. H. Janzen (eds.), *Herbivores: Their Interactions with Secondary Plant Metabolites.* New York: Academic, pp. 3–54.

Rhoades, D. F. 1983. Responses of alder and willow to attack by tent caterpillars and webworms: evidence for pheromonal sensitivity of willows. In P. E. Hedin (ed.), *Plant Resistance to Insects.* Washington, DC: American Chemical Society, pp. 55–68.

Ryan, C. A., and T. R. Green. 1974. Proteinase inhibitors in natural plant protection. *Rec. Adv. Phytochem.* 8:123–140

Tuomi, J., and T. Vuorisalo. 1989. What are the units of selection in modular organisms? *Oikos* 54:229–233.

Tuomi, J., J. Salo, E. Haukioja, P. Niemelä, T. Hakala, and R. Mannila. 1983. The existential game of self-maintaining units: selection and defense tactics of trees. *Oikos* 40:369–376.

Tuomi, J., P. Niemelä, E. Haukioja, S. Sirén, and S. Neuvonen. 1984. Nutrient stress: an explanation for anti-herbivore responses to defoliation? *Oecologia (Berlin)* 61:208–210.

Tuomi, J., P. Niemelä, F. S. Chapin III, J. P. Bryant, and S. Sirén. 1988a. Defensive responses of trees in relation to their carbon/nutrient balance. In W. J. Mattson, J. Levieux, and C. Bernard-Dagan (eds.), *Mechanisms of Woody Plant Defenses against Insects: Search for Pattern.* New York: Springer, pp. 57–72.

Tuomi, J., P. Niemelä, M. Rousi, S. Sirén, and T. Vuorisalo. 1988b. Induced accumulation of foliage phenols in mountain birch: branch response to defoliation? *Am. Natur.* 132:602–608.

Tuomi, J., P. Niemelä, I. Jussila, T. Vuorisalo, and V. Jormalainen. 1989. Delayed bud-break: a defensive response of mountain birch to early-season defoliation? *Oikos* 54:87–91.

Vuorisalo, T., and J. Tuomi. 1986. Unitary and modular organisms: criteria for ecological division. *Oikos* 47:382–385.

Watson, M. A. 1986. Integrated physiological units in plants. *Trends Ecol. Evol.* 1:119–123.

Watson, M. A., and B. B. Casper. 1984. Morphogenetic constraints on patterns of carbon distribution in plants. *Annu. Rev. Ecol. Syst.* 15:233–258.

Whitham, T. G., and C. N. Slobodchikoff. 1981. Evolution by individuals, plant-herbivore interactions, and mosaics of genetic variability: the adaptive significance of somatic mutations. *Oecologia (Berlin)* 49:287–292.

Zucker, W. V. 1983. Tannins: does structure determine function? An ecological perspective. *Am. Natur.* 121:335–365.

CHAPTER 5

BIOCHEMICAL ASPECTS AND SIGNIFICANCE OF THE RAPIDLY INDUCED ACCUMULATION OF PHENOLICS IN BIRCH FOLIAGE

S. E. HARTLEY

Institute of Terrestrial Ecology, Banchory Research Station, Banchory, Kincardineshire AB 31 4BY, United Kingdom; Department of Biology, University of York, Heslington, York Y01 5DD, England

J. H. LAWTON

NERC Centre for Population Biology, Imperial College, Silwood Park, Ascot SL57PY, United Kingdom; Department of Biology, University of York, Heslington, York Y01 5DD, England

1   Introduction
2   Rapid phytochemical induction in damaged birch leaves
   2.1   Initial experiments
   2.2   Is insect attack the same as artificial damage?
   2.3   Problems of interpretation
   2.4   The induction of phenolic biosynthesis: phenylalanine ammonia lyase
   2.5   Summary of rapidly induced changes in birch foliage
3   The effect of wound-induced changes on insects
   3.1   Preference tests with mature leaves and late instar caterpillars
   3.2   Tests with young leaves and early instars
   3.3   *Spodoptera*, *Apocheima*, and manipulated phenolics
   3.4   Links between laboratory and field
   3.5   Field experiments
   3.6   Summary of effects on insects
4   Conclusion
   References

1   INTRODUCTION

The idea that induction of phytochemicals by herbivores defends plants against further damage is both attractive and plausible. As the present volume makes plain, phytochemical induction is now well documented in a number of plant systems. The problem, as we see it, is that much of the evidence for a defensive role of induced chemical changes is not particularly rigorous (see Fowler and Lawton 1985 for discussion). Nevertheless, it creates the impression that such responses are widespread and common, if not universal, components of herbivore–plant interactions (Rhoades 1985). In this chapter, we are concerned only with rapid phytochemical induction (plant responses measured over a few days or weeks, rather than over months or years), and with one particular system, namely the interaction between birch trees (*Betula*) and their insect herbivores. The birch system clearly highlights the dangers of making sweeping generalizations about the role of rapidly induced chemical changes in damaged plant foliage on the basis of limited evidence.

There is no doubt that rapid, wound-induced changes in the foliage of both *B. pendula* and *B. pubescens* can influence food preferences, larval performance, and ultimately the population dynamics of at least some birch-feeding insects (Neuvonen and Haukioja, this volume, Chapter 12, Edwards et al. this volume, Chapter 9). Haukioja and Niemela (1979) showed that the growth of *Epirrita autumnata* caterpillars was slower on previously grazed leaves, or leaves adjacent to them, than on undamaged leaves; effects on caterpillars were also correlated with increases in the level of phenolic compounds in damaged and neighboring foliage. Other species of Lepidoptera and sawfly larvae were also affected, sometimes being smaller at pupation and in one case having lower survival (Neuvonen and Haukioja, this volume, Chapter 12). However, such effects are far from universal, and one species of sawfly (*Dineura viridorsata*) actually performed better on the damaged leaves (Haukioja and Niemela 1979). In another laboratory study (Wratten et al. 1984), both species of birch showed a wound-induced reduction in the palatability of damaged and neighboring leaves to *Spodoptera littoralis* caterpillars, again correlated with an increase in phenolic levels. But an attempt to repeat these results for *Spodoptera*, using a different population of trees and caterpillars from a different laboratory stock was unsuccessful (Hartley 1988). Indeed, in a second more recent experiment, Wratten and his colleagues (Edwards et al. 1986) found that only damaged *B. pubescens* foliage was unpalatable to *Spodoptera*; caterpillars were apparently unaffected by damage to leaves of *B. pendula* and in one case actually preferred the foliage adjacent to damaged leaves.

In a field experiment, *Apocheima pilosaria* caterpillars caged on experimentally damaged leaves suffered a fourfold increase in mortality relative to controls (Fowler and MacGarvin 1986). In contrast, in this same experiment *Coleophora serratella* larvae showed no increase in mortality, and the larvae actually ate more on the damaged branches. But in a second experiment (Bergelson et al. 1986) *Coleophora* yielded results consistent with a defensive role for induced chemical changes by moving away from damaged leaves, and taking longer to pupate when reared on damaged trees.

This brief review illustrates some of the problems in studying rapid phytochemical induction in birch, particularly in field situations, and in attempting to draw simple general conclusions: The effects are often small, unclear, and variable both within and between trees, and could also be different in different populations of birch (see also Neuvonen and Haukioja this volume, Chapter 12). Furthermore, interpretation of particular experimental results involves a hierarchy of events (Fowler and Lawton 1985). Chemical changes in foliage need not lead to changes in insect performance, and even if they do, these may not produce significant changes in insect population dynamics, or subsequent benefits to the plant. As we shall see, it is dangerous to extrapolate from one level in the hierarchy to a higher level without sufficient grounds.

This chapter is predominantly concerned with the first step in the hierarchy, namely with damaged-induced chemical changes in *Betula pendula* focusing upon changes in phenolics and tannins. Phenolics are a major group of secondary compounds in birch, and although it would be desirable to investigate other aspects of leaf chemistry, this has not yet been attempted. Hence, we do not know if other classes of secondary compounds are involved in the insects' response to foliar damage. Our results should be interpreted in this light. We are concerned with wound-induced responses to both artificial (mechanical) and herbivore (caterpillar) damage, and we approach the problem in a more detailed way than in any previous studies on birch, using an enzyme assay to measure induction of the phenolic biosynthetic pathway as well as looking at changes in phenolics themselves. In later parts of the chapter, we briefly consider higher levels in the hierarchy, particularly the effects of leaf damage on the performance and population dynamics of birch feeding insects. A more comprehensive account of the effects of phytochemical induction in birch on herbivore populations is given by Neuvonen and Haukioja (this volume, Chapter 12).

Our study site is Skipwith Common, an acid lowland heath 15 km south of York. Some of our study trees were transplanted from here to the University of York campus; they were moved as small saplings at least four years before being used in experiments and were maintained free of insect attack in a tree nursery. (Few birch-feeding insects occur on the campus and occasional infestations were sprayed.) Other experimental trees from Skipwith were transferred to the tree nursery in 32-cm pots, 18 months prior to use in experiments. We have never found any difference in response to damage between transplanted trees on campus and trees at Skipwith. *B. pendula* at Skipwith grow to over 12 m tall, but only one experiment used mature trees. We worked mainly with saplings 1.5–3 m in height. We do not know whether the response of *B. pendula* to leaf damage differs between very old and very young trees. However, we have been unable to find differences between saplings and mature trees over the size and age range used in these experiments.

Table 1 summarizes the taxonomy and relevant biology of our study insects. Approximately 330 species of phytophagous insects feed on birch in

TABLE 1   The Taxonomy and Relevant Biology of the Insect Species Referred to in This Chapter

Species	Authority	Family
Native free-living folivores		
Apocheima pilosaria	Denis and Schiffermüller	Geometridae
Epirrita dilutata	Denis and Schiffermüller	Geometridae
Erranis defoliaria	Clerck	Geometridae
Euproctis similis	Fuessly	Lymantriidae
Nonnative free-living folivore used under laboratory conditions		
Spodoptera littoralis	Boisduval	Noctuidae
Leaf miners		
Eriocrania spp.	—	Eriocraniidae
Case-bearing leaf miner living externally on leaf		
Coleophora serratella	Linnaeus	Coleopheridae

Britain (Southwood et al. 1982), with over 120 species recorded at Skipwith, more than half of which are Lepidoptera (Fowler 1983). We have studied only a tiny fraction of these species and already the results are complicated.

2   RAPID PHYTOCHEMICAL INDUCTION IN DAMAGED BIRCH LEAVES

2.1   Initial Experiments

Initial studies on the changes in phenolic levels following damage used potted trees in the tree nursery on the university campus. Five such trees were selected and 6-mm holes were punched (with a metal hole-punch) in some of their leaves. Phenolics were monitored over the subsequent 8-day period using a simple colormetric assay, the Folin Denis method (see Bergelson et al. 1986 and Hartley and Firn 1989 for experimental details). The damaged leaves showed a small increase in phenolic levels (Fig. 1) relative to control leaves (undamaged leaves on the same tree, which showed no change). No increase occurred in adjacent undamaged trees. A similar experiment showed that the increase in phenolics after such damage is larger in young leaves than in mature leaves on the same trees (Table 2), and that young leaves had the highest constituent levels of phenolics.

The wound-induced chemical changes were also measured using a protein precipitation technique (Bradford method; see Hartley and Lawton 1987 for details). Polyphenolic compounds are able to precipitate protein, and may

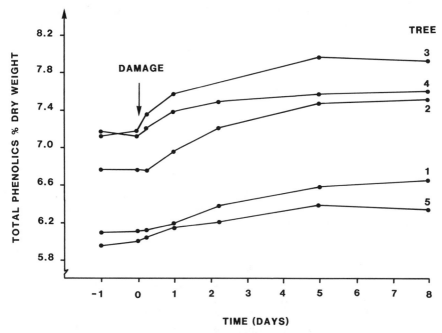

Figure 1 The effect of artificial damage (hole-punching) on the phenolic content of leaves from five birch trees over the 8 days subsequent to damage. Phenolics are significantly increased in damaged leaves within 24 h ($t = 3.96$, df $= 4$, $p < .01$).

TABLE 2   Folin Denis Phenolic Levels (% dry wt) of Young and Mature Birch Leaves from Five Different Trees, before (day 0) and 5 Days after Damage (day 5)[a]

	Young Leaves		Mature Leaves	
Tree	Day 0	Day 5	Day 0	Day 5
1	8.23	9.20	7.06	7.73
2	5.82	6.82	5.11	5.98
3	7.07	7.91	6.23	6.93
4	7.33	8.63	7.18	8.05
5	7.18	7.93	6.59	7.06
Mean % increase	13.8%		11.4%	

[a] The increase in phenolic content after damage was significant in both young (paired t test: $t = 10.45$, df $= 4$, $p < .001$) and mature ($t = 9.68$, df $= 4$, $p < .001$) leaves. The increase is larger in young leaves (paired t test: $t = 5.20$, df $= 4$, $p < .01$).

TABLE 3   The Protein-Precipitating Abilities of Artificially Damaged Foliage from Five Different Birch Trees, 5 Days after Damage[a]

	Protein Precipitated (SE)	
Tree	Undamaged	Damaged
1	67.46 (8.35)	103.0 (13.7)
2	79.70 (3.56)	104.0 (5.80)
3	13.92 (1.75)	29.17 (3.69)
4	28.77 (1.60)	50.12 (4.24)
5	23.83 (3.13)	20.64 (2.32)

[a] The results are expressed as regression coefficients of micrograms protein precipitated vs. milligrams leaf extract, together with the SE of the fitted slope. Each estimated slope is based on four sample points and large slopes imply high protein-precipitating ability. The increase in protein-precipitating ability after damage is significant (paired t test: $t = 2.93$, df $= 4$, $p < .05$).

do so in the guts of herbivores (e.g., Feeny 1970, Rhoades and Cates 1976, Klocke and Chan 1982), thereby reducing foliage digestibility by precipitating herbivore gut enzymes and/or plant proteins (although recent work challenges this classical view of their mode of action; e.g., Berenbaum 1980, Bernays 1981, Mole and Waterman 1985, Martin et al. 1987). Results of the protein precipitation assays were highly variable: Some trees showed no significant increase in protein-precipitating ability in damaged leaves, while others responded with increases of over 100% (Table 3).

These early results demonstrate some of the difficulties with interpreting damage-induced chemical changes. Increases in phenolics are small compared with the variation in constituent phenolic levels that exists between trees; in fact, the highest levels of phenolics in the first experiment were found in some of the undamaged control trees rather than in trees with hole-punched leaves. Variability between trees is even greater in the case of protein-precipitating ability, and there is no correspondence between protein precipitation and the same tree's total Folin Denis phenolic content (other studies show a similar lack of correlation between different measures on foliage phenolics; e.g., Martin and Martin 1982, Wilson 1984, Mole and Waterman 1987a,b). There is also considerable within-tree variation; young leaves have the highest levels of phenolics and the largest increases following wounding. This last result is consistent with the hypothesis that the most valuable young foliage in a quick-growing species like birch should be the best defended (Fowler 1984, Edwards and Wratten 1987, Coleman and Jones this volume, Chapter 1) but, of course, other explanations are also possible. For example, the physiological state of the leaf may determine the degree to which damage induces phenolic synthesis. In other words, herbivores may encounter much greater differences in phenolic levels among undamaged foliage than occurs subsequent to damage. It is difficult to see the significance for a foliage-feeding caterpillar of a 1–2% increase in phenolics in damaged leaves, when undamaged leaves similar in age and condition may vary by five or even ten times this amount on the same or adjacent trees; it is possible, however, that changes in types of phenolics are more important than amounts.

2.2   Is Insect Attack the Same as Artificial Damage?

A second possibility is that chemical induction may be much greater and more conclusive after insect attack than after artificial damage. Differences in the magnitude of phytochemical induction following natural and artificial damage have been observed by a number of workers (e.g., Woodhead 1981, Shain and Hillis 1972, Baldwin 1988). Usually (but not always, e.g., Baldwin 1988) insects induce greater responses than artificial damage, lending support to the hypothesis that induced chemicals are involved in defense against herbivores (e.g., Cole 1985, Edwards and Wratten 1985, Haukioja and Neuvonen 1985). Thus, we asked how the chemical changes induced in birch after damage differ with type of damage.

Our initial experiment used 15 trees at Skipwith. Five were left undamaged as controls and five were grazed overnight by *A. pilosaria* caterpillars caged on three branches of each tree. The five remaining trees had the same amount and distribution of damage inflicted on three of their branches using sterilized scissors. Before damage and at intervals afterward, damaged leaves (where appropriate) and undamaged leaves from a branch adjacent to the treatment branch were removed and analyzed for phenolics by the Folin Denis and Bradford methods.

Artificial damage caused an increase in Folin Denis phenolics (though again this was small compared to between-tree variation), and no increase in phenolics in adjacent undamaged leaves. In contrast, insect attack caused a much larger increase in total phenolics, and there was also a small but significant increase in the phenolics in adjacent undamaged leaves (Fig. 2).

The results of the Bradford assay were more difficult to interpret. In fact, if anything, in the grazed leaves the trend (although not statistically significant) is in the opposite direction (Table 4) with a slight decrease in protein-precipitating ability in damaged leaves. In contrast to the Folin Denis results, cut leaves had a larger change in protein precipitation than grazed leaves, with up to a 60% increase following damage (Table 4). However, despite this the trees were highly variable, and the overall change was not statistically significant.

The influence of both the type of damage and the technique used to measure wound-induced chemical changes on the interpretation of results was confirmed by a second experiment. From five mature trees at Skipwith, 50 leaves of each of three different types were collected: undamaged leaves, leaves freshly damaged by leaf chewers, and leaves damaged by recently vacated eriocraniid leaf miners (Lawton 1987, Hartley and Lawton 1987). The Folin Denis phenolics and protein-precipitating ability of the leaves were measured (for the mined leaves, the mine itself was excluded from the analysis). The total phenolic content was significantly increased in the mined

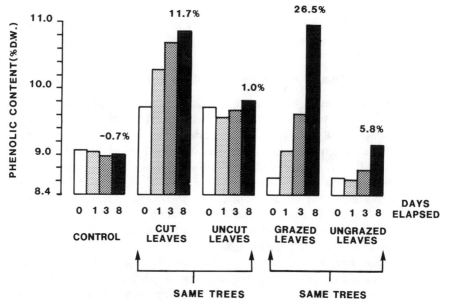

Figure 2   The effect of leaf damage on the phenolic content of birch leaves. Each value is the mean of measurements of five trees, each tree sampled on the days shown. The percentage change in phenolic content after 8 days is also shown. The phenol contents of cut leaves and both damaged and undamaged leaves from insect-grazed trees were significantly increased after 8 days ($t = 5.91$, df $= 4$, $p < .01$; $t = 7.57$, df $= 4$, $p < .002$; $t = 2.99$, df $= 4$, $p < .05$, respectively).

leaves (compared to undamaged leaves from the same tree), but even larger increases in leaf phenolic levels occurred following insect grazing (Table 5). Results for the protein precipitation assay were different. In a two-way ANOVA there was a significant effect of both tree and damage type on protein precipitation (Table 6). Mined leaves were significantly better at precipitating protein than both chewed and undamaged leaves, while chewed leaves (which had the highest levels of total phenolics) were not different from undamaged leaves.

2.3   Problems of Interpretation

Clearly birch responds differently to different types of damage. It is equally clear that it is difficult to draw conclusions about the roles of induced birch chemicals given the variable and conflicting results thus far observed and the relatively crude methods used to obtain them. One possibility is that the Folin Denis reagent detects phenolic hydroxyl groups nonstoichiometrically on both mono- and polyphenolics, while the Bradford assay detects only phenolics that can precipitate proteins, that is, high molecular weight polyphenolics such as tannins. If this interpretation is correct, it follows that

TABLE 4   The Protein-Precipitating Abilities (expressed as a slope of micrograms protein precipitated vs. milligrams leaf extract, together with the SE of the fitted slope) for Birch Leaves from Undamaged, Artificially Damaged, and Insect-Grazed Trees, before and 8 days after Treatment[a]

Trees	Protein Precipitated (SE)	
	Before	After
Control		
Tree 1	277 (12)	257 (49)
Tree 2	181 (7)	181 (13)
Tree 3	279 (7)	251 (24)
Artificially damaged		
Tree 1	256 (15)	250 (14)
Tree 2	193 (15)	309 (66)
Tree 3	217 (32)	285 (5)
Tree 4	201 (17)	215 (23)
Tree 5	178 (6)	224 (10)
Insect-grazed		
Tree 1	183 (4)	140 (16)
Tree 2	259 (6)	237 (38)
Tree 3	237 (22)	214 (38)
Tree 4	233 (38)	196 (4)
Tree 5	204 (16)	240 (23)

[a] Each estimated slope is based on four sample points. Paired t tests on the regression coefficients showed no significant increase in protein-precipitating ability in control ($t = 1.56$, df $= 2$, $p > .1$), artificially damaged ($t = 2.24$, df $= 4$, $p > .05$), or insect-grazed ($t = -1.29$, df $= 4$, $p > .1$) trees.

TABLE 5   Folin Denis Phenolic Levels (% dry wt) of Bulked Samples of 50 Mined, Insect-Grazed, and Undamaged Leaves from Five Trees at Skipwith[a]

Tree	Undamaged	Insect-Grazed	Mined
1	6.11	7.01	7.01
2	6.38	7.82	6.77
3	5.96	7.81	6.60
4	6.83	8.52	7.58
5	4.99	6.35	5.62
Mean increase above undamaged level	—	24.6%	12.2%

[a] There is a significant increase in phenolics after damage in both insect-grazed (paired t test: $t = 8.94$, df $= 4$, $p < .001$) and mined ($t = 7.88$, df $= 4$, $p < .002$) leaves.

TABLE 6   Protein-Precipitating Abilities (slope of microgram protein precipitated vs. milligrams leaf extract) and SE (of the fitted slope) of Mined, Insect-Grazed and Undamaged Foliage from Five Trees at Skipwith[a]

Tree	Protein-Precipitating Ability		
	Undamaged	Insect-Grazed	Mined
1	118.7 (2.57)	101.7 (2.32)	120.2 (11.2)
2	199.9 (18.7)	184.7 (21.8)	212.3 (11.7)
3	102.0 (0.58)	127.7 (5.51)	133.0 (5.08)
4	171.7 (10.2)	181.1 (4.39)	192.2 (1.73)
5	84.7 (1.86)	105.8 (9.89)	127.7 (1.73)

[a] Two-way ANOVAs were performed on the regression coefficients and showed there is a significant effect of both treatment ($F = 5.17$, df $= 2,8$, $p < .05$) and tree ($F = 53.66$, df $= 4,8$, $p < .001$). Least significant difference tests showed that chewed leaves are not significantly better at precipitating protein than undamaged leaves, while mined leaves are significantly better than both.

different types of damage not only induce different quantities of phenolics, but may also give rise to the synthesis of compounds with different molecular weights. Why this should be so is unclear, but it is plausible that the differences are in some way related to different defensive needs posed by different types of damage. We have been investigating these ideas using high-performance liquid chromatography (HPLC). Although our studies are at a preliminary stage, HPLC has confirmed the quantitative differences observed using the Folin Denis technique. Insect-grazed leaves contain more phenolics than mined leaves, which have substantially more phenolics than undamaged foliage. However, qualitative changes have proved more difficult to establish and, in the case of artificial damage, no new phenolic compounds have so far been found (Prof. P. G. Waterman, personal communication).

2.4   The Induction of Phenolic Biosynthesis: Phenylalanine Ammonia Lyase

Given the problems outlined above we decided to avoid measuring the end-products and to concentrate instead on characterizing the induction of the phenolic pathway in terms of the activation of biosynthetic enzymes. This approach has both advantages and disadvantages. The disadvantages are that studying phenolic induction clearly cannot provide any information about the endproducts or their specific ecological effects. What it can do is tell us more about the plant's overall response to wounding. For example, if phenolics appear at sites remote from insect attack, is this due to transport of phenolics from damaged leaves, or to the induction of phenolic synthesis in undamaged tissue? If phenolic synthesis increases in undamaged, adjacent tissue, some sort of message is involved. What is the message? How does the plant recognize the damage? Can plants distinguish between different sorts of damage? Assaying enzymes of the phenolic pathway can lead to answers to these important questions.

We used a modified version of the intact cell assay of Amrhein et al. (1976) (see Hartley and Firn 1989) to study phenylalanine ammonia lyase (PAL). PAL catalyses the conversion of phenylalanine to cinnamic acid, the committed step in phenolic biosynthesis. PAL levels in the plant can be assessed by incubating leaf disks with radioactive phenylalanine in phosphate buffer (pH 5.5) for 3 h, during which time the labeled phenylalanine is taken up by the cells and the action of PAL liberates ^{3}HOH as a side-product. Thus, a measurement of the production of labeled water is a measure of enzyme activity and, since this is an intact cell assay, a measure of phenolic production in the leaf.

Twelve trees in the tree nursery on the campus were selected for the experiment. Four were left undamaged as controls and four were grazed overnight by *A. pilosaria* caged on three different branches of each tree as in our previous experiment. The four remaining trees were artificially damaged with a pair of sterile scissors. Damaged and immediately adjacent undamaged leaves were collected before and at intervals after treatment for analysis for PAL activity and total phenolics (Folin Denis method).

The results for total phenolics were similar to our earlier experiment comparing insect attack and artificial damage. Insect grazing caused a much larger induction of phenolics than did the same amount of artificial damage (Fig. 3). This time, however, the induction in the undamaged leaves was

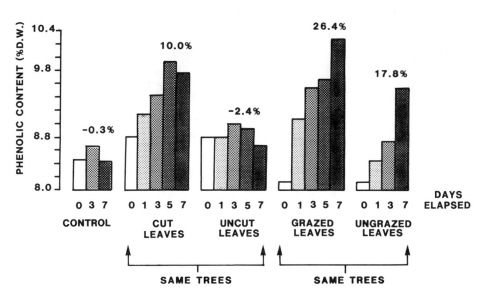

Figure 3 The effect of leaf damage on phenolic levels in birch leaves. The values are measurements made on four trees and the same trees were used on each sampling day. The percentage change in phenolic content after seven days is also shown. Cut leaves and both damaged and undamaged leaves on insect-grazed trees had increased phenolic levels after 7 days ($t = 6.64$, df $= 3$, $p < .001$; $t = 13.6$, df $= 3$, $p < .001$; $t = 5.79$, df $= 3$, $p < .02$, respectively).

larger than in the previous experiment. Again, there was no increase in the phenolic levels of undamaged leaves from artificially damaged trees. These differences were even more apparent when phytochemical induction was measured as changes in enzyme activity (Fig. 4). The results were expressed as mean percentage change from the start of the experiment to reduce the effect of between-tree variation in initial PAL activity, although there was still considerable variability. However, it is clear that there was no change in enzyme activity in control trees, or in undamaged leaves on artificially damaged trees; there was, nevertheless, a small increase in PAL activity in artificially damaged leaves. The induction of PAL activity was much greater, and lasted longer, following insect attack (confirming the results of the Folin Denis assay). Intriguingly, the largest induction of PAL activity took place in the *undamaged* leaves on insect-grazed trees. Thus, a signal appears to pass from the site of damage to undamaged leaves and induces an increase in PAL activity. There were no signs of any such signal from artificially damaged leaves.

Not surprisingly, this experiment raises more questions than it answers. What is the nature of the signal that passes to the undamaged leaves, and why does this not occur in the case of artificial damage? How does the plant

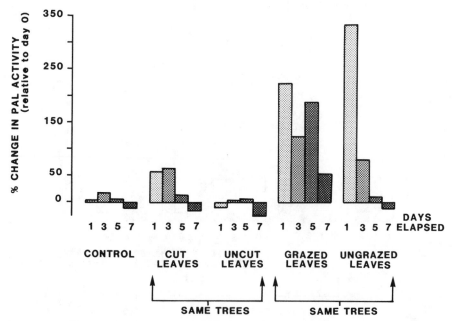

Figure 4 The percentage change in PAL activity in damaged and adjacent undamaged birch leaves over the 7 days subsequent to damage, and in control trees. Each value is the mean of measurements of four trees (the same four trees for each sample). There is a significant effect of treatment on days 1 ($F = 10.64$; df $= 4,15$; $p = .0003$) and 5 ($F = 5.46$; df $= 4,14$; $p = .0084$).

recognize insect attack? There are many possibilities; wound-induced signals such as ethylene and action potentials are well known (Yang and Pratt 1978; Van Seembeck et al. 1976) and the idea of "wound hormones" is also well established (Ryan 1974, Davies and Schuster 1981). Interest in such signals has recently been generated by two much studied phenomena: the production of phytoalexins in response to fungal attack (Mansfield 1982, Dixon 1986), and the wound-induced production of proteinase inhibitors (e.g., Green and Ryan 1973, Ryan 1974), a process that also involves the transport of inducing factors away from the site of damage to undamaged leaves (Pena-Cortes et al. 1988). The induction of proteinase inhibitors and PAL can both be stimulated (in some systems) by oligosaccharide cell wall fragments (Hahlbrock et al. 1981, Bishop et al. 1981).

It is interesting that both of these processes involve induction of chemical changes in response to fungal and/or plant cell wall fragments. This is a possible mechanism for the larger response to insect attack than to artificial damage. Insects may have fungal spores of hyphae associated with their mouthparts, leading to a larger plant reaction than that associated for example, with sterile scissors. The importance of fungi in insect-induced chemical changes has been noted before (Shain and Hillis 1972, Grisham et al. 1987), as has insect saliva (Miles 1969). Similar studies in the birch system are at an early stage, but are encouraging, in that both the application of droplets of liquid regurgitated by *Apocheima* to wounds made with sterile scissors, and the use of nonsterile scissors produce changes in phenolic levels similar in magnitude to insect grazing (Fig. 5). Clearly much more work is required in this area before these processes are fully understood.

2.5   Summary of Rapidly Induced Changes in Birch Foliage

Wounding triggers a complex series of reactions in birch leaves. The chemical changes following damage are small when compared to the within- and between-tree variation in levels of phenolics already present in undamaged leaves, and the magnitude of induction also varies within and between trees. Different amounts, and possibly different types of phenolics, are induced depending upon the type of damage inflicted (i.e., artificial mechanical damage, leaf miners, or caterpillar grazing). Most intriguing is the observation that grazing by *Apocheima* caterpillars triggers a larger induction of PAL activity, and hence phenolic levels, than an equivalent amount of artificial damage. Insect grazing also causes an increase in PAL activity and phenolic levels at sites remote from damage; this does not occur in the case of artificial damage. These chemical changes are summarized in Table 7.

The fact that insects are more effective than artificial damage in increasing phenolic levels, and that they do so in adjacent, undamaged foliage could be taken as support for a defensive role by phenolics against insect herbivores. While we await more detailed information on the nature of the phenolic compounds induced and the mechanisms of induction in undamaged

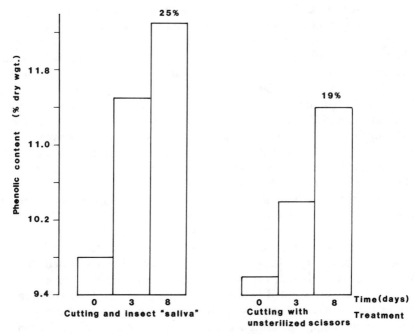

Figure 5   The effect of two types of artificial damage on the phenolic levels of *Betula pendula* leaves. For each treatment, each value is the mean of measurements taken from the same four trees at three times after damage. The percentage change in phenolic levels after 8 days is also shown and in both treatments this was a significant increase (caterpillar saliva $t = 13.25$, df $= 3$, $p < .001$; nonsterile scissors $t = 5.52$, df $= 3$, $p < .02$).

foliage, one way to assess the likelihood of a defensive role is to test for adverse effects of the induced phenolics against birch-feeding insects. In the next section we briefly describe some of the experiments we have carried out at York in search of evidence for such effects (see Neuvonen and Haukioja, this volume, Chapter 12, for a more extensive review of work carried out on birch).

3   THE EFFECT OF WOUND-INDUCED CHANGES ON INSECTS

3.1   Preference Tests with Mature Leaves and Late Instar Caterpillars

One way in which chemical changes induced by damage could exert an effect on insect herbivores is by modifying herbivore feeding behavior. Insects could avoid feeding on damaged foliage and the plant may then benefit from reduced herbivory, particularly if increased movement to search for palat-

TABLE 7   Summary of the Mean Induction (as a percentage of change) of Phenolics, Protein-Precipitating Ability, and PAL Activity in Different Types of Damaged Birch Leaves[a]

Type of Leaf	Chemical Change Induced (%)		
	Total Phenolics	Protein Precipitation	PAL Activity
Artificial damage	13.8[b]	50.7	63
	11.4	(24.4)	
	11.7		
	10.0		
Adjacent to	(1.0)	—	(6)
artificial damage	(−2.4)		
Insect-grazed	24.6	(3.0)	222
	26.4	(8.1)	
	26.5		
Adjacent to	5.8	—	338
insect-grazed	17.8		
Mined leaves	12.2	(16.1)	

[a] The results represent a summary of all the experiments described in this chapter, therefore several values can appear in each category. Nonsignificant results are in parentheses. The results are for induction 5–8 days after damage except in the case of PAL induction, which is shown after 1–3 days.

[b] Youngest leaves only.

able food leads to greater exposure to natural enemies and hence to increased herbivore mortality (Schultz 1983, Bergelson and Lawton 1988).

The effects of three types of damage (hole-punched, insect-grazed, and mined) on herbivore behavior were examined in laboratory preference tests (Lawton 1987, Hartley and Lawton 1987). As we have seen, all three classes of damaged leaves have increased phenolic levels compared with undamaged leaves from the same tree; feeding behavior in response to these increases was monitored with caterpillars of four birch-feeding Lepidoptera: *A. pilosaria*, *Erannis defolaria*, *Epirrita dilutata*, and *Euproctis similis* (Table 1). For each trial up to 30 replicates were established using one final or penultimate instar caterpillar and a pair of freshly collected mature birch leaves, one damaged and one undamaged, from the same tree. Trials used leaves from at least 5 different trees. Preferences were expressed as amount eaten from the damaged leaf/total amount consumed; equal amounts eaten from the two types of leaf (no preference) therefore gives a value of .5.

The results were rather surprising. In marked contrast to earlier reports in the literature on rapid phytochemical induction in birch (Edwards and Wratten 1982, Wratten et al. 1984; Edwards et al. this volume, Chapter 9), most caterpillars were indifferent to chewing damage, in spite of the increased phenolic levels, and one species (*Erannis*) even preferred hole-punched leaves (Table 8). The mined leaves were avoided, despite not having

TABLE 8   Feeding Preferences of Final- and Penultimate-Instar Birch Caterpillars for Mined, Chewed, and Artificially Damaged Leaves (damaged 5 days previously)[a]

Insect Species	Sample Size	Mean Preference (standard error)	p Value Wilcoxon	t Test	Result
Mined Damage					
Apocheima pilosaria	18	0.29 (0.08)	= .06	<.05	Marginally prefers undamaged
Epirrita dilutata	27	0.31 (0.08)	= .04	<.01	Prefers undamaged
Erannis defolaria	19	0.23 (0.08)	= .01	<.01	Prefers undamaged
Euproctis similis	24	0.21 (0.07)	= .003	<.001	Prefers undamaged
Chewed Damage					
Apocheima pilosaria	19	0.53 (0.07)	= .96	>.5	Not significant
Epirrita dilutata	35	0.54 (0.05)	= .66	>.4	Not significant
Erannis defolaria	19	0.69 (0.08)	= .1	<.05	Prefers undamaged
Artificial Damage					
Apocheima pilosaria	29	0.51 (0.06)	= .73	>.05	Not significant
Epirrita dilutata	21	0.67 (0.05)	= .07	<.05	Marginally prefers damaged
Erannis defolaria	30	0.70 (0.06)	= .03	<.01	Prefers damaged
Euproctis similis	21	0.52 (0.07)	= .68	>.5	Not significant

[a] Preference was expressed as amount eaten from the previously damaged leaf vs. total amount eaten. A mean preference of 0.5 indicates equal amounts consumed from damaged and undamaged leaves. Replicates refer to individual trials with single caterpillars; each caterpillar was used only once. The significance of the preference was assessed using both a paired comparison *t* test and a Wilcoxon matched-pairs signed rank test.

the largest induced changes in phenolics (see earlier results). Thus, the feeding behavior reflected damage type and herbivore species, rather than changes in phenolic compounds. However, we recognize that the Folin Denis method is a crude assay and may not be a good estimate of the chemical signals to which the insects are responding; specific phenolic compounds, or other types of compounds altogether may be much more important in determining preference.

Although conclusions about the reasons for the observed behavior must await more detailed chemical information, these preliminary results have interesting implications for the field. In particular, the avoidance of mined leaves by other herbivores raises the prospect of indirect competition between different guilds of herbivores via the host plant (Strong et al. 1984, Neuvonen and Haukioja, this volume, Chapter 12). Such effects have been observed by several workers (Edwards and Wratten 1985, West 1985, Faeth 1986). Paradoxically, studies during the spring of 1987 at Skipwith (G. Valladares, J. H. Lawton, and S. E. Hartley, unpublished results) failed to find any evidence that chewing caterpillars avoid eriocraniid-mined leaves in the field (Fig. 6) (although the miner *Coleophora* did feed on leaves mined by *Eriocrania* less often than expected by chance; we have not tested *Coleophora*'s behavior in laboratory choice tests). Failure to detect any avoidance of eriocraniid-mined leaves by free-living caterpillars in the field, despite very large sample sizes and despite very clear results in the laboratory assays (Table 8) is remarkable. Indeed, these data have forced us to question the usefulness of results based on simple laboratory choice tests with excised leaves (Risch 1985). At the very least, they must be interpreted cautiously. Similar remarks obviously apply to the results indicating no preference (Table 8). The situation in the field may be very different.

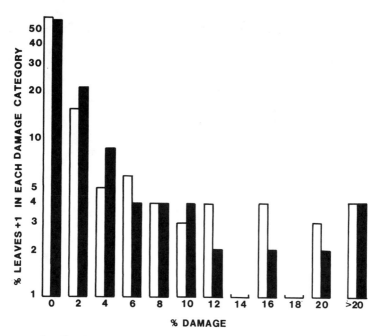

Figure 6 The distribution of damage on previously undamaged (closed bars) or mined (open bars) birch leaves, accumulating over a 3-week period. There is no significant difference between the two distributions (Kolmogorov-Smirnov test, $D = 0.138, p > .05$).

3.2   Tests with Young Leaves and Early Instars

Another possibility for differences between phytochemical analyses and feeding behaviors is that responses may have been different if we had used different ages of leaves and younger caterpillars. For example, it has been suggested that in the case of a quick-growing species like birch, maximum benefit to the tree may occur if the youngest, most valuable foliage is better protected than older foliage, so herbivores are driven onto the less valuable leaves (Edwards and Wratten 1986). Furthermore, young caterpillars may be more vulnerable to plant secondary compounds.

These two ideas were tested using laboratory preference tests as described above, but carried out before we knew the results of the 1987 field trials. We do not know whether the same results would be obtained in the field. The feeding trials compared young damaged leaves against undamaged ones at intervals after the damage was inflicted, and did the same for more mature leaves on the same trees. The results (Table 9) show that the only statistically significant preference was for *damaged* leaves, and this occurred even in the youngest leaves, despite the fact that they show greater induction of phenolics. The results for feeding trials with first and second instar *Apocheima* and *Epirrita* caterpillars comparing three types of young damaged leaves (hole-punched, chewed, and mined) against undamaged leaves from the same trees were similar to those with the later instars: No preferences were expressed except in the case of mined leaves, which were again avoided (Table 10).

Taken at their face value, these laboratory trials suggest that birch-feeding caterpillars are certainly not consistently repelled by damaged leaves, and that sometimes they even prefer them. Moreover, laboratory preferences were not correlated in any obvious way with levels of wound-induced phe-

TABLE 9   Feeding Preferences of Final-Instar *Apocheima pilosaria* for Artificially Damaged Leaves 1, 3, and 8 days after Damage[a]

Day	Sample Size	Mean Preference (standard error)	p Value Wilcoxon	t Test	Result
		Young Leaves			
1	26	0.66 (0.07)	= .08	>.05	Not significant
3	27	0.70 (0.05)	= .002	<.005	Prefers damaged
8	24	0.42 (0.08)	= .5	>.4	Not significant
		Mature Leaves			
1	26	0.69 (0.06)	= .01	<.01	Prefers damaged
3	24	0.62 (0.06)	= .058	>.05	Not significant
8	27	0.37 (0.07)	= .3	>.1	Not significant

[a] Preferences were calculated and tested for significance as described in Table 8.

TABLE 10   The Feeding Preferences of 2nd-Instar Caterpillars of Two Birch-Feeding Herbivores for Three Different Damage Types[a]

Species	n	Mean Preference	p	Result
		Mined Damage		
Apocheima pilosaria	23	0.27	<.01	Prefers undamaged
Epirrita dilutata	21	0.28	<.05	Prefers undamaged
		Insect-Grazed Leaves		
Apocheima pilosaria	27	0.43	>.1	Not significant
Epirrita dilutata	23	0.5	>.1	Not significant
		Artificial Damage		
Apocheima pilosaria	21	0.47	>.1	Not significant
Epirrita dilutata	22	0.67	>.05	Not significant

[a] Mean preferences were calculated as in Table 8, and tested for significance with a Wilcoxon matched-pairs signed rank test.

nolics in the leaves. Yet similar experiments with *Spodoptera littoralis* (Wratten et al. 1984), a broadly polyphagous lepidopteran caterpillar that does not normally feed on birch, have produced evidence that damaged birch leaves are avoided in laboratory preference tests. Perhaps there is a difference between adapted caterpillars that normally feed on birch and nonadapted herbivores such as *Spodoptera*.

3.3 *Spodoptera, Apocheima,* and Manipulated Phenolics

This experiment was designed to see whether the effects of damage were due to phenolics or to other chemicals by manipulating phenolics independently of damage. Full experimental details are in Hartley (1988). We blocked phenolic synthesis using an inhibitor of PAL, amino(oxyacetic) acid (AOA), which when sprayed onto trees causes up to 80% inhibition of the enzyme and a marked reduction in leaf phenolics (Amrhein 1979). (AOA itself has no discernable effects on feeding preferences, Hartley 1988.) Both insect-grazed leaves and those damaged with scissors were compared in laboratory preference tests with undamaged leaves from the same trees. In addition, some leaves were taken from trees that had been sprayed with inhibitors, so they were damaged but had no increase in phenolics.

Neither species of caterpillar showed any significant preference for any type of leaf in pairwise tests (Table 11); nor was there any significant effect of treatment on preference (Kruskal-Wallis nonparametric one-way ANOVA $H = 7.23$, df $= 5$, $p = .2$). These results suggest that total phenolic content

TABLE 11   Feeding Preferences of Final- and Penultimate-Instar *Apocheima pilosaria* **and 2nd- and 3rd-Instar** *Spodoptera littoralis* **for Damaged Leaves from Artificially Damaged or Insect-Grazed Trees, Some of Which were Treated with an Inhibitor of Phenolic Biosynthesis**[a]

Time after Damage	Type of Leaf Damage			
	Cut	Grazed	Cut + Inhibited	Grazed + Inhibited
Apocheima pilosaria				
3 days				
n	21	21	27	22
Mean preference	0.54	0.58	0.38	0.44
p	>.1	>.1	>.1	>.1
8 days				
n	23	23	22	25
Mean preference	0.53	0.40	0.49	0.41
p	>.1	>.1	>.1	>.1
Spodoptera littoralis				
3 days				
n	30	Not tested	28	Not tested
Mean preference	0.59		0.61	
p	>.1		=.051	
8 days				
n	26		27	
Mean preference	0.49		0.51	
p	>.1		>.1	

[a] Preferences were expressed as amount eaten from damaged leaves vs. total amount eaten, so a mean preference of 0.5 indicates equal amounts consumed from damaged and undamaged leaves. The significance of the preference was assessed using a Wilcoxon matched-pairs signed rank test.

as measured by the Folin Denis method has no effect on leaf preference in these two herbivores.

Most puzzling of all is why *Spodoptera* failed to show a preference for undamaged leaves in these trials, when other experiments done under similar conditions (but without the elaboration of using a phenolic blocker) show that damaged leaves from *B. pendula* and *B. pubescens* are avoided (Wratten et al. 1984). Coleman and Jones (this volume, Chapter 1) discuss a number of causes of such disparities to which we add that different laboratory strains of *Spodoptera*, reared under different conditions, may give different results (we used cabbage and Wratten et al. used dandelion prior to testing on birch). Also, our birch trees were younger than the trees tested by Wratten et al. Interestingly, in a later experiment, Edwards et al. (1986) found that *Spodoptera* avoided damaged leaves of *B. pubescens*, but was either indifferent to damage in the case of *B. pendula*, or actually preferred leaves adjacent to damaged ones, two days after damage. At the very least, we are forced to conclude that phytochemical induction in damaged birch foliage has variable effects on *Spodoptera*, but specific reasons for this variation are unknown (see also Karban et al. 1987).

3.4 Links between Laboratory and Field

As we have tried to emphasize, laboratory feeding trials with excised leaves are difficult to interpret. A negative result in a feeding trial does not mean that rapidly induced chemical changes in damaged foliage have no adverse effects on herbivores; it simply means they do not respond behaviorally to these changes under laboratory conditions. For example, *Apocheima*, an insect that has consistently shown no response to damage in laboratory feeding trials suffered a fourfold increase in mortality when caged on damaged branches in the field at Skipwith (Fowler and MacGarvin 1986). Viewed in this light, its consistent failure to reject damaged leaves in the laboratory is bizarre. Furthermore, the main effect of damage-induced changes may be indirect, via an increase in development time (e.g., Haukioja and Niemela 1977, 1979) leading to increased mortality via increased exposure to enemies. We believe that field experiments, or at least experiments involving whole trees and not excised leaves, are much more informative than laboratory bioassays for testing the role of rapidly induced chemical changes as defenses against insects. (See Neuvonen and Haukioja, this volume, Chapter 12, for further discussions on the advantages and disadvantages of examining rapidly induced chemical changes via field experiments.)

3.5 Field Experiments

Some of the field experiments performed on the birch system at Skipwith have already been briefly reviewed. Here we briefly discuss two other experiments carried out at Skipwith and designed to test whether deterioration in food quality following foliage damage causes insects to move more, and so suffer increased exposure to parasites and predators (Schultz 1983, Faeth 1985).

A. *pilosaria* was reared on damaged or undamaged birch trees in open-ended cages that allowed free access by winged parasitoids but prevented the escape of the caterpillars (Fowler and MacGarvin 1986). No significant difference in mortality due to parasitoids was found between the two treatments, which does not support the hypothesis of enhanced damage-induced herbivore mortality via increased rates of parasitism. However, the results are equivocal because rates of parasitism were low, and some of the larvae were parasitized in closed cages where parasitoids were supposed to have been excluded (such larvae were probably attacked before the experiment was started). We would like to see this sort of field experiment repeated in other systems.

An experiment to test whether increased movement influenced predation rates has also been carried out on the birch system at Skipwith (Lawton 1987, Bergelson and Lawton 1988). Three sorts of trees were used: untreated controls, trees with every alternate leaf on every alternate branch mechanically damaged with a hole-punch; and fertilized trees, which increased leaf nitrogen content and hence leaf quality for herbivores, potentially reducing movement and exposure to enemies. These treatments were crossed with predator treatments: trees to which the predators had free access, trees from which insectivorous birds were excluded by coarse plant-netting, and trees from which aggressive predatory ants were excluded with sticky barriers. A significant interaction between foliage treatments and predator treatments on herbivore numbers would support the idea that herbivores moving in response to foliage quality were more vulnerable to predation. Although there were direct effects of predator exclusion on herbivore abundance, there was no significant interaction between predator treatments and foliage treatments either for *A. pilosaria* or *C. serratella* (Fig. 7), the two most common insect herbivores on the experimental trees. This is particularly surprising for *Coleophora*, because this herbivore has previously been shown to move away from damaged foliage (Bergelson et al. 1986). Perhaps this increase in movement was undetectable by the predators in this particular experiment. Again, parallel studies in other systems would be valuable, but have not yet been attempted.

3.6 Summary of Effects on Insects

The results of our experiments on the responses of insect herbivores to damage-induced chemical changes in birch are mainly negative. Laboratory feeding trials with several species of herbivores and several types of damaged leaves with various amounts of wound-induced chemicals (including leaves with phenolic production inhibited) generally revealed negligible effects on feeding behavior. We were unable to verify previous reports that *S. littoralis* caterpillars avoid damaged birch foliage. In fact, there were instances of damaged leaves actually being preferred by some birch-feeding caterpillars. The only leaves consistently avoided in the laboratory were mined ones; we were unable to obtain the same results in a field experiment. Paradoxically, one species (*A. pilosaria*) that consistently failed to select undamaged leaves in the laboratory suffered enhanced mortality when forced to develop on damaged leaves in the field. Lastly, other field experiments provide no support for the hypothesis that wound-induced changes in the foliage of birch trees enhance herbivore mortality via natural enemies, either parasitoids or predators.

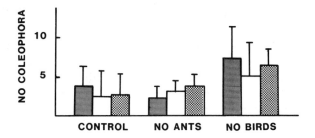

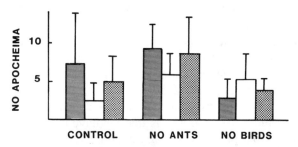

Figure 7 The abundance of *Apocheima pilosaria* and *Coleophora serratella* after 1 month, expressed as a mean (± 2 SD) of four saplings. The saplings were either foliar fed (lines), damaged (dots), or left as controls (white), and had either ants, birds or no predators excluded. Both *Apocheima* ($F = 4.59$; df $= 2,27$; $p < 0.05$) and *Coleophora* ($F = 5.74$, df $= 2,27$, $p < 0.01$) showed a significant effect of predators, but neither showed a significant effect of leaf treatment, and there was no significant interaction between predator treatments and leaf treatments.

4 CONCLUSION

Rapid phytochemical induction of phenolics in damaged birch foliage has proven to be complex. The chemical changes vary quantitatively with the type of damage inflicted, and possibly qualitatively as well, since different methods of analysis produce different results. This complexity, together with the fact that induced changes are usually swamped by inherent between- and within-tree variation in phytochemicals, makes results difficult to interpret. More precise chemical information on the nature of induced compounds may help to resolve these difficulties; assays of total phenolics are probably no longer useful for this type of work.

One of the most intriguing results to emerge from an otherwise messy and generally negative picture is that birch trees apparently recognize different types of damage and respond differently to each. Several features of these different damage-induced responses are consistent with their being defenses against insects. The response to insects is larger than to artificial damage and is associated with induction of phenolics in undamaged leaves, a process that involves in situ synthesis via an increase in PAL activity rather than passive transport of phenolics from the site of damage. These responses do not occur in the case of artificial damage, but the application of secretions from caterpillar mouthparts to artificial wounds induces chemical changes similar in magnitude to genuine insect attack. Somehow birch trees can distinguish between sterile scissors and caterpillar mouthparts.

Despite these thought-provoking results, we have so far found few definite effects on birch herbivores. There is no clear link between the induced chemical changes in birch phenolics and herbivore feeding behavior, and unequivocal effects on insect population dynamics in the field are even harder to demonstrate. Nevertheless, small effects from induced chemical changes, though difficult to detect, may be ecologically significant over the lifetime of a tree. We often lack the statistical power to detect weak, but biologically significant effects of phytochemical induction on herbivore population dynamics in the field. A second important reason why our results, at least superficially, appear to be at variance with other related work, particularly that carried out in Finland by Haukioja and his colleagues (Neuvonen and Haukioja, this volume, Chapter 12), is that we have not, in the main, done the same kind of experiments. Until we do, comparisons between these two bodies of data are very difficult. It is also clear that the case for phenolics being involved in any insect responses is now weak. We do not know what chemicals the insects are responding to when and if they do respond.

Against this confusing background, it may be appropriate to consider an alternative hypothesis for the adaptive significance of damage-induced phenolics in birch foliage. Our work on the means by which birch distinguishes insect attack from artificial damage and communicates the message of injury to sites remote from the damage itself is only just beginning, but it is likely that the mechanisms involved will be similar to those well known to plant pathologists. Phytoalexin production is known to be stimulated by plant and fungal cell well fragments (Darvill and Albersheim 1984, Dixon 1986); fungal contamination on the mouthparts of Hemiptera stimulates wound-induced ethylene production (Grisham et al. 1987); elicitors capable of inducing PAL in plant cells are found in some fungal–plant interactions (Hahlbrock et al. 1981); and cotton's response to attack by both fungi and mites appears to be identical, or at least similar enough for organisms from these two different kingdoms to compete indirectly, via the host plant (Karban et al. 1987). Speculating, we believe that fungi and other microorganisms associated with insects may be the reason why birch's reaction to herbivores and to artificial damage is so different; indeed, the primary role of damage-induced changes in birch phenolics may be antifungal or antimicrobial, rather than directed principally at insect herbivores. Entomologists would undoubtedly benefit from a closer consideration of the role of plant pathogens in their studies of rapid phytochemical induction.

REFERENCES

Amrhein, N. 1979. Biosynthesis of cyanidin in buckwheat hypocotyls. *Phytochemistry* 18:585–589.

Amrhein, N., K. H. Goedeke, and J. Gerhardt. 1976. The estimation of phenylalanine ammonia lyase (PAL) activity in intact cells of higher plant tissues. *Planta* 131:33–40.

Baldwin, I. T. 1988. The alkaloidal responses of wild tobacco to real and simulated herbivory. *Oecologia* 77:378–381.

Berenbaum, M. R. 1980. Adaptive significance of midgut pH in larval Lepidoptera. *Am. Natur.* 115:138–146.

Bergelson, J., S. Fowler, and S. Hartley. 1986. The effects of foliage damage on casebearing moth larvae, *Coleophora serratella*, feeding on birch. *Ecol. Entomol.* 11:241–250.

Bergelson, J. M., and J. H. Lawton. 1988. The influence of foliage damage on interactions between the insect herbivores of birch, and their predators. *Ecology* 69:434–445.

Bernays, E. A. 1981. Plant tannins and insect herbivores: an appraisal. *Ecol. Entomol.* 6:353–360.

Bishop, P. D., D. J. Makus, G. Pearce, and C. A. Ryan. 1981. Proteinase inhibitor-inducing factor activity in tomato leaves resides in oligosaccharides enzymically released from cell walls. *P.N.A.S.* 78:3526–3540.

Cole, R. A. 1985. Relationship between the concentration of chlorogenic acid and the incidence of carrot-fly larvae. *Ann. Appl. Biol.* 106:211–217.

Coleman, J. S., and C. Jones. 1991. A phytocentric perspective of phytochemical induction by herbivores. In D. W. Tallamy and M. J. Raupp (eds.), *Phytochemical Induction by Herbivores*. New York: Wiley, pp. 3–46.

Darvill, A. P., and P. Albersheim. 1984. Phytoalexins and their elicitors: a defense against microbial infection in plants. *Annu. Rev. Plant Physiol.* 35:243–275.

Davies, E., and A. Schuster. 1981. Intercellular communication in plants: evidence for a rapidly generated, bidirectionally transmitted wound signal. *P.N.A.S.* 78:2422–2426.

Dixon, R. A. 1986. The phytoalexin response: elicitation, signalling and control of host gene expression. *Biol. Rev.* 61:239–291.

Edwards, P. J., and S. D. Wratten. 1982. Wound-induced changes in palatability in birch (*Betula pubescens* Ehrh. ssp. *pubescens*). *Am. Natur.* 120:816–818.

Edwards, P. J., and S. D. Wratten. 1985. Induced plant defenses against insect grazing: fact or artefact? *Oikos* 44:70–74.

Edwards, P. J., and S. D. Wratten. 1987. Ecological significance of wound-induced changes in plant chemistry. In V. Labeyne, E. Fabres, and D. Lachaise (eds.),

Proceedings of the Sixth International Symposium on Insect-Plant Relationships. Dordrecht: Junk, pp. 213–218.

Edwards, P. J., S. D. Wratten, and S. Greenwood. 1986. Palatability of British trees to insects: constitutive and induced defenses. *Oecologia* 69:316–319.

Edwards, P. J., S. D. Wratten, and R. M. Gibberd. 1991. The impact of inducible phytochemicals on food selection by insect herbivores and its consequences for the distribution of grazing damage. In D. W. Tallamy and M. J. Raupp (eds.), *Phytochemical Induction by Herbivores.* New York: Wiley, pp. 205–222.

Faeth, S. H. 1985. Quantitative defense theory and patterns of feeding by oak insects. *Oecologia* 68:34–40.

Faeth, S. H. 1986. Indirect interactions between temporally separated herbivores, mediated by the host plant. *Ecology* 67:474–494.

Feeny, P. 1970. Seasonal changes in oak leaf tannin and nutrients as a cause of spring feeding by winter moth caterpillars. *Ecology* 51:565–585.

Fowler, S. V. 1983. The foliage feeding insects on birch: plant fitness loss, apparency, and the levels of anti-herbivore defenses. Ph.D. thesis, University of York.

Fowler, S. V. 1984. Foliage value, apparency and defense investment in birch seedlings and trees. *Oecologia* 62:387–392.

Fowler, S. V., and J. H. Lawton. 1985. Rapidly induced defenses and talking trees: the devil's advocate position. *Am. Natur.* 126:181–195.

Fowler, S. V., and M. MacGarvin. 1986. The effects of leaf damage on the performance of insect herbivores on birch, *Betula pubescens. J. Anim. Ecol.* 55:565–574.

Green, T. R., and C. A. Ryan. 1973. Wound-induced proteinase inhibitor in tomato leaves. *Plant Physiol.* 51:19–21.

Grisham, M. P., W. L. Sterling, R. D. Powell, and P. W. Morgan. 1987. Characterization of the induction of stress ethylene synthesis in cotton caused by the cotton fleahopper (Hemiptera: Miridae) and its microorganisms. *Ann. Entomol. Soc. Am.* 80:411–416.

Hahlbrock, K., C. J. Lamb, C. Purwin, J. Ebel, E. Fautz, and E. Schaffer. 1981. Rapid response of suspension-cultured parsley cells to the elicitor from *Photophthora megasperma* var. *sojae.* Induction of the enzymes of general phenylpropanoid metabolism. *Plant Physiol.* 67:768–773.

Hartley, S. E. 1988. The inhibition of phenolic biosynthesis in damaged and undamaged birch foliage and its effect on insect herbivores. *Oecologia* 76:65–70.

Hartley, S. E., and R. D. Firn. 1989. Phenolic biosynthesis, leaf damage and insect herbivory in birch (*Betula pendula*). *J. Chem. Ecol.* 15:275–283.

Hartley, S. E., and J. H. Lawton. 1987. The effects of different types of damage on the chemistry of birch foliage and the responses of birch feeding insects. *Oecologia* 74:432–437.

Haukioja, E., and S. Neuvonen. 1985. Induced long-term resistance of birch foliage against defoliators: defensive or incidental? *Ecology* 66:1303-1308.

Haukioja, E., and P. Niemela. 1977. Retarded growth of a geometrid larva after mechanical damage to leaves of its host tree. *Ann. Zool. Fenn.* 14:48–52.

Haukioja, E., and P. Niemela. 1979. Birch leaves as a resource for herbivores: seasonal occurrence of increased resistance in foliage after mechanical damage of adjacent leaves. *Oecologia* 39:151–159.

Karban, R., R. Adamchak, and W. C. Schnathorst. 1987. Induced resistance and interspecific competition between spider mites and a vascular wilt fungus. *Science* 235:678–680.

Klocke, J. A., and B. G. Chan. 1982. Effects of cotton condensed tannin on feeding and disgestion in *Heliothus zea*. *J. Insect Physiol.* 28:911–915.

Lawton, J. H. 1987. Food-shortage in the midst of apparent plenty: the case for birch feeding insects. In H. W. Velthius (ed.), *Proceedings of the Third European Congress of Entomology*. Amsterdam: Nederlandse Entomolgische Verenining, pp. 219–228.

Mansfield, J. W. 1982. Role of phytoalexins in disease reistance. In J. A. Bailey and J. W. Mansfield (eds.), *Phytoalexins*. Glasgow: Blackie, pp. 253–288.

Martin, J. S. and M. M. Martin. 1982. Tannin assays in ecological studies: lack of correlation between phenolic, proanthocyanidin and protein-precipitating constituents in mature foliage of six oak species. *Oecologia* 54:205–211.

Martin, J. S., M. M. Martin, and E. A. Bernays. 1987. Failure of tannic acid to inhibit digestion or reduce digestibility of plant protein in gut fluids of insect herbivores: implications for theory of plant defense. *J. Chem. Ecol.* 13:605–621.

Miles, P. W. 1969. Interaction of plant phenols and salivary phenolases in the relationship between plants and Hemiptera. *Entomol. Exp. Applic.* 12:736–744.

Mole, S., and P. G. Waterman. 1985. Stimulatory effects of tannins and cholic acid on tryptic hyrolysis of proteins: ecological implications. *J. Chem. Ecol.* 11:1323–1331.

Mole, S., and P. Waterman. 1987a. A critical analysis of techniques for measuring tannins in ecological studies. I. Techniques for chemically defining tannins. *Oecologia* 72:137–147.

Mole, S., and P. Waterman. 1987b. A critical analysis of techniques for measuring tannins in ecological studies. II. Techniques for biochemically defining tannins. *Oecologia* 72:148–156.

Neuvonen, S., and E. Haukioja. 1991. The effects of inducible responses in host foliage on birch-feeding herbivores. In D. W. Tallamy and M. J. Raupp (eds.), *Phytochemical Induction by Herbivores*. New York: Wiley, pp. 277–291.

Pena-Cortes, H., J. Sanchez-Serrano, M. Rocha-sosa, and L. Willmitzer. 1988. Systemic induction of proteinase inhibitor-II gene expression in potato plants by wounding. *Planta* 174:84–89.

Rhoades, D. F. 1985. Pheromonal communication between plants. *Rec. Adv. Phytochem.* 19:195–218.

Rhoades, D. F., and R. G. Cates. 1976. Towards a general theory of plant antiherbivore chemistry. *Rec. Adv. Phytochem.* 10:168–213.

Risch, S. J. 1985. Effects of induced chemical changes on feeding preference tests. *Entomol. Exp. Applic.* 39:81–84.

Ryan, C. A. 1974. Assay and biochemical properties of the proteinase inhibitor inducing factor: a wound hormone. *Plant Physiol.* 54:328–332.

Schultz, J. C. 1983. Impact of variable plant defensive chemistry on susceptibility of insects to natural enemies. In P. A. Hedin (ed.). *ACS Symposium Series,* Vol.

208, *Plant Resistance to Insects*. Washington, DC: American Chemical Society, pp. 37–54.

Shain, L., and W. E. Hillis. 1972. Ethylene production in *Pinus radiata* in response to *Sirex-Amylostereum* attack. *Phytopathology* 62:1407–1409.

Southwood, T. R. E., V. C. Moran, and C. E. J. Kennedy. 1982. The richness, abundance and biomass of arthropod communities on trees. *J. Anim. Ecol.* 51:635–649.

Strong, D. R., J. H. Lawton, and R. Southwood. 1984. *Insects on Plants: Community Patterns and Mechanisms*. Oxford: Blackwell Scientific.

Van Sambeek, J. W., B. G. Pickard, and C. E. Ulbright. 1976. Mediation of rapid electrical, metabolic, transpirational and photosynthetic changes by factors released from wounds. II. Mediation of the variation potential by Ricca's factor. *Can. J. Bot.* 54:2651–2661.

West, C. 1985. Factors underlying the late seasonal appearance of the lepidopterous leaf-mining guild on oak. *Ecol. Entomol.* 10:111–120.

Wilson, M. F. 1984. Comparison of tannin levels in developing fruit buds of two orchard pear varieties using two techniques: Folin Denis and protein precipitation assays. *J. Chem. Ecol.* 10:493–498.

Woodhead, S. 1981. Environmental and biotic factors affecting the phenolic content of different cultivars of *Sorghum bicolour*. *J. Chem. Ecol.* 7:1035–1047.

Wratten, S. D., P. J. Edwards, and I. Dunn. 1984. Wound-induced changes in the palatability of *Betula pubescens* and *B. pendula*. *Oecologia* 61:372-375.

Yang, S. F. and H. K. Pratt. 1978. The physiology of ethylene in wounded plant tissues. In G. Kahl (ed.), *Biochemistry of Wounded Plant Tissues*. Berlin: de Gruyter, pp. 595–625.

PART II

IMPACT OF INDUCIBLE PHYTOCHEMICALS ON HERBIVORE FITNESS, BEHAVIOR, POPULATION DYNAMICS, AND COMMUNITIES

CHAPTER 6

EFFECTS OF MAMMAL BROWSING ON THE CHEMISTRY OF DECIDUOUS WOODY PLANTS

J. P. BRYANT
Institute of Arctic Biology, University of Alaska, Fairbanks, AK 99775-0180

K. DANELL
The Swedish University of Agricultural Sciences, Faculty of Forestry,
Department of Wildlife Ecology, UMEA, Sweden H 5-90183

F. PROVENZA
Department of Range Science, College of Natural Resources, UMC 52, Utah
State University, Logan, UT 84322

P. B. REICHARDT and T. A. CLAUSEN
Department of Chemistry, University of Alaska, Fairbanks, AK 99775-0180

R. A. WERNER
United States Forest Service, Institute of Northern Forestry, University of
Alaska, Fairbanks, AK 99775-0180

1   Introduction
2   Long-term chemical responses of deciduous woody plants to herbivory
    2.1   Leaves
    2.2   Twigs
3   Physiological mechanisms of long-term responses to herbivory
    3.1   Leaves
    3.2   Twigs
4   Implications for plant defense theory
    4.1   Leaf responses to herbivory
    4.2   Twig responses to herbivory
5   Summary
    References

1   INTRODUCTION

Browsing by mammals can affect the food value of woody plants. In some
cases, palatability and/or nutritional quality decrease in response to browsing

135

(Bryant 1981a,b, Bryant et al. 1983, 1985a,b, Provenza and Malechek 1983, Miquelle 1983, Basey 1987, Danell et al. 1987). Such negative changes have attracted attention because they imply that plants actively defend themselves against herbivory (Rhoades 1979, Haukioja 1980). However, browsing does not always decrease the food value of woody plants (Provenza and Malechek 1984, Danell et al. 1985a, Danell and Huss-Danell 1985, Danell and Neuvonen 1987). In this chapter we attempt to resolve these conflicting observations with respect to *deciduous* woody species.

Previous studies have emphasized evolutionary explanations of plant chemical responses to herbivory (Haukioja and Hakala 1975, Rhoades 1979, Haukioja 1980, Haukioja et al. 1983). However, the physiological mechanisms causing such responses must be elucidated before evolutionary explanations can be evaluated. Thus, we first construct physiological models that can account for the chemical responses of deciduous woody plants to defoliation and pruning by herbivores. Then we use these models to test the validity of evolutionary explanations of long-term induced defense (LTI) sensu Haukioja (1980). We conclude by discussing the relevance of LTI to the population dynamics of browsing mammals.

2   LONG-TERM CHEMICAL RESPONSES OF DECIDUOUS WOODY PLANTS TO HERBIVORY

Interactions between woody plants and browsing mammals differ from plant–insect interactions. Unlike most insects, individual mammals feed in a single day upon several plant parts (leaves, twigs, bark, fruits, roots) from a variety of woody species of different inherent growth rates (slow vs. fast) and physiological ages (mature vs. juvenile) (Freeland and Janzen 1974, Bryant and Kuropat 1980, Robbins 1983). Thus, effects of defoliation and pruning upon plant growth and chemistry, as modified by plant inherent growth rate and physiological age, must be considered when attempting to understand the impact mammalian herbivory has upon woody plant chemistry. Furthermore, in ecosystems such as boreal forests insect herbivores are not active in winter. However, browsing by mammals such as the snowshoe hare (*Lepus americanus*) is usually more severe in winter than in summer (Grange 1949, Keith 1963, Wolff 1978). Seasonality is important, because summer defoliation and winter pruning have different effects on woody plant growth and chemistry (Kozlowski 1971, Kramer and Kozlowski 1979).

2.1   Leaves

Severe defoliation of deciduous species early in the growing season can reduce the food value of leaves in subsequent years (Benz 1974, Baltensweiler et al. 1977, Haukioja and Niemela 1977, Werner 1979, Myers and

TABLE 1   Characteristics of Forest Defoliating Insect Outbreaks Where Reduced Nutritional Value of Reflushed Foliage to the Defoliating Insect is Well Documented

	Defoliator				
Characteristic	*Zeiraphera diniana*	*Epirrita autumnata*	*Choristoneura conflictana*	*Rheumaptera hastata*	*Malacosoma californicum pluviale*
Defoliation event					
Defoliation phenology[a]	E	E	E	E–M	E
Duration of severe defoliation (years)	1–2	1–2	1–3	1–3	4–5
Host					
Species[b]	*L. decidua*	*B. pubescens*	*P. tremuloides*	*B. papyrifera*	*A. rubra*
Growth form	Deciduous	Deciduous	Deciduous	Deciduous	Deciduous
Reflushed foliage					
Leaf size	Decrease	Decrease	Decrease	Decrease	?
Nitrogen concentration	Decrease	Decrease	Decrease	Decrease	Decrease
Phenolic concentration	Increase	Increase	Increase	Increase	?
Food value	Decrease	Decrease	Decrease	Decrease	Decrease
References[c]	1–3	4–8	9,10	10,11	12

[a] E, early growing season; M, middle growing season.

[b] *L., Larix* (larch); *B., Betula* (birch); *P., Populus* (aspen); *A., Alnus* (alder).

[c] (1) Benz 1974; (2) Baltensweiler et al. 1977; (3) Fischlin and Baltensweiler 1979; (4) Haukioja and Niemela 1976; (5) Haukioja 1980; (6) Haukioja et al. 1983; (7) Kallio and Lehtonen 1973; (8) Niemela, unpublished results; (9) Beckwith 1968, 1970; (10) Werner, unpublished results; (11) Werner 1977; (12) Myers and Williams 1987.

Williams 1987). This response is called long-term induction (LTI) and has been attributed to an evolved, active defense targeted directly at herbivores (Rhoades 1979, Haukioja 1980, Haukioja and Neuvonen 1985).

Long-term induction can be elicited by either insect defoliation (Benz 1974, Baltensweiler et al. 1977, Haukioja and Neuvonen 1985, Myers and Williams 1987) or manual defoliation (Haukioja and Niemela 1977, Werner 1979, Haukioja and Neuvonen 1985). Irrespective of the cause, LTI is characterized by several attributes (Table 1). First, defoliation is complete before a significant fraction of mineral nutrients in mature leaves at midsummer have been removed by preabscission retranslocation (Bryant et al. 1987b). For example, defoliation of Alaska paper birch (*Betula resinifera* = *papyrifera* sbsp. *humilis*) (Dugle 1966, Hulten 1968) by outbreaking spear-marked black moths (*Rheumaptera hastata*) occurs comparatively late in the growing season (Table 1). However, by the time *R. hastata* pupates in August (Werner 1977), only 25% of the nitrogen contained in Alaska paper birch leaves has been removed by preabscission retranslocation (Chapin and Kedrowski 1983). Leaves of induced plants are small (Fig. 1) in comparison to leaves of undefoliated conspecifics (Baltensweiler 1985, Werner, unpublished results, Danell and Bergstrom unpublished results) and often have comparatively low nitrogen concentrations (Baltensweiler et al. 1977, Tuomi et al. 1984, Myers and Williams 1987, Werner, unpublished results) and comparatively high phenol or fiber concentrations (Baltensweiler et al. 1977, Tuomi et al. 1984, Werner, unpublished results).

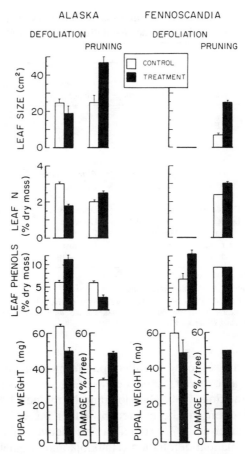

Figure 1   Effects of defoliation and pruning on leaves of birches in Alaska and Fen-
noscandia and on pupal weight of *Rheumaptera hastata*. Alaskan data provided by
Bryant (unpublished results). Fennoscandian data are from Tuomi et al. (1984; Leaf
N and Phenols), Haukioja and Niemela (1976; Pupal Weight), and Danell and Huss-
Danell (1985; Effects of Pruning). Means ± SE presented.

These changes in leaf chemistry are related to the intensity and duration
of defoliation. In the case of Alaska paper birch (Werner 1979, unpublished
results), 100% defoliation reduced leaf nitrogen more than did 50% defoliation. Conversely, 100% defoliation increased leaf phenols more than did
50% defoliation. Successive defoliations result in progressive changes in leaf
chemistry (Fig. 2) and progressive deterioration of leaf growth and food value
(Werner 1979).

Winter pruning of deciduous species by mammals has effects on leaves
opposite those described above for severe early season defoliation by in-

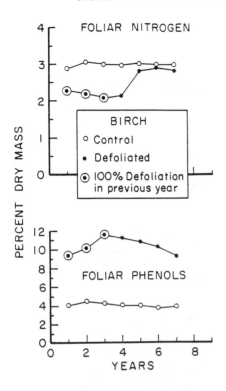

Figure 2 Effects of repetitive 100% defoliation upon Alaska paper birch foliar nitrogen and total phenolics (Folin-Dennis assay tannic acid standard). Calculated from Werner (unpublished results).

sects. For example, moose pruning of birch (Fig. 1) results in increased leaf size, increased leaf nitrogen, decreased leaf phenols and protein precipitating capacity of leaf methanol extract, and increased herbivory of leaves in the following growing season (Danell and Huss-Danell 1985, Bryant, unpublished results, Danell unpublished results).

2.2 Twigs

Twigs of juvenile-form stump sprouts regenerated by heavily pruned or ring-barked mature-form woody plants (Kozlowski 1971) are less palatable to a variety of small mammals including hares, microtine rodents, beavers and domestic goats than similar diameter twigs from mature-form conspecifics (Libby and Hood 1976, Klein 1977, Bryant 1981a,b, Pehrson 1981, 1983, Bryant et al. 1983, 1985a,b, 1987b, Provenza and Malechek 1984, Reichardt et al. 1984, 1990, Sinclair and Smith 1984, Danell et al. 1985b, Tahvanainen et al. 1985, Clausen et al. 1986, Basey 1987). In the case of snowshoe hares (*Lepus americanus*) (Reichardt et al. 1984, Bryant et al. 1985a) and mountain hares (*L. timidus*) (Pehrson 1981, 1983), unpalatable stump sprout twigs (Fig. 3) are also of low nutritional quality in comparison to similar diameter twigs from mature-form plants (Fig. 4). In the case of small ru-

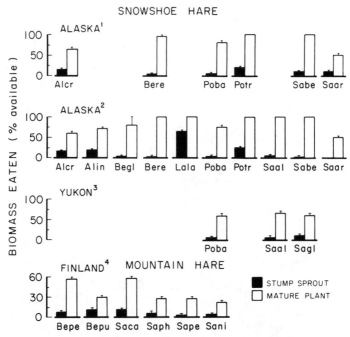

Figure 3   Palatability to hares of juvenile-form stump sprouts and similar diameter twigs from mature-form conspecifics. Means ± 1 SE presented. *References:* 1, Klein 1977; 2, Bryant unpublished results; 3, Sinclair and Smith 1984; 4, Tahvanainen et al. 1985, Tahvanainen, unpublished results. Alcr, *Alnus crispa*; Alin, *A. incana*; Begl, *Betula glandulosa*; Begl, *B. pendula*; Bepu, *B. pubescens*; Bere, *B. resinifera*; Lala, *Larix laricina*; Poba, *Populus balsamifera*; Potr, *P. tremuloides*; Saal, *Salix alaxensis*; Saar, *S. arbusculoides*; Saca, *S. caprea*; Sabe, *S. bebbiana*; Sagl, *S. glauca*; Sani, *S. nigricans*; Sape, *S. pentandra*; Saph, *S. phyllicifolia*.

minants there appears to be no consistent correlation between stump sprout palatability and nutritional quality (Provenza and Malechek 1983, 1984). Larger ruminants such as moose feed preferentially upon stump sprouts of some species such as willow (*Salix* spp.) (Machida 1979).

Chemical analyses and preference bioassays (Figs. 3–5) indicate that feeding deterrent secondary metabolites are the most likely cause of the low palatability of juvenile-form stump sprouts to small mammals. The low palatability of juvenile-form stump sprouts cannot be explained by a lack of mineral nutrients such as nitrogen (Klein 1977, Bryant 1981a,b, Pehrson 1981, 1983, Bryant et al. 1983, 1985a,b, 1988, Fox and Bryant 1984, Provenza and Malechek 1984, Reichardt et al. 1984, 1987, 1990, Sinclair and Smith 1984, Tahvanainen et al. 1985, Clausen et al. 1986, Basey 1987). Bryant et al. (1985a) and Danell et al. (1985b) have suggested that low concentrations

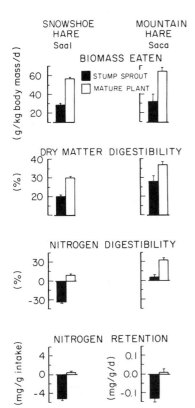

Figure 4 Effects of growth stage upon consumption, apparent dry matter digestibility, apparent nitrogen digestibility, and nitrogen retention by hares fed willow stump sprout twigs or twigs of mature-form plants. Means ± 1 SE presented. Results for snowshoe hares and *Salix alaxensis* (Saal) from Bryant et al. (1985a). Results for mountain hares and *S. caprea* (Saca) from Pehrson (1981, 1983).

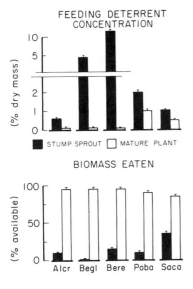

Figure 5 Concentrations of feeding deterrent secondary metabolites in winter-dormant, juvenile-form, stump sprout, current year twigs and similar diameter current year twigs of mature-form conspecifics, and their effects upon hare consumption of oats. Alcr, *Alnus crispa*—pinosylvin, pinosylvin methyl ether (Clausen et al. 1986); Begl, *Betula glandulosa*—resin containing papyriferic acid and greenic acid (Bryant et al., in preparation); Bere, *B. resinifera*—papyriferic acid (Reichardt et al. 1984); Poba, *Populus balsamifera*—salicaldehyde plus 6-hydroxyclyclohexene-one (Reichardt et al., 1990); Saca, *Salix caprea*—phenolic glycoside mixture (Tahvanainen et al. 1985). Means ± 1 SE presented.

BIOMASS EATEN (4mm twig)

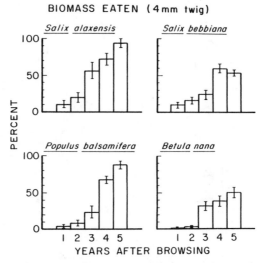

Figure 6   Consumption of winter-dormant stump sprouts by snowshoe hares. Means ± 1 SE presented. Data from Bryant et al. (1985a), Fox and Bryant (1984), and Bryant (unpublished results).

of soluble carbohydrate may contribute to the low palatability of willow stump sprouts. However, this is not true of balsam poplar (*Populus balsamifera*) (Reichardt et al., 1990). Juvenile-form balsam poplar stump sprouts and equal-diameter flowering twigs from mature-form balsam poplar have similar concentrations of fructose, sucrose, and glucose, the major sugars of woody plants (Kramer and Kozlowski 1979).

The low palatability of juvenile-form stump sprouts persists for several years (Provenza and Malechek 1983, Fox and Bryant 1984, Bryant et al. 1985a). In the case of preferred rapidly growing deciduous species (Fig. 6) such as some willows (*Salix*), dwarf birch (*B. glandulosa*) and balsam poplar (Bryant and Kuropat 1980), low palatability persists for 2–3 years (Bryant 1981a, Bryant et al. 1983, 1985a, Fox and Bryant 1984). As plant inherent growth rate and palatability decline (Bryant and Kuropat 1980, Bryant et al. 1983, Coley et al. 1985), low palatability persists for an increasing number of years (Fox and Bryant 1984), as does expression of other juvenile traits (Kozlowski 1971).

Conversely, pruning of juvenile-form plants can increase their palatability (Bryant et al. 1983, Danell et al. 1985a). Associated with this increase in palatability is a decrease in concentrations of feeding deterrent secondary metabolites in twigs (Chapin et al. 1985, Danell et al. 1985a). Such chemical changes are related to the carbohydrate reserves of the plant. Plants with comparatively large carbohydrate reserves are less affected by repetitive pruning than plants with smaller carbohydrate reserves (Garrison 1972, Bryant et al. 1983).

3 PHYSIOLOGICAL MECHANISMS OF LONG-TERM RESPONSES TO HERBIVORY

3.1 Leaves

As woody plants progress from the seedling stage through juvenility toward maturity they age physiologically (Waring 1959, Kozlowski 1971, Borchert 1976, Kramer and Kozlowski 1979). The apical dominance characteristic of juvenility relaxes and lateral buds develop into shoots, thereby increasing the size and complexity of the crown. Although the root system increases in size and complexity, this increase is not commensurate with that of the crown: There is a proportionately greater increase in the mass of above ground parts in comparison to the mass of roots (Kozlowski 1971, Borchert 1976). Thus, sooner or later in the life of a woody plant there is a reduction in its root–shoot ratio and a compensation point is reached whereby nutrient demands of shoots exceed the absorption capacity of roots (Borchert 1976). This change in supply and demand results in the gradual decrease in shoot growth caused by increased competition for nutrients between shoots (Moorby and Waring 1963); it is the cause of physiological aging (Waring 1959, Kozlowski 1971, Borchert 1976, Kramer and Kozlowski 1979). It is no coincidence that nutrient stress associated with heavy fruiting and/or growth in infertile soil increases physiological aging (Moorby and Waring 1963).

Long-term induction and physiological aging are remarkably similar in several respects. Severe defoliation results in increased fine root mortality (Kozlowski 1971). Fine root biomass and thus nutrient absorption declines. Consequently, leaf nutrient concentrations decline (Baltensweiler et al. 1977, Tuomi et al. 1984, Werner unpublished results, Myers and Williams 1987) as does shoot and leaf growth (Kulman 1971, Baltensweiler 1985, Werner unpublished results). An obvious explanation of reduced growth is nutrient stress (Tuomi et al. 1984, 1987, Bryant et al. 1987a, b). This speculation is supported by two observations. First, like aging, LTI appears to be most pronounced in plants growing in infertile soil (Tuomi et al. 1984, Haukioja and Neuvonen 1985, Bryant, unpublished results). Second, root damage elicits symptoms of LTI, such as reduced leaf nitrogen and increased leaf phenols (Tuomi et al. 1984).

Pruning releases apical dominance and reduces the competition among growing points for mineral nutrients that causes the reduced shoot growth associated with aging (Moorby and Waring 1963). Similarly, winter pruning of birch by moose (Fig. 1) releases apical dominance and reduces bud numbers, thereby increasing nutrition (Danell and Huss-Danell 1985, Danell et al. 1985a, Bergstrom and Danell 1987).

The analogies between aging and LTI and between reversal of aging and leaf responses to winter browsing are not coincidental. It is our hypothesis that LTI, like aging in woody plants, is caused by nutrient stress and, fur-

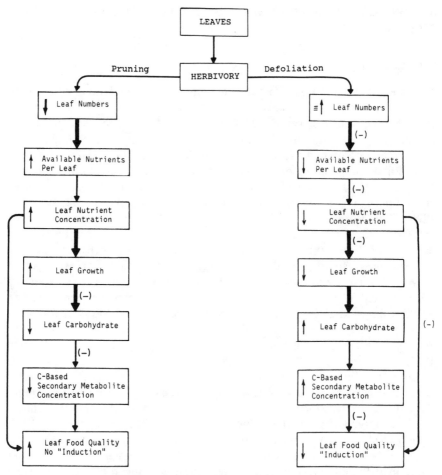

Figure 7   Effects of defoliation and pruning upon leaves and twigs of juvenile-form and mature-form deciduous woody plants. Vertical arrows within boxes indicate increase (↑) or decrease (↓) in the parameter. Arrows between boxes indicate positive effect unless otherwise (−) indicated. Thickness of arrow indicates magnitude of effect.

thermore, that winter pruning by mammals relaxes this stress in remaining shoots. Thus, we propose the following physiological mechanism for LTI (Fig. 7, *right*). Severe and repeated early growing season defoliation as occurs in insect outbreaks results in fine root mortality, a decline in nutrient absorption, and, in the event of no bud death, reduced leaf nutrient concentrations. If carbohydrate is not limiting to leaf growth, as is often the case (Moorby and Waring 1963, Chapin 1980, Bryant et al. 1983), more carbohydrate is available for carbon-based secondary metabolite synthesis so that concentrations of substances such as defensive phenolics rise in leaves (Bryant et al. 1983). Thus, leaf food value declines because of reduced nutrient concentrations and/or increased concentrations of defensive carbon-based secondary metabolite in leaves. Winter pruning by mammals reverses this process (Fig. 7, *left*) because pruning releases apical dominance, removes buds competing for nutrients, and therefore decreases competition for growth-limiting mineral nutrients among remaining buds (Moorby and Waring 1963).

3.2 Twigs

Bryant et al. (1983, 1985a) proposed that increased defense of stump sprouts as compared to twigs of mature trees and shrubs is a consequence of a reversion to the juvenile form. Whereas physiological aging involves a gradual reduction in vigor associated with increasing crown size, maturation or phase change involves rapid and predictable changes that occur during a restricted period early in the life of a woody plant (Kozlowski 1971). Although the physiological mechanism of maturation is complex and poorly understood, it involves hormonal activation of genes within individual cells (Kozlowski 1971, Kramer and Kozlowski 1979). Thus, maturation is not a consequence of simple nutrient stress as is aging (Kozlowski 1971).

After an individual cell matures, its lineage continues to express mature traits when propagated vegetatively (Kozlowski 1971). For example, cuttings from the mature-zone in the crown of a mature tree exhibit mature traits when propagated in a garden, whereas cuttings from the juvenile-zone at the base of the same tree exhibit juvenile traits when propagated in the same garden (Schaffalitzky de Muckadell 1969, Kozlowski 1971, Libby and Hood 1976). Such experiments indicate that juvenile traits are relatively stable (Moorby and Waring 1963) and, in the case of tall trees, related to height. It is possible that relaxation of juvenile defensive traits during maturation is a result of escape from browsing by vertical growth (Klein 1977, Janzen and Martin 1981, Sinclair and Smith 1984). However, experiments with the shrub, Alaska feltleaf willow (*S. alaxensis*), have demonstrated that phase change can occur within the height range used for feeding in winter by hares (Bryant et al. 1985a). Thus, juvenile-phase defense is a result of age-specific selection rather than escape in space.

Severe pruning of juvenile-phase plants cannot cause them to revert to the juvenile form (Bryant et al. 1983). Rather, severe and repeated pruning of juvenile-phase plants eventually causes carbon stress (Bryant et al. 1983) and an associated reduction in allocation of carbon to defensive substances (Bryant et al. 1983, Chapin et al. 1985, Danell et al. 1985a). The severity and duration of pruning required to significantly lower juvenile-phase defenses depends on the plant's carbohydrate reserves (Bryant et al. 1983). In general, young saplings established from seed and evergreens will be more susceptible to carbon stress caused by pruning than will be stump sprouts of well-established deciduous species that have larger below-ground carbohydrate reserves (Bryant et al. 1983).

Severe browsing must result in an increase in the proportion of small twig biomass composed of current year growth (CAG), because regrowth after pruning is, by definition, CAG (Aldous 1952, Garrison 1972, Krefting et al. 1966, Willard and McKell 1978, Wolff 1978, Provenza et al. 1983, Fox and Bryant 1984, Bergstrom and Danell 1987). Irrespective of physiological age (juvenile or mature), such a shift in the internode age distribution of some woody plants will negatively affect their palatability in winter to small mammals such as hares, goats, and sheep, because these mammals feed preferentially upon larger, older current-year internodes and reject smaller-diameter current-year internodes (Currie and Goodwin 1966, Fox and Bryant 1984, Provenza and Malechek 1984, Bryant et al. 1987a, Yabann et al. 1987, Reichardt et al., 1990). In the case of birch and hares, small-diameter, unpalatable current-year internodes are toxic in comparison to older more palatable internodes (Pehrson 1981, 1983, Reichardt et al. 1984, Palo 1987a,b).

Current-year internodes are less palatable than older larger-diameter internodes, because most secondary metabolites in twigs are in bark rather than wood (McKey 1979, Palo 1987a,b, Sunnerheim et al. 1988): Small-diameter current-year internodes have a greater bark–wood ratio than larger-diameter older internodes (Danell and Bergstrom 1987) and, thus, higher concentrations of feeding deterrent secondary metabolites. However, epidermal cells producing bark secondary metabolites such as resin glands are often active only in the first year of internode growth, because subsequent radial growth of the cambium results in their destruction (Esau 1965). Thus, internode palatability increases with time until internodes are of such a large diameter that the amount of wood in them again depresses their palatability and nutritional quality (Pehrson 1981, 1983, Fox and Bryant 1984, Palo 1987a,b). The winter food supply of small browsing mammals such as hares that, because of small mouth parts, can reject current year internodes is restricted to a window of availability bounded by internode age and large diameter.

Results presented in Figure 6 indicate a defensive time lag in the palatability and nutritional quality of stump sprouts to hares. In some species, for example, Alaska feltleaf willow, change in palatability with time occurs within small-diameter current-year internodes (Fox and Bryant 1984, Bryant et al. 1985a). In other species such as dwarf birch (*B. glandulosa*) change in palatability with age is caused by change in the plant's internode age distribution (Bryant unpublished results). Following severe pruning, a geometric increase in current-year shoot numbers (Moorby and Waring 1963) results in an increase in the 2-year-old or older internode biomass preferred by hares after 2–3 years (Fox and Bryant 1984, Bryant unpublished results). In species such as balsam poplar (Reichardt et al., 1990), change in palatability with time occurs both as a result of changes within current-year internodes and as a result of changes in the plant's internode age distribution.

Ongoing studies in boreal forests and the Great Basin indicate that changes in shoot palatability after severe browsing as occurs during hare outbreaks are primarily caused by change in the age distribution of internodes (Bryant and Provenza, unpublished results).

4   IMPLICATIONS FOR PLANT DEFENSE THEORY

4.1   Leaf Responses to Herbivory

We propose that LTI is *not* an active defensive response. Rather, its similarity to aging strongly suggests that it is a consequence of a passive defoliation-caused nutrient stress as proposed by Tuomi et al. (1984). Such an explanation for LTI (Fig. 7) is appealing for several reasons. First, it provides a plausible physiological mechanism for LTI. Second, it reveals a congruity between the Tuomi carbon–nutrient balance model of LTI and the passive deterioration explanation of LTI proposed by Myers (Myers and Williams, 1984, 1987, Williams and Myers 1984, Myers 1988). The model presented in Figure 7 predicts that LTI is caused by passive deterioration in a woody plant's ability to meet nutritional demands of the shoot. Third, our model relates the opposing responses of deciduous woody plants to severe early season defoliation and pruning to the same mechanisms: shoot mineral nutrition.

Our model provides several easily tested predictions. If LTI is caused by nutrient stress, then fertilization with growth-limiting nutrients should minimize LTI and damage to roots should enhance LTI. Results of Tuomi et al. (1984) and Haukioja and Neuvonen (1985) are consistent with this prediction. However, results of Haukioja and Niemela (1976) are not. They found that root damage of mountain birch (*B. pubescens* ssp. *tortuosa*) in the previous year resulted in increased growth of *Oporinia autumnata* larvae and fertilization had little effect upon larval growth. Furthermore, Haukioja and Neuvonen (1985) found that LTI was elicited in mountain birch by insect defoliation even in plants grown in fertile soil. However, ongoing studies in Alaska have failed to confirm the generality of this result. In the case of Alaska paper birch and quaking aspen (*P. tremuloides*) (Bryant, unpublished results), fertilization with nitrogen plus phosphorus stopped LTI even in plants undergoing severe defoliation by *R. hastata* and the large aspen tortrix (*Choristoneura conflictana*), respectively. Our model assumes that early season defoliation is an important component of LTI. Thus, we predict that early season defoliation will elicit a stronger induced response than late season defoliation after preabscission retranslocation of nutrients from leaves has begun. Furthermore, aging can affect individual branches in a tree, and its reversal by pruning can take place locally within individual branches (Moorby and Waring 1963). By analogy we predict that LTI can be initiated or reversed in individual branches by defoliation early in the growing season or pruning, respectively.

We conclude this section by pointing out that mammals rarely, if ever, defoliate woody plants severely enough to elicit widespread LTI in their food supply. Furthermore, most summer browsing removes buds and twig ends as well as leaves. Such bud death will counter the nutrient stress we propose as the cause of LTI. Thus, we predict that LTI does not commonly affect browsing mammal population dynamics.

4.2   Twig Responses to Herbivory

Juvenile-phase physiological and morphological traits are under genetic control and are thus subject to natural selection (Kozlowski 1971). It is therefore reasonable to conclude that juvenile defensive traits are a consequence of age-specific selection for defense against herbivory and pathogen attack (Schaffalitzky de Muckadell 1969, Libby and Hood 1976, Bryant et al. 1983, Amo et al. 1986). This conclusion does not, however, imply that reversion to the juvenile-form is an active defense response. Rather, the chemical defenses of juvenile-phase stump sprouts are best viewed as a passive expression of juvenile-phase defenses following a reversion to the juvenile form (Bryant et al. 1983, Chapin et al. 1985). Similarly, a decline in palatability resulting from a browsing-caused change in the ratio of current-year internode biomass versus older internode biomass within a woody plant's small-diameter twig population is probably not an active defense response. In short, the only active aspect of internode defense is the active rejection of small-diameter current-year internodes by mammals such as hares. This conclusion implies that lowered palatability in browsed twigs caused by change in internode age distribution is a consequence of the animal's rejection of the plant's constitutive defenses rather than an induced defense on the part of the plant.

Even though we propose that browsing-induced increases in the chemical defense of internodes are passive responses, we do not wish to imply that such changes are irrelevant to the population dynamics of small mammals, especially hares. Irrespective of the cause (juvenile reversion or internode age distribution), a 2- to 3-year induced deterioration in the palatability and nutritional quality of winter-dormant internodes (Figs. 3–6) can, if widespread, generate a 10-year hare population cycle (May 1972, 1974, Bryant 1981a,b, Bryant et al. 1983, 1985a,b, Fox and Bryant 1984). Moreover, such a food-related time lag is highly probable in regions where species such as rapidly growing willows, dwarf birch, and poplar are important winter foods of hares. Such regions include Alaska (Wolff and Zasada 1979, Bryant and Kuropat 1980, Bryant 1981a,b, Bryant et al. 1983, 1985a,b), the Yukon Territory of Canada (Sinclair et al. 1982, Sinclair and Smith 1984) and Alberta, Canada (Pease et al. 1979, Keith et al. 1984). All that is required to generate such a time lag is the normal, rapid regeneration of current-year internodes observed in deciduous species after heavy browsing ceases (Aldous 1952, Garrison 1972, Krefting et al. 1966, Willard and McKell 1978, Wolff 1978,

Provenza et al. 1983, Fox and Bryant 1984, Bergstrom and Danell 1987). On the other hand, such an effect of internode age is not likely to affect the population dynamics of large browsers such as moose that do not selectively feed on parts of small-diameter twigs.

5 SUMMARY

We have provided physiological mechanisms that can account for long-term chemical responses of deciduous woody plants to defoliation and twig removal by herbivores. We suggest that these responses are not active defense responses in the accepted sense of the word (Rhoades 1979, Haukioja 1980, Haukioja and Neuvonen 1985). Rather, they are caused by passive nutrient stress, juvenile reversion, and passive changes in internode age distributions. We further suggest that herbivory-induced changes in leaf chemistry do not affect the population dynamics of mammals. However, a browsing-causing reversion to the juvenile form in concert with a browsing-caused increase in the ratio of current-year internode biomass versus older internode biomass in the winter food supply of hares can destabilize their population dynamics and generate a 10-year hare cycle. Such a destabilization of population dynamics is less likely for browsing mammals with larger mouth parts, for example, moose. Further research is required to determine the importance of chemical defense to browsing mammal population dynamics.

ACKNOWLEDGMENTS

We thank F. S. Chapin III, P. Coley, P. Morrow, and J. Thompson for reviewing this manuscript. NSF Grants BSR-8416461 and BSR-8614856 and the National Swedish Environment Protection Board, the Swedish Sportsmen's Association, and the Swedish National Research Council funded the research that led to this paper.

REFERENCES

Aldous, S. E. 1952. Deer browse clipping study in the Lake States Region. *J. Wildl. Manage.* 16:401–409.

Amo, R. S. del, J. G. Ramirez, and O. Espejo. 1986. Variation in some seedling secondary metabolites in juvenile stages of three plant species from tropical rain forest. *J. Chem. Ecol.* 12:2021–2038.

Baltensweiler, W. 1985. On the mechanisms of the outbreaks of the larch bud moth (*Zeiraphera diniana* Gn., Lepidoptera, Tortricidae) and its impact on the sub-alpine larch-Cembran pine ecosystem. In J. Turner and W. Tranquillini (eds.), *Establishment and Tending of Subalpine Forest: Research and Management.*

Proc. 3rd Int. Workshop IUFRO Project Group P1.07–00 Ecology of Subalpine Zones, Sept. 3–5, 1984. CH-3981 Riederalp, Switzerland, pp. 215–219.

Baltensweiler, W., G. Benz, P. Bovey, and V. Delucchi. 1977. Dynamics of larch bud moth populations. *Annu. Rev. Entomol.* 22:79–100.

Basey, J. M. 1987. Central place foraging by beaver in the Sierra Nevada: tree-size selection and inducible defenses in quaking aspen. MS Thesis, University of Nevada, Reno.

Beckwith, R. C. 1968. The large aspen tortrix, *Choristoneura conflictana* (Wlkr.), in interior Alaska. *USDA Forest Serv. Res. Note PNW-81.*

Beckwith, R. C. 1970. Influence of host on larval survival and adult fecundity of *Chorostoneura conflictana* (Lepidoptera: Torticidae). *Can. Entomol.* 102:1474–1480.

Benz, G. 1974. Negative Ruckkoppelung durch Raumund Nahrungskonkurrenz sowie zyklische Veranderung der Nahrungsgrundlage als Regelprinzip in der Populationsdynamik des Grauen Larchenwicklers, *Zeiraphera diniana* (Guenoe). *Z. Angew. Entomol.* 76:196–228.

Bergstrom, R., and K. Danell. 1987. Effects of simulated winter browsing by moose on morphology and biomass of two birch species. *J. Ecol.* 75:533–544.

Borchert, R. 1976. The concept of juvenility in woody plants. *Acta Hortic. (The Hague)* 56:21–35.

Bryant, J. P. 1981a. The regulation of snowshoe hare feeding behavior during winter by plant antiherbivore chemistry. In K. Myers and D. C. MacInnes (eds.), *Proceedings 1st International Lagomorph Conference.* Guelph University, Canada, pp. 720–731.

Bryant, J. P. 1981b. Phytochemical deterrence of snowshoe hare browsing by adventitious shoots of four Alaskan trees. *Science* 313:889–890.

Bryant, J. P., and P. J. Kuropat. 1980. Selection of winter forage by subarctic browsing vertebrates: the role of plant chemistry. *Annu. Rev. Ecol. Syst.* 11:261–285.

Bryant, J. P., F. S. Chapin III, and D. R. Klein. 1983. Carbon/nutrient balance of boreal plants in relation to vertebrate herbivory. *Oikos* 40:357–368.

Bryant, J. P., G. D. Wieland, T. Clausen, and P. Kuropat. 1985a. Interactions of snowshoe hares and feltleaf willow (*Salix alaxensis*) in Alaska. *Ecology* 66:1564–1573.

Bryant, J. P., F. S. Chapin III, P. Reichardt, and T. Clausen. 1985b. Adaptation to resource availability as a determinant of chemical defense strategies in woody plants. In G. A. Cooper-Driver and T. Swain (eds.). *Recent Advances in Phytochemistry, Vol. 19, Chemically Mediated Interactions between Plants and Other Organisms,* pp. 219–237.

Bryant, J. P., F. D. Provenza, and A. Gobina. 1987a. Environmental controls over woody plant chemical defenses: implications for goat management. In W. C. Foote, A. Gabriel de Silva, and O. P. Santana (eds.). *Proceedings of the IV International Conference on Goats.* Brasilia, Brazil: Departmento de Difusao de Tecnologia.

Bryant, J. P., J. Tuomi, and P. Niemela. 1988. Environmental constraint of constitutive and long-term inducible defenses in woody plants. In K. Spencer (ed.), *Chemical Mediation of Coevolution.* New York: Academic Press, pp 367–389.

Chapin, F. S., III. 1980. The mineral nutrition of wild plants. *Annu. Rev. Ecol. Syst.* 11:233–260.

Chapin, F. S., III, and R. A. Kedrowski. 1983. Seasonal changes in nitrogen and phosphorus fractions and autumn retranslocation in evergreen and deciduous taiga trees. *Ecology* 64:376–391.

Chapin, F. S., III, J. P. Bryant, and J. F. Fox. 1985. Lack of induced chemical defense in juvenile Alaskan woody plants in response to simulated browsing. *Oecologia (Berlin)* 67:457–459.

Clausen, T. P., J. P. Bryant, and P. B. Reichardt. 1986. Defense of winter-dormant green alder against snowshoe hares. *J. Chem. Ecol.* 12:2117–2131.

Coley, P. D., J. P. Bryant, and F. S. Chapin III. 1985. Resource availability and plant antiherbivore defense. *Science* 230:895–899.

Currie, P. O., and D. L. Goodwin. 1966. Consumption of forage by black-tailed jackrabbits on salt-desert ranges of Utah. *J. Wildl. Manage.* 30:304–311.

Danell, K., and R. Bergstrom. 1987. Studies on interactions between moose and two species of birch in Sweden. In F. D. Provenza, J. Flinders, and E. D. McArthur (eds.), *Proceedings, Symposium on Plant–Herbivore Interactions. USDA Forest Service Intermountain Research Station General Technical Report INT-222*, pp. 48–57.

Danell, K., and K. Huss-Danell. 1985. Feeding by insects and hares on birches earlier affected by moose browsing. *Oikos* 44:75–81.

Danell, K., and S. Neuvonen. 1987. Does browsing modify the quality of birch foliage for *Epirrita* larvae? *Oikos* 49:156–160.

Danell, K., K. Huss-Danell, and R. Bergstrom. 1985a. Interactions between browsing moose and two species of birch in Sweden. *Ecology* 66:1867–1878.

Danell, K., T. Elmqvist, L. Ericson, and A. Salmonson. 1985b. Sexuality in willows and preference by bark-eating voles: Defense or not. *Oikos* 44:82–90.

Danell, K., T. Elmqvist, L. Ericson, and A. Salomonson. 1987. Are there general patterns in bark-eating by voles on different shoot types from woody plants. *Oikos* 50:396–402.

Dugle, J. R. 1966. A taxonomic study of western Canadian species in the genus *Betula*. *Can. J. Bot.* 44:929–1007.

Esau, K. 1965. *Plant Anatomy*. New York: Wiley.

Fischlin, A., and W. Baltensweiler. 1979. Systems analysis of the larch bud moth system. I. The larch–larch bud moth relationship. *Bull. Soc. Entomol. Suisse* 52:273–289.

Fox, J. F., and J. P. Bryant. 1984. Instability of the snowshoe hare and woody plant interaction. *Oecologia (Berlin)* 63:128–135.

Freeland, W. J., and D. H. Janzen. 1974. Strategies in herbivory by mammals: the role of plant secondary compounds. *Am. Natur.* 108:269–289.

Garrison, G. A. 1972. Carbohydrate reserves and response to use. In *Wildland Shrubs: Their Biology and Utlization. USDA Forest Service General Technical Report T-1*. Logan, UT: Utah State University, pp. 271–286.

Grange, W. B. 1949. *The Way to Game Abundance*. New York: Scribner's.

Haukioja, E. 1980. On the role of plant defenses in the fluctuation of herbivore populations. *Oikos* 35:202–213.

Haukioja, E., and T. Hakala. 1975. Herbivore cycles and periodic outbreaks. Formulation of a general hypothesis. *Rep. Kevo Subarctic Res. Sta.* 12:1–9.

Haukioja E., and S. Neuvonen. 1985. Induced long-term resistance of birch foliage against defoliators: defensive or incidental? *Ecology* 66:1303–1308.

Haukioja, E., and P. Niemela. 1976. Does birch defend itself actively against herbivores? *Rep. Kevo Subarctic Res. Sta.* 13:44–47.

Haukioja, E., and P. Niemela. 1977. Retarded growth of a geometrid larva after mechanical damage to leaves of its host tree. *Ann. Zool. Fenn.* 14:48–52.

Haukioja, E., K. Kapiainen, P. Niemela, and J. Tuomi. 1983. Plant availability hypothesis and other explanations of herbivore cycles: complementary or exclusive alternatives. *Oikos* 40:419–432.

Hulten, E. 1968. *Flora of Alaska and Neighboring Territories.* Stanford, CA: Stanford University Press.

Janzen, D. H., and P. S. Martin. 1981. Neotropical anachronisms: the fruits the *Gomphotheres* ate. *Science* 215:19–27.

Kallio, P., and J. Lehtonen. 1973. Birch forest damage caused by *Oporinia autumnata* (Bkh.) in 1965–1966 in Utsjoki, N. Finland. *Rep. Kevo Subarctic Res. Sta.* 10:55–69.

Keith, L. B. 1963. *Wildlife's Ten-Year Cycle.* Madison, WI: University of Wisconsin Press.

Keith, L. B., J. R. Cary, O. J. Rongstad, and M. C. Brittingham. 1984. Demography and ecology of a declining snowshoe hare population. *Wildl. Monogr.* 90:1–43.

Klein, D. R. 1977. Winter food preferences of snowshoe hares (*Lepus americanus*) in interior Alaska. *Proc. Int. Congr. Game Biol., 13th, Atlanta,* pp. 266–75.

Kozlowski, T. T. 1971. *Growth and Development of Trees,* Vol. 1. New York: Academic.

Kramer, P. J., and T. T. Kozlowski. 1979. *Physiology of Woody Plants.* New York: Academic.

Krefting, L. W., M. H. Stenlund, and R. K. Seemel. 1966. Effect of simulated and deer natural browsing on mountain maple. *J. Wildl. Manage.* 30:481–488.

Kulman, H. M. 1971. Effects of insect defoliation on growth and mortality of trees. *Annu. Rev. Entomol.* 16:289–324.

Libby, W. J., and J. V. Hood. 1976. Juvenility in hedged radiata pine. *Acta Hort. (The Hague)* 56:91–98.

Machida, S. 1979. Differential use of willow species by moose in Alaska. Thesis, University of Alaska, Fairbanks.

May, R. M. 1972. Limit cycles in predator–prey communities. *Science* 177:900–902.

May, R. M. 1974. *Stability and Complexity in Model Ecosystems,* 2d ed. Princeton, NJ: Princeton University Press.

McKey, D. 1979. The distribution of secondary compounds within plants. In G. A. Rosenthal and D. H. Janzen (eds.), *Herbivores: Their Interaction with Secondary Plant Metabolites.* New York: Academic, pp. 55–133.

Miquell, D. G. 1983. Browse regrowth and consumption following summer defoliation by moose. *J. Wildl. Manage.* 47:17–24.

Moorby, J., and P. F. Waring. 1963. Aging in woody plants. *Ann. Bot. (N.S.)* 106:291–309.

Myers, J. 1988. The induced defense hypothesis: does it apply to the population dynamics of insects? In K. Spencer (ed.), *Chemical Mediation of Coevolution.* New York: Academic Press, pp. 345–365.

Myers, J. S., and K. H. Williams. 1984. Does tent caterpillar attack reduce the food quality of red alder foliage? *Oecologia (Berlin)* 62:74–79.

Myers, J. S., and K. H. Williams. 1987. Lack of short or long term inducible defenses in the red alder-western tent caterpillar system. *Oikos* 48:73–78.

Palo, R. T. 1987a. Chemical defense in a woody plant and the role of digestive systems of herbivores. In F. D. Provenza, J. T. Flinders, and E. D. McArthur (eds.), *Proceedings, Symposium on Plant–Herbivore Interactions. USDA Forest Service Intermountain Research Station General Techical Report INT-222.*

Palo, R. T. 1987b. Phenols as defensive compounds in birch (*Betula* spp.): implications for digestion and metabolism in browsing mammals. Ph.D. Thesis, Department of Animal Physiology, Swedish University of Agricultural Sciences, Uppsala.

Pease, J. L., R. H. Vowles, and L. B. Keith. 1979. Interaction of snowshoe hares and woody vegetation. *J. Wildl. Manage.* 43:43–60.

Pehrson, A. 1981. Winter food consumption and digestibility in caged mountain hares. In K. Myers and C. D. MacInnes (eds.), *Proceedings, 1st International Lagomorph Conference.* Guelph University, Canada.

Pehrson, A. 1983. Digestibility and retention of food components in caged mountain hares (*Lepus timidus* L.) during the winter. *Holarctic Ecol.* 6:395–403.

Provenza, F. D., and J. C. Malechek. 1983. Tannin allocation in blackbrush (*Coleogyne ramosissima*). *Biochem. Syst. Ecol.* 3:233–238.

Provenza, F. D., and J. C. Malechek. 1984. Diet selection by domestic goats in relation to blackbrush twig chemistry. *J. Appl. Ecol.* 21:831–841.

Provenza, F. D., J. C. Malechek, P. J. Urness, and J. E. Bowns. 1983. Some factors affecting twig growth in blackbrush. *J. Range Manage.* 36:518–520.

Reichardt, P. B., J. P. Bryant, T. P. Clausen, and G. D. Wieland. 1984. Defense of winter-dormant Alaska paper birch against snowshoe hare. *Oecologia (Berlin)* 65:58–69.

Reichardt, P. B., T. P. Clausen, and J. P. Bryant. 1987. Plant secondary metabolites as feeding deterrents to vertebrates. In. F. D. Provenza, J. T. Flinders, and E. D. MacArthur (eds.), *Proceedings, Symposium of Plant–Herbivore Interactions. USDA Forest Service Intermountain Research Station General Technical Report INT-222.*

Reichardt, P. B., J. P. Bryant, B. R. Mattes, T. P. Clausen, F. S. Chapin III, and M. Meyer. 1990. The winter chemical defense of balsam poplar against snowshoe hares, 16:1961–1970.

Rhoades, D. F. 1979. Evolution of plant chemical defense against herbivores. In G. A. Rosenthal, D. H. Janzen (eds.), *Herbivores: Their Interaction with Secondary Plant Metabolites.* New York: Academic, pp. 3–54.

Robbins, C. T. 1983. *Wildlife Feeding and Nutrition.* New York: Academic.

Schaffalitzky de Muckadell, M. 1969. Environmental factors in developmental stages of trees. In T. T. Kozlowski (ed.), *Tree Growth*. New York: Ronald.

Sinclair, A. R. E., and J. N. M. Smith. 1984. Do plant secondary compounds determine feeding preferences of snowshoe hares? *Oecologia (Berlin)* 61:403–410.

Sinclair, A. R. E., C. J. Krebs, and J. N. M. Smith. 1982. Diet quality and food limitation in herbivores: The case of the snowshoe hare. *Can. J. Zool.* 60:889–897.

Sunnerheim, K., R. T. Palo, O. Theander, and P. G. Knutsson. 1988. Chemical defense in birch. II. Platyphylloside, a phenol from *Betula pendula* inhibiting digestibility. *J. Chem. Ecol.* 14:549–560.

Tahvanainen, J., E. Helle, R. Julkunen-Tiitto, and A. Lavola. 1985. Phenolic compounds of willow bark as deterrents against feeding by mountain hare. *Oecologia (Berlin)* 65:319–323.

Tuomi, J., P. Niemela, E. Haukioja, S. Siren, and S. Neuvonen. 1984. Nutrient stress: an explanation for plant anti-herbivore responses to defoliation. *Oecologia (Berlin)* 61:208–210.

Tuomi, J., P. Niemela, F. Stuart Chapin III, J. P. Bryant, and S. Siren. 1987. Defensive responses of trees in relation to their carbon/nutrient balance. In W. Mattson, J. Levieus, and C. Bernard-Dagan (eds.), *Mechanisms of Woody Plant Defenses against Insects: Search for a Pattern*. IUFRO Symposium. New York: Springer, pp. 55–70.

Waring, P. F. 1959. Problems of juvenility and flowering in trees. *J. Linn. Soc., Bot.* 56:282–289.

Werner, R. A. 1977. Biology and behavior of the spear-marked black moth, *Rheumaptera hastata*, in interior Alaska. *Ann. Entomol. Soc. Am.* 70:328–336.

Werner, R. A. 1979. Influence of host foliage on development, survival, fecundity, and oviposition of the spear-marked black moth, *Rheumaptera hastata* (Lepidoptera: Geometridae). *Can. Entomol.* 111:317–322.

Willard, E. E., and C. M. McKell. 1978. Response of shrubs to simulated browsing. *J. Wildl. Manage.* 42:514–519.

Williams, K. S., and J. H. Myers. 1984. Previous herbivore attack of red alder may improve food quality for fall webworm larvae. *Oecologia (Berlin)* 63:166–170.

Wolff, J. O. 1978. Burning and browsing effects on willow growth in interior Alaska. *J. Wildl. Manage.* 42:135–140.

Wolff, J. O., and J. Zasada. 1979. Moose habitat and forest succession on the Tanana River floodplain and Yukon-Tanana upland. *Proc. N. Am. Moose Conf.* 15:213–244.

Yabann, W. K., E. A. Burritt, and J. C. Malechek. 1987. Sagebrush (*Artemisia tridentata*) monoterpenoid concentrations as factors in diet selection by free-grazing sheep. In F. D. Provenza, J. T. Flinders, and E. D. McArthur (eds.), *Proceedings, Symposium on Plant–Herbivore Interactions. USDA Forest Service General Technical Report INT-222*, pp. 64–70.

CHAPTER 7

SQUASH BEETLES, CUCUMBER BEETLES, AND INDUCIBLE CUCURBIT RESPONSES

DOUGLAS W. TALLAMY
Department of Entomology and Applied Ecology, University of Delaware, Newark, DE 19717-1303

ERIC S. McCLOUD
Department of Entomology, University of Illinois, Urbana, IL 61801

1   Introduction
2   Cucurbitacin induction in *Cucurbita*
3   *Cucurbita*–squash beetle interactions
4   *Cucurbita*–cucumber beetle interactions
5   Cucurbitacins and the assumption of defense
6   Conclusion
    References

1   INTRODUCTION

A decade ago Carroll and Hoffman (1980) described the unusual feeding behavior of *Epilachna* squash beetles as a counteradaptation against bitter cucurbitacins induced within cucurbit foliage by tissue damage. Working with *E. tredecimnotata* (Latreille), they noted that before feeding, beetles chew a circular trench in their leaf, severing all but sporadic epidermal connective tissues, apparently isolating the area on which they are about to feed from the active vascular network of the plant. Data from bioassays prompted Carroll and Hoffman to conclude that *Epilachna* trenching behavior prevents the rapid (≤40 min) accumulation of cucurbitacins within the isolated tissues, thus preserving the palatability and quality of the leaf resource while it is consumed. The increase of cucurbitacins in an attacked leaf was the most rapid inducible phytochemical response to herbivory yet reported, and it helped promote interest in the hypothesis that inducible plant changes are

155

antiherbivore adaptations rather than chemical components of wound repair mechanisms.

Tallamy (1985) expanded on Carroll and Hoffman's studies using different species of *Cucurbita* and *Epilachna* in hopes of critically testing their hypothesis. Based on quantification of cucurbitacin responses to foliar injury, leaf disk bioassays with *Epilachna* and *Diabrotica* beetles, and studies of the deleterious effects on *Epilachna* fitness of ingesting induced foliage, Tallamy concluded that the hypothesis was valid; *Epilachna* trenching behavior did indeed prevent an accumulation of cucurbitacins within the isolated tissues and thus represented a behavioral counteradaptation to the inducible chemical defenses of cucurbits. The evidence was compelling, the reasoning was seductively intuitive, but the conclusion was incorrect.

In this chapter we propose an alternative interpretation of *Epilachna* feeding behavior based on further studies of the interactions between *Epilachna* and their cucurbit hosts. We then compare *Epilachna* interactions with cucurbits to those of a very different guild of cucurbit beetle herbivores, the diabroticite cucumber beetles. Diabroticites have placed cucurbits in an ironic evolutionary bind: The bitter compounds usually regarded as defensive in origin are phagostimulants to these beetles, triggering compulsive feeding and destruction of bitter plant tissues (reviewed by Metcalf and Lampman 1989). If inducible cucurbit responses are, in fact, adaptations against herbivores, diabroticite attacks should elicit plant responses that differ from those triggered by physiologically nonadapted herbivores. The veracity of this hypothesis will also be discussed. We conclude by questioning the evolutionary role of cucurbitacins in cucurbit defense.

2   CUCURBITACIN INDUCTION IN *CUCURBITA*

Members of the genus *Cucurbita* (Cucurbitaceae) are nonwoody tropical and subtropical plants of the Americas that are characterized by long trailing vines and large fleshy fruits (Whitaker and Davis 1962). The genus divides ecologically into xerophytic perennials with thick, pubescent leaves and large tap roots, and mesophytic annuals bearing broad leaves and a more fibrous root system (Whitaker and Bemis 1975). There is evidence from both groups that the known oxygenated tetracyclic triterpenes (Fig. 1), commonly labeled cucurbitacins A-S (Rehm et al. 1957, Lavie and Glotter 1971), function in defense against herbivory. These compounds are responsible for the extreme bitterness of wild cucurbit fruits and roots and effectively deter many invertebrate and vertebrate herbivores (DaCosta and Jones 1971, Nielson et al. 1977, Ferguson et al. 1983, Stroesand et al. 1985, reviewed by Guha and Sen 1975, Metcalf 1985, Metcalf and Lampman 1989).

Although foliar concentrations of cucurbitacins are typically far lower than those of fruits and roots (Metcalf et al. 1982), they fluctuate dramatically as the plant responds to even small levels of tissue damage. Standardized

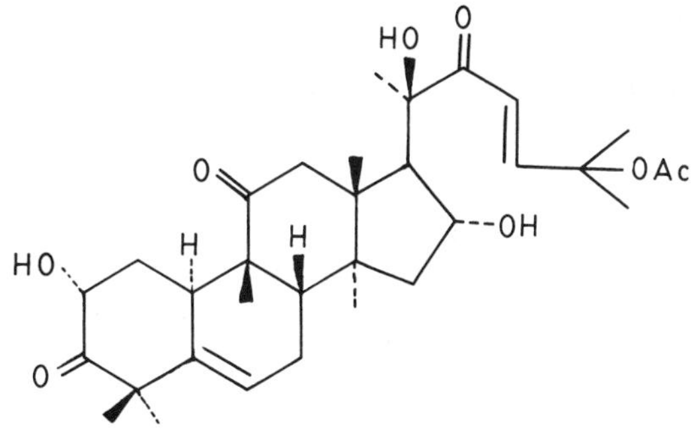

B

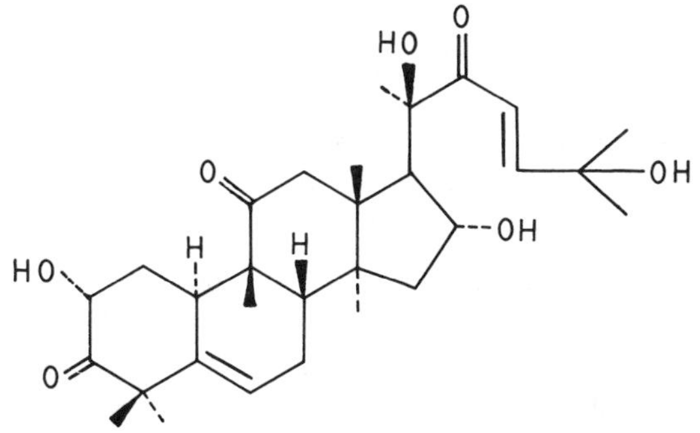

D

Figure 1   Structures of cucurbitacins B and D.

mechanical abrasions of less than 1% of a mature *C. pepo* L. leaf surface cause quantities of cucurbitacin D to rise from undetectable concentrations to a mean of 8.8 μg/g fresh wt within 3 h of injury (Tallamy 1985). The concentration of cucurbitacin B increases similarly, more than doubling after tissue damage (Fig. 2). Carroll and Hoffman (1980) reported that two parallel cuts in *C. moschata* leaves trigger within 40 min foliar responses (presumably the accumulation of cucurbitacins) that are detectable by squash and cucumber beetles. Recent data, however, suggest that the response is not always as dramatic or as rapid (McCloud 1989, McCloud unpublished results). Time-course analyses of the effect of mechanical abrasions on foliar cucurbitacins in *C. andreana* and *C. texana* have revealed extreme variability in the response. A comparatively small increase in the concentration of cu-

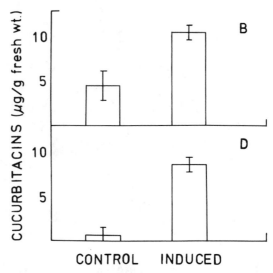

Figure 2   Comparison of the concentration (μg/g fresh wt) of cucurbitacins B and D in *Cucurbita pepo* "Black" leaves 3 h after mechanical injury versus concentrations in uninjured leaves. Error bars show ±1 SD. (From Tallamy 1985.)

curbitacin B peaked in *C. andreana* approximately 2.6 h post injury and relaxed to control levels within 11 h (Fig. 3). In contrast, induction in *C. texana* was more powerful and more punctuated: A compound coeluting with cucurbitacin I peaked 10.5 h following injury and relaxed within 21 h (Fig. 4). This compound has the same thin-layer chromotography R_f as by-products of cucurbitacin I after exposure to air during extraction procedures.

The mechanisms regulating cucurbitacin induction are not yet understood. Carroll and Hoffman (1980) proposed a translocation model whereby cucurbitacins are mobilized within the vascular system of the plant and transported to the source of injury. Their argument is supported by bioassays from several experiments suggesting that local biosynthesis of cucurbitacins does not occur in the absence of turgor pressure: Neither *Epilachna* nor diabroticite beetles change their preference over time for damaged leaf tissue once it has been detached from a turgid plant (Carroll and Hoffman 1980, Tallamy 1985, Tallamy 1986, data below).

The mechanisms by which cucurbitacins may be mobilized in the translocation model are also unknown, but several hypotheses offer direction for future studies. Plant tissue damage can trigger an electrical stimulus having many of the properties of an action potential (Van Sambeek and Pickard 1976, reviewed by Davies 1987). These potentials can propagate basipetally or acropetally throughout a shoot. The progress of the potential is thought to be the result of the concomitant spread of various phytochemicals through

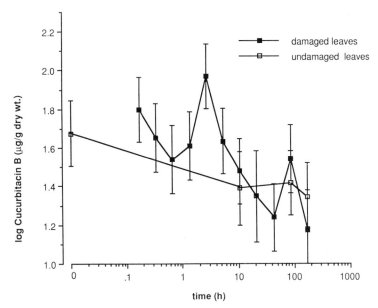

Figure 3 Time-course analysis of the induction of cucurbitacin B in damaged and undamaged *Cucurbita andreana* leaves. Three standardized wounds were applied with a rat-tailed file to one leaf of each treated plant. Cucurbitacin amounts are expressed as least-squares means of log transformed values ± 1 SE.

the vascular system. As the potential enters undamaged tissues, photosynthesis and transpiration rapidly decrease, while catabolism increases. Such a response has been reported for a variety of plants, including *C. pepo* "Jack-O-Lantern" (Van Sambeek and Pickard 1976), and may be responsible for triggering cucurbitacin transport via the xylem (as are other steroidal compounds of comparatively high molecular weight; Dixon 1975) from local or remote storage tissues, such as the roots, to the area being attacked. Induced xylem translocation of toxic compounds from root storage sites does occur in plants and is particularly well understood in tobacco (Dawson 1941, Baldwin, this volume, Chapter 2). Alternatively, leaf wounding may release a water-soluble damage cue, such as lipopolysaccharide components of ruptured cell walls (Ryan 1983), which is then transported by phloem bundles to the root system or other possible cucurbitacin storage or synthesis sites.

Because the rapid responses described by Carroll and Hoffman seem to defy the physics of water movement in phloem tissue, it is important to note that the Cucurbitaceae are morphologically specialized for rapid phloem transport. Despite an uncommonly complex vascular course (Fukuda 1967), cucurbits boast the largest sieve tubes of all angiosperms, and their sieve plates bear unusually large pores (Esau and Cheadle 1959, Esau 1965).

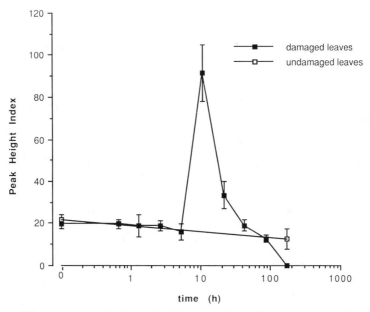

Figure 4   Time-course analysis of the induction of an unknown compound coeluting with cucurbitacin I in damaged and undamaged *Cucurbita texana* leaves. Three standardized wounds were applied with a rat-tailed file to one leaf of each treated plant. Error bars show ±1 SE. Peak height index is a linear transformation of observed peak height on the HPLC chromatogram using cucurbitacin I as an internal standard.

Phloem velocities as high as 290 cm/h have been reported in cucurbits (Craft and Crisp 1971).

Regardless of how cucurbitacins are mobilized, only indirect evidence supports the translocation model regarding cucurbitacin induction. If induction triggers the intra- and interleaf transport of cucurbitacins, these compounds should be detectable by taste and chemical assays within stem tissue and the vascular system of leaves. Cucurbitacins have not as yet been detected in stems, even following foliar injury known to induce these compounds (Metcalf et al. 1982, McCloud, unpublished results). Cheesman (1989), however, noted that the primary veins of previously injured *C. pepo* leaves were preferentially attacked by *Diabrotica undecimpunctata howardi* beetles over leaf lamina or veins of uninjured leaves from a control plant. Though chemical analyses were not conducted, he suggested that diabroticites were attracted to the vascular system because it contained cucurbitacins mobilized by the original tissue damage.

An alternative possibility is that leaf wounding systemically triggers *de novo* production of cucurbitacins in undamaged foliage via the rapid vascular transport of a damage cue or electrical stimulation. But perhaps the most plausible mechanism for cucurbitacin induction follows from the suggestion that relatively nonbitter cucurbitacin glycosides are rapidly converted to bitter aglycones by β-glucosidase (elaterase) released from storage vacuoles when plant cells are crushed (Enslin et al. 1956). By this model, the strength of the induced response will be a function of the constitutive level of cucurbitacin glycosides present prior to tissue damage, the amount of tissue damaged, and the degree to which elaterase is transported to undamaged tissues.

3 *CUCURBITA*–SQUASH BEETLE INTERACTIONS

Squash beetles are herbivores in the subfamily Epilachninae (Coccinellidae) that feed only on members of the Cucurbitaceae. Not all Epilachninae are cucurbit feeders; many specialize on legumes or solanaceous hosts (Schaefer 1983). All available evidence indicates that feeding trenches are employed only by species that consume cucurbits (Table 1). Curiously, however, trenching behavior is not universal among epilachnine cucurbit specialists (Table 1): *E. septima*, for example, feeds only on *Momordica* spp. but does not utilize feeding trenches (Nakano, personal communication, Tallamy, unpublished results). Even more curious is the feeding behavior of *E. admirabilis*. This species cuts circular trenches in the leaves of *Schizopepon bryoniaefolius* prior to feeding, but does not exhibit trenching behavior when feeding on *Gynostemma pentaphyllum*.

Though data are limited, mandibular morphology of *Epilachna* trench-makers suggests considerable convergence relating to trench-snipping behavior. Trench-feeding species have relatively large mandibles bearing a long apical tooth well separated from two simple major teeth (Fig. 5A). The apical teeth function like scissors, slicing cleanly through leaf tissue with little or no tearing (Tallamy, personal observation). In comparison, the mandibles of nontrenching *Epilachna* that skeletonize solanaceous plants are typically smaller, with all three teeth more uniform in size (Fig. 5B; data from Gordon 1975). We predict, then, that squash-feeding *Epilachna* that do not trench do not have the produced apical tooth characteristic of trenching epilachnines.

It is evident that epilachnine beetles associated with particular members of the Cucurbitaceae have developed a specialized behavior that prevents the tissues about to be consumed from interacting with the rest of the leaf and plant. Aspects of this behavior provoke two questions of considerable interest, especially regarding the role of inducible cucurbit "defenses": Why is such behavior necessary when hosts are cucurbits but not legumes or solanaceous plants, and why must *Epilachna* specializing on certain cucurbit species trench before feeding, while species dependent on other cucurbits can feed without trenches? To answer these questions we have studied the interaction of *Epilachna borealis* and its cucurbit hosts in detail (Tallamy

TABLE 1   Feeding Behavior and Host Records of *Epilachna* Species

Species That Trench Prior to Feeding	Host Plant	Location	Reference
E. borealis Fabricius	*Cucurbita*	North America	Brannon 1937
	Lagenaria		Tallamy 1985, personal observation
E. tumida Gorham	*Cucurbita*		
	Wild cucubit	Central America	Tallamy, personal observation
E. tredecimnotata (Latreille)	*Cucurbita*	Central America	Carroll and Hoffman 1980
	Lagenaria		Tallamy personal observation
E. pocohantae Gordon	*Cucurbita*	Central America	Tallamy, personal observation
E. paenulata (Germar)	*Cucurbita*	South America	Schaefer 1983
	Echinocystis		
E. dodecastigma (Wiedemann)	*Benincasa*	West Sumatra	
	Cucumis		Katakura, personal communication
	Luffa		Nakamura, personal communication
E. admirabilis Crotch	*Citrullus*		Katakura, personal communication,
	Schizopepon		Schaefer 1983
	Trichosanthes		
	Momordica		
	Luffa		
	Melothria		
Epilachna sp. 1	*Sechium*	West Sumatra	Nakano, personal communication
Epilachna sp. 2	*Sechium*	West Sumatra	Nakano, personal communication

Species That Do Not Trench Prior to Feeding	Host Plant	Location	Reference
E. cucurbitae Richards	Domestic *Cucurbita* (wild hosts unknown)	Australia	Richards 1983 Richards, personal communication
E. septima Dieke	*Momordica*	West Sumatra	Nakano, personal communication
E. admirabilis Crotch	*Gynostemma*	Japan	Katakura, personal communication
E. vigintioctomaculata Motschulsky	*Schizopepon*	Japan	Katakura, personal communication

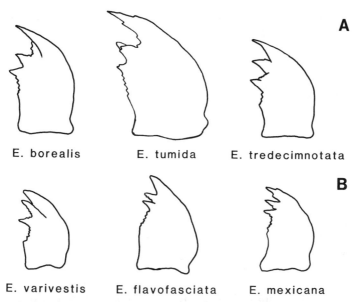

E. borealis   E. tumida   E. tredecimnotata

E. varivestis   E. flavofasciata   E. mexicana

Figure 5 Lateral view of the right mandible of (*A*) trenching *Epilachna* species that feed only on Cucurbitaceae and (*B*) nontrenching *Epilachna* species feeding on solanaceous or leguminous host plants.

1985, Tallamy and Krischik 1989, McCloud 1989, McCloud and Tallamy, unpublished).

Of primary importance is whether cucurbitacins negatively affect *Epilachna* fitness and, if so, whether beetles can detect and appropriately avoid these compounds. Two experiments have confirmed that *E. borealis* fitness is reduced by their consumption of cucurbit tissues previously wounded with a bastard file (Tallamy 1985, Tallamy and Krischik 1989). First, the survivorship of larvae reared on induced foliage is less than half that of larvae reared on controls (Student's *t* test: $N = 10$ cohorts; $p < .01$). Second, adult female fitness parameters, particularly fecundity, also suffer significantly when beetles are reared on induced foliage (Table 2). The assumption in these studies was that losses in fitness were due to increases in cucurbitacins within induced plants (Tallamy and Krischik 1989); deleterious effects from other compounds that may have been simultaneously induced have not been seriously considered to date.

Although induced foliage is responsible for reducing *Epilachna* fitness, a recent experiment casts serious doubt on the role of cucurbitacins as agents detrimental to these beetles (McCloud and Tallamy, unpublished). *E. borealis* adults were individually presented with two 3-cm^2 leaf disks. One disk had exogenously received a 25-μL solution bearing increasing doses of cucurbitacin B or I, while the other disk had received only methanol. After 2 h, the difference between the area remaining of the treated and control disks

TABLE 2   Fitness Traits of *Epilachna borealis* Females Reared from Egg until Death on Induced (Damaged) and Control Zucchini Leaves

Trait	Control Leaves			Induced Leaves			
	X	SD	N	X	SD	N	p^a
Dry mass (g)	0.0233	0.0021	27	0.02008	0.0016	16	.03
Age at first reproduction (days)	30.3	1.0	31	31.4	0.8	18	.04
Age at last reproduction (days)	62.1	12.7	24	46.1	5.2	17	.009
Number of eggs	598.4	186.0	24	332.4	111.0	17	.007

Source: From Tallamy 1985.

[a] Student's *t* test.

demonstrated that squash beetles overwhelmingly preferred disks treated with cucurbitacins, even at concentrations higher than those found in the most bitter cucurbit foliage (Table 3).

These data suggest that,

1. Contrary to previous suggestions (Carroll and Hoffman 1980, Tallamy 1985, Tallamy and Krischik 1989), cucurbitacins do not deter squash beetles from feeding on bitter foliage.
2. Some compound(s) other than cucurbitacins is responsible for reducing growth rate, survivorship and fecundity of *Epilachna* squash beetles that consume plants subjected to previous injury.

Because of such vulnerability to cucurbit responses, squash beetles have been likened to physiologically nonadapted herbivores that are deterred or killed by cucurbit chemistry (Tallamy and Krischik 1989). The ability of squash beetles to specialize on cucurbits, therefore, has been attributed to trenching, a unique adaptation that presumably counters the defensive effects of inducible cucurbit responses (Carroll and Hoffman 1980, Tallamy 1985, Tallamy 1986).

It was this reasoning that led to the prediction that the benefits of overcoming inducible cucurbit defenses through specialized feeding behavior may be balanced by costs from feeding constraints and increased foraging movement (Carroll and Hoffman 1980, Tallamy 1986). To the extent that tissue damage elicits a systemic response in cucurbits (Tallamy 1985), squash beetles should be limited to single feeding bouts on any given leaf. Deleterious compounds elevated by the initial feeding damage within the attacked leaf and uninjured adjacent leaves should force beetles to forage some dis-

TABLE 3   Preference of *Epilachna borealis* Adults for *Cucurbita maxima* Leaf Disks Treated with a Methanol Solution of Cucurbitacin I or B [a]

Concentration Applied (mg/mL^{-1})	Dose/disk wt (mg/g^{-1})	Preference (least squares mean)	Probability least squares mean = 0
Cucurbitacin I			
0.25	.08	29.06	.0204*
0.5	.16	32.50	.0154*
1.0	.32	35.63	.0047*
2.0	.64	47.81	.0002*
Cucurbitacin B			
0.25	.08	− 12.56	.3117
0.5	.16	− 41.44	.0011*
1.0	.32	− 26.56	.0337*
2.0	.64	− 37.81	.0028*

Source: From McCloud and Tallamy, unpublished.

[a] Control disks were treated with solvent only. Control disk area remaining was subtracted from that of the treated disk to determine the direction and intensity of the preference ($N = 16$).

* Significantly different at $p < .05$.

tance for uncontaminated leaves before feeding again. Indeed, such forays have been reported for *E. tredecimnotata* adults, which move about 6 m along host vines between feeding bouts (Carroll and Hoffman 1980).

Nevertheless, a detailed study of *E. borealis* feeding behavior (McCloud and Tallamy, unpublihed) failed to detect this predicted pattern of movement following feeding bouts. Marked beetles on *C. texana* within 1.8-m^3 saran field cages fed once in the morning and once again in the evening (Fig. 6) with very little movement in the interim. Though caged, the plants available to beetles were long and viney; beetles had the option of moving from leaf to leaf within and among five different plants throughout the study, yet they made no effort to avoid leaves bearing their own feeding damage (Table 4). Typically, beetles attacked a fresh leaf in the morning, rested at the base of that leaf throughout the day, and then fed again on that leaf at dusk. Changes from leaf to leaf occurred at night or as a result of movement motivated by diurnal oviposition patterns (Fig. 6). Movement between plants occurred only once every 3 days, in spite of the fact that plants were thoroughly intertwined with each other.

We then wondered whether *E. borealis* also defied predicted movement patterns when it encountered artificially damaged leaves. After caging beetles as before, we manipulated foliar damage by applying six file wounds to those *C. texana* leaves on which beetles were feeding, repeating the wounding every 24 h for as long as a beetle remained on that leaf (McCloud and

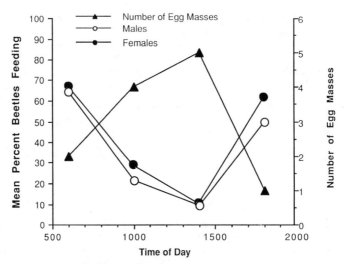

Figure 6 Diurnal feeding and oviposition activity of *Epilachna borealis*. The right ordinate indicates the number of egg masses accumulated in a study arena within a given observation time. (From McCloud and Tallamy, unpublished.)

Tallamy, unpublished). Such wounding did modify beetle feeding behavior in two significant ways (Table 5): The feeding rate of beetles in treatment cages decreased relative to that of controls, and beetles fed fewer consecutive times upon leaves that had previously received file wounds (i.e., moved more often to undamaged leaves). Although the increased movement observed in beetles encountering previously wounded leaves did not compare in magnitude to that reported by Carroll and Hoffman (1980), it did indicate that *E. borealis* detects and responds to changes in foliar quality triggered by mechanical tissue damage, changes that are not triggered by its own feeding damage.

TABLE 4 Feeding Activity of Adult *E. borealis* on *C. texana*

Activity	Males	Females
Feeds/day (SD)[a]	1.77 (0.31)	2.2 (0.56)*
New leaves/day (SD)	0.875 (0.32)	1.17 (0.21)*
Plant changes/day (SD)	0.352 (0.16)	0.365 (0.20)
Consecutive feeds per leaf (SD)	1.82 (0.24)	1.75 (0.37)
Mean percent feeds[b] on damaged leaves (SD)	60.69 (2.7)	50.47 (3.7)*

Source: From McCloud and Tallamy, unpublished.
[a] Means ($n = 12$) of within-cage means (1.8-m^3 field cages).
[b] Feeding bouts on leaves damaged by feeding within the preceding 24 h.
* Significant, $p < .05$.

TABLE 5   Feeding of *E. borealis* on Mechanically Damaged (Induced) and Nondamaged *C. texana*[a] Leaves

Activity	Control	Induced
Feeds/day (SD)[a]	2.06 (0.34)	1.64 (0.54)[b]
New leaves per day (SD)	0.80 (0.30)	0.89 (0.28)
Consecutive feeds/leaf (SD)	2.28 (0.18)	1.16 (0.18)*

Source: From McCloud and Tallamy, in press.
[a] Figures given are means of within-cage means; $n = 8$ cages.
[b] $0.1 > p < .05$, $n = 8$.
* $p < .05$, $n = 8$.

The results of these experiments hardly clarified the interaction between squash beetles and their cucurbit hosts. Previously, Tallamy (1985) had demonstrated that cucumber beetles (for which cucurbitacins are phagostimulants) prefer *C. pepo* leaf disks taken from outside of (proximal to) *E. borealis* feeding trenches over disks taken from within (distal to) trenches, while *E. borealis*, as expected, prefer disks from inside over those from outside of feeding trenches. Unfortunately, chemical analyses were not performed to confirm that beetle preferences were based on cucurbitacin content of the leaf disks, but that was the assumption. It now seems likely that, in addition to cucurbitacins, a powerful compound that deters squash beetles but not cucumber beetles is triggered by tissue damage to *C. pepo*. However, the foraging experiments described above do not corroborate these findings. Rather, they suggest that 1. the phytochemical response of *C. texana* to squash beetle feeding damage may be quite different from the response of *C. pepo* (Tallamy 1985) and *C. moschata* (Carroll and Hoffman 1980), or 2. Tallamy's (1985) preference studies reflect the induction of a deterent compound other than cucurbitacins which is blocked by squash beetle trenches.

In an attempt to reconcile these discrepancies, we compared the induced response of *C. texana* to artificial file damage with its response to *E. borealis* feeding damage (McCloud 1989, McCloud, unpublished results). A time-course analysis of the response to file wounding did show the expected induction of Cucurbitacins E and I (or their breakdown products; see section 2) 10.5 h following wounding (Fig. 4); no measurable response of cucurbitacins was triggered, however, by squash beetle feeding. Thus, it is possible that *C. texana* and *C. pepo* do respond in distinctly different ways, and that squash beetle trenches prevent the induction of compounds in *C. texana* that, if induced by other types of tissue damage, squash beetles attempt to avoid. If squash beetle trenches prevent rather than elicit the induction of noxious cucurbit compounds, beetles could feed comfortably on a single leaf as long as nutritious tissue remained available. *E. borealis* larvae behave in this manner, frequently consuming an entire leaf via a dozen or more trenches before undertaking the long and possibly dangerous journey in search of a fresh leaf. Adults are more mobile than larvae, yet our data indicate that multiple trenching on a single leaf is the rule rather than the exception, and that the interleaf movement that does occur is more likely to result from mate seeking or ovipositional demands than from the need to avoid induced phytochemicals (McCloud and Tallamy, unpublished).

We can only speculate at this time as to how squash beetle trenching might prevent the induction of deleterious phytochemicals. If, as Green and Ryan (1972) suggest, induction is a response to wound messages in the form of cell wall fragments released by injury into the vascular system of the plant, the strength of induction might depend upon the degree to which cells are crudely damaged. Green and Ryan found that clean cuts from a razor blade failed to elicit the induction of proteinase inhibitors in tomato. Mandibular specializations of squash beetles suggest that trenching might be analogous to razor blade damage, severing tissue cleanly and with minimal release of cell wall fragments. Once a trench is completed, messages of injury are trapped within the trench and do not communicate an alarm to the rest of the plant. If this is the case, squash beetles could be considered classic "stealth" feeders (sensu Rhoades 1985).

Underlying the arguments presented above is the assumption that epilachnine trenching behavior is an adaptation countering the flexibility and effectiveness of inducible cucurbitacins as a defense mechanism in the Cucurbitaceae. If this is so, however, we would not expect squash beetles to invest time or energy in trenching behavior while consuming cucurbits with little or no foliar cucurbitacins. Observations of lab colonies confirm that trenching behavior is indeed facultative in *E. borealis*, yet beetles always trench prior to feeding on cucurbitacin-free cultivars such as *C. pepo* "Patty Green Tint" and *Cucumis sativus* "Marketmore Non-Bitter" when these plants are field grown. One explanation yet to be critically examined is that cucurbitacins are only one component in a suite of defensive compounds that are induced by tissue damage in the Cucurbitaceae. For example, in addition to Cucurbitacins, cucurbits produce potentially noxious protease inhibitors (MacGibbon and Mann 1986) and saponins (Iwamoto et al. 1985a,b), to name just two candidates. Nothing is known about the inducibility of these compounds in cucurbits or their effect on cucurbit herbivores. Conceivably, these or other deleterious compounds could present a formidable defense against squash beetles in cultivars from which cucurbitacins have been selectively removed. Under such conditions, feeding trenches would remain a useful means of preserving the palatability and quality of host plant tissue.

Recent evidence, however, suggests a more parsimonious explanation for trenching in the absence of foliar cucurbitacins. Phloem sap of many species of Cucurbitaceae congeals upon exposure to air, effectively sealing wounds from water loss and bacterial invasion (Walker and Thaine 1971, Kleinig 1975, McEuen and Hill 1982, Richardson and Baker 1982). Feeding behaviors such as trenching, vein snipping, and stem girdling are mechanisms that have arisen in insects specifically for avoiding the copious flow of latex, resins, or mucilaginous sap (Dussourd and Eisner 1987, Dussourd and Denno, in press). In most, if not all cases, such substances directly impede mandibular processes when released by herbivory from specialized ducts or the vascular system itself. Clearly, the stimulus to avoid these plant products differs greatly from that prompting a squash beetle to block a relatively slow influx of minute quantities of a toxic compound (Carroll and Hoffman 1980, Tallamy 1985, 1986). If squash beetle trenching is an adaptation to avoid the congealing action of cucurbit sap, interpretations of the interactions between squash beetles and cucurbits must be revisited.

We have conducted several experiments to test this possibility (McCloud and Tallamy, unpublished). To determine whether exposure to cucurbit sap can interfere with squash beetle mandibular functions we compared the rate at which starved female beetles recovered sufficiently to feed after being dipped head-first into either water, snap-bean sap, or squash (*C. texana*) sap. Treatment with bean sap was included in the design for comparison because *Epilachna varivestis*, the Mexican bean beetle, skeletonizes bean foliage without the aid of feeding trenches. The results of this experiment (Fig. 7) demonstrate a clear advantage to avoiding exposure to squash sap: Beetles whose mandibles were dipped in squash sap groomed on average more than 5 h longer before feeding than beetles dipped in bean sap (Wilcoxin two-sample test: $Z = -2.2721, p = .0231$) and more than 8 h longer than beetles dipped in water (Wilcoxin two-sample test: $Z = -4.1221, p < .0000$). In fact, 8.5% of the beetles dipped in squash sap ($N = 35$) never fed following treatment and eventually died. Though impedance from squash sap was far stronger, bean sap also significantly delayed beetle feeding when compared to water (Wilcoxin two-sample test: $Z = -1.9820; p = .0457$). These results may provide an explanation for Mexican bean beetle feeding behavior: By skeletonizing, these beetles avoid exposure to sap by leaving major veins in the vascular system of the plant intact.

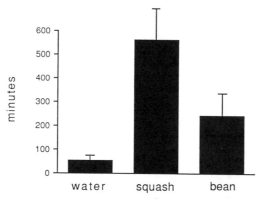

Figure 7 Comparison of the time elapsed before first feeding after *Epilachna borealis* adult females were dipped head-first into water, *Cucurbita texana* sap, and snap bean sap. (From McCloud and Tallamy, unpublished.)

Next we asked whether the probability of exposure to sap as mediated by plant turgor pressure affects squash beetle trenching "decisions" and whether this effect has an influence greater than foliar cucurbitacin content. Using *C. texana* (a plant with little foliar cucurbitacins) and *C. andreana* (a cucurbit with relatively bitter leaves), we quantified squash beetle propensity to trench as a function of leaf turgor pressure in the following way. Leaves of both species were severed from the parent plant with a razor, held varying lengths of time to manipulate turgor, and then placed stem first in vials of water. A squash beetle was then placed on each leaf and observed. Before actually feeding, beetles always chewed a small hole in the leaf; the "decision" to then snip a feeding trench or to feed without first trenching was apparently based upon information obtained from this initial feeding hole. Immediately after it had elected to feed with or without a trench we removed the beetle, sliced a major leaf vein 2 mm in diameter with a razor blade proximal to the feeding damage, and placed a micropipette flush against the severed vein. The distance the squash sap traveled up the pipette in 15 s provided an index of turgor pressure.

Beetles feeding on both plant species reached a decision to trench or not when turgor pressures were between 1 and 4 mm (Fig. 8). Pressures below 1 mm always resulted in a no trench decision, while pressures exceeding 4 mm always triggered trenching behavior. Thus, trenching frequency responds dramatically to changes in turgor pressure. To consider the possibility that the turgor threshold that activates trenching is lower on bitter *C. andreana* than on nonbitter *C. texana*, we used a 2 × 2 contingency table to compare the percentage of beetles electing to trench on *C. texana* within that 1- to 4-mm range of pressures with beetles trenching under similar pres-

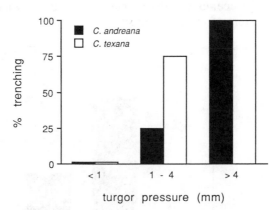

Figure 8   The percentage of *Epilachna borealis* adults that utilized a feeding trench on *Cucurbita andreana* and *C. texana* leaves of varying turgor pressure, as measured by the distance (mm) phloem sap traveled up a microcapillary pipette placed flush to the severed end of a midrib 2 mm in diameter for 15 s. (From McCloud and Tallamy, in press.)

sures on *C. andreana*. Beetles feeding on *C. texana* trenched 70% of the time even though its foliage is low in cucurbitacin content. In contrast, significantly fewer (25%) of the beetles feeding on the more bitter *C. andreana* foliage trenched ($\chi^2 = 4.45$, $p = .035$). It is apparent that under these conditions, the decision to trench is a function of turgor pressure and the consequent likelihood of being doused with gummy sap rather than a decision based on foliage cucurbitacin content.

We conclude this section by summarizing observations that are or may be in conflict with the suggestion that trenching behavior in epilachnines constitutes an adaptation countering inducible cucurbitacins:

1. Cucurbitacins do not deter squash beetles from attacking cucurbit foliage.
2. Not all cucurbit-feeding epilachnines trench before feeding.
3. Preliminary chemical analyses indicate that trenching may not always result in cucurbitacin induction within leaf tissues proximad to trenches.
4. Behavioral studies of *E. borealis* in field cages suggest that beetles do not forage as the induction hypothesis predicts.
5. Manipulation of leaf turgor dramatically alters trenching frequency and trenching is not correlated with foliar cucurbitacin content.
6. Exposure to mucilaginous squash sap impedes *Epilachna* feeding activities.

It is important to recognize, however, that although sap avoidance provides a more satisfactory explanation for the evolution of squash beetle trenching behavior, it does not preclude other benefits associated with trenching; blocking the influx of induced cucurbit defenses or preventing the induction process itself may have equally profound effects on beetle fitness, even if such effects are serendipitous.

4 *CUCURBITA*–CUCUMBER BEETLE INTERACTIONS

In addition to squash beetles, the Cucurbitaceae serve completely or in some capacity as hosts to a large group of galerucine chrysomelid beetles in the tribe Luperini. The Old World Aulacophorina and New World Diabroticina (hereafter diabroticites) comprise hundreds of species that share a long evolutionary history (at least 30 million years) with cucurbits (reviewed by Metcalf 1985, Metcalf and Lampman 1989). Cucurbitacins stimulate compulsive feeding behavior in diabroticites and thus function as kairomones rather than allomones to these beetles. Even diabroticites not dependent upon cucurbits for food (i.e., the spotted cucumber beetle, *Diabrotica undecimpunctata howardi*) regularly seek out and consume these sources of cucurbitacins, often in spectacular feeding frenzies (Metcalf et al. 1982). The likelihood that a particular plant part will be attacked by diabroticites is a function of its cucurbitacin content; constitutively bitter tissues such as cotyledons (Gould 1944, Ferguson et al. 1983, Lewis et al 1990, Tallamy and Krischik 1989), roots (Merrick, personal communication), fruits (Metcalf et al. 1982, Tallamy, personal observation), and stem galls formed on *C. foetidissima* by the western squash vine borer, *Melittia snowii* (Smith and Binder, personal communication), are all highly attractive cucurbitacin resources for diabroticites and, as such, are subject to decimation when beetle populations are high. Undamaged leaves, in contrast, are usually the least bitter components of a cucurbit and suffer little herbivory from diabroticites. Recent wound sites on foliage, however, are attractive to these beetles, possibly because of the local induction of cucurbitacins (Tallamy, unpublished results).

It is obvious that induction of cucurbitacins as an antiherbivore defense is less than effective against diabroticites. Tallamy and Krischik (1989) demonstrated that induced concentrations of cucurbitacins had no effect on the spotted cucumber beetle fitness (Fig. 9). In fact, most species of diabroticites thus far examined can consume cucurbitacins in quantities exceeding those occurring in nature with no ill effects (Metcalf 1985). The bulk of ingested cucurbitacins is rapidly excreted (Ferguson et al. 1985), but diabroticites also readily metabolize these compounds through glucosylation, hydrogenation, desaturation, and acetylation (Andersen et al. 1988). Excretion of cucurbitacins prior to energy-demanding detoxification, as occurs in the diabroticite *Acalymma vittata*, is obviously advantageous. Females of this species can consume as much as 277 mg cucurbitacins/g body w with no detectable reduction in fitness (Ferguson et al. 1985). This is remarkable given that the intraperitoneal LD_{50} of cucurbitacin B for mice is 0.03 mg/g body wt (David and Vallance 1955). Cucurbitacins are thought to be general cytotoxins that can disrupt membranes or act as Michael acceptors for biological nucleophiles (Kupchan 1970). Nevertheless, cucurbitacins not only fail to reduce diabroticite fitness, but Ferguson and Metcalf (1985) provide evidence that the consumption of bitter tissues actually increases beetle fitness by rendering them unpalatable to invertebrate predators. Praying mantids rejected up to 72% of diabroticites previously fed bitter *C. andreana* fruits. Analyses for cucurbitacins revealed the accumulation of bitter glycoside conjugates in the hemolymph and elytra of these beetles.

Interactions between diabroticites and the inducible properties of cucurbit foliage remains unclear. Attempts to quantify fluctuations in foliar cucurbitacins that may be triggered by diabroticite herbivory have been frustrated by the reluctance of these beetles to attack mature, undamaged leaves (Cheesman 1989, McCloud, unpublished results). Observations of *D. undecimpunctata howardi* confined on individual *Cucurbita* leaves suggest that adults have difficulty penetrating intact foliar epidermis with their relatively small mandibles; consequently, they usually restrict their feeding on mature

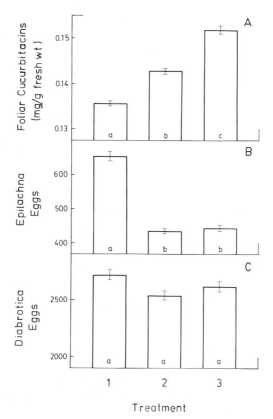

Figure 9   Effects of *Cucurbita pepo* "Black" cotyledons grown without ultraviolet-B irradiance (treatment 1), with UV-B irradiance (treatment 2), and with UV-B irradiance plus two standardized wounds from a rat-tail file (treatment 3): (*A*) on total cucurbitacin content, and on fecundity of (*B*) *Epilachna borealis* and (*C*) *Diabrotica undecimpuntata howardi*. Error bars show ± 1 SE. Histograms with different letters are significantly different (Duncan's multiple range, $p < .05$). (From Tallamy and Krischik 1989.)

foliage to sites where the epidermis has been previously damaged by some other source. Curiously, the arrestant and stimulatory effects on diabroticites of fruit and cotyledon cucurbitacins are not observed when these beetles feed on mature foliage. Instead, diabroticites that feed on foliage at all, generally do so in short, ephemeral episodes, moving frequently within and among plants between feeding bouts. It is this flighty feeding behavior that makes diabroticites such insidious vectors of plant pathogens.

This feeding behavior is not consistent with the indiscriminate induction of cucurbitacins. If diabroticite herbivory triggers the rapid accumulation of cucurbitacins within attacked foliage, beetles should remain for long periods at those feeding sites, receiving cucurbitacin rewards with every bite. Furthermore, leaf damage from one beetle should increase the attractiveness of that leaf for other beetles, resulting in an ever-increasing aggregation of diabroticites. The absence of these phenomena circumstantially suggests that cucurbits, through differential induced responses, may distinguish between two opposing herbivore responses to cucurbitacin induction: that of the generalist herbivore, deterred by increasing foliar cucurbitacin content, and that of the specialist diabroticite, incited to potential plant destruction by the same chemical.

Though it has not been critically tested, plausible arguments can be mounted in defense of this hypothesis (Tallamy and Krischik 1989). There is an incredible discrepancy between the minute quantity of cucurbitacins required to negatively affect physiologically nonadapted herbivores and the huge (but as yet unknown) amounts required to reduce attractiveness to diabroticites. (Diabroticites may become temporarily or permanently satiated with cucurbitacins after prolonged ingestion of high concentrations; Tallamy, unpublished results). Since increased commitments to the production of cucurbitacins do not readily translate into increased costs for diabroticites, selection may favor a reduction in foliar cucurbitacins that is constrained only by the minimum requirements for defense against nonadapted herbivores. Maintaining relatively small commitments to defense that are conservatively deployed through discriminating inducible responses should minimize any metabolic costs associated with defense and, perhaps more importantly, reduce cucurbit attractiveness to diabroticite beetles. Thus, the dynamic properties of cucurbitacin induction may permit cucurbits to chemically hide from their adapted enemies.

Patterns of constitutive bitterness in leaves, fruits, and roots support this hypothesis (Fig. 10). Of these plant parts, only the relatively nonbitter leaves are exposed to diabroticite attack. Roots, which are subject to attack from a variety of soil inhabiting arthropods, nematodes, and mammals, are collectively highest in cucurbitacin concentration, but are physically concealed from most adult diabroticites. Bitter fruits are protected by tough rinds that are difficult for beetles to penetrate. If diabroticites do succeed in entering a fruit, they feed voraciously on the fleshy endocarp which contains high levels of cucurbitacins (Sharma and Hall 1973, Rhodes et al. 1980), but virtually ignore the seeds which are as low in cucurbitacin content as foliage (Tallamy, personal observation).

Comparatively greater bitterness in cotyledons may reflect the absolute necessity of preventing even casual grazing by generalist herbivores on this vulnerable stage of development. Whereas the more permanent foliage of mature plants may escape diabroticite attack through low levels of bitterness and, hence, nonattractance, cotyledons may escape destruction from these specialists by virtue of being an inconspicuous and ephemeral stage of development (see Feeny 1976). This, of course, is not the case in the artificial setting of agricultural fields. Here unnaturally high diabroticite populations

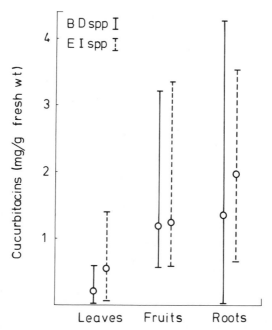

Figure 10 Comparison of total cucurbitacins among leaves, fruits, and roots of *Cucurbita* species producing cucurbitacins B and D and of species producing cucurbitacins E and I: circles = means; lines = ranges. (From Tallamy and Krischik 1989.)

and unnaturally dense cucurbit dispersion encourage crop destruction (Ferguson et al. 1983).

Specificity in inducible cucurbit responses to invading organisms is not without precedence. While cucumbers achieve cross resistance to several phytopathogens by a generalized induction of lignification and increased peroxidase activity (Hammerschmidt and Kuc 1980, 1982, Metraux and Boller 1986), cucurbitacins are apparently mobilized only by nonpathogenic tissue damage (Kuc, personnel communication; but see Bar-Nun and Mayer 1989). In addition, mechanical injuries to leaf tissues fail to evoke systemic immunization against cucurbit pathogens (Kuc 1982), but rapidly trigger systemic increases in foliar bitterness (Carroll and Hoffman 1980, Tallamy 1985, McCloud 1989).

5 CUCURBITACINS AND THE ASSUMPTION OF DEFENSE

Because of their extreme bitterness to human taste buds and their demonstrable effectiveness as antifeedants against vertebrate and invertebrate herbivores (reviewed by Metcalf and Lampman 1989), cucurbitacins have been considered antiherbivore defenses since they were first characterized (Enslin 1954). That these compounds are stockpiled within the vital plant parts of most cucurbits such as cotyledons, roots, and fruits supports this assumption. Yet, to label cucurbitacins plant "defenses" is to imply that they are chemical adaptations derived and maintained by natural selection for the singular purpose of defending their producers from herbivores. There is growing evidence that this view is oversimplified, if not entirely incorrect. Alternative causes of the evolution and dynamic properties of cucurbitacins must be considered carefully before the defense hypothesis can be exclusively accepted.

As attractive as it has been, the assumption that cucurbitacins evolved for defensive purposes is not without imperfections. Problems arise when interpreting the inducible response itself. Several seemingly unrelated types of stress have been shown to induce foliar cucurbitacins. Folklore from the cucurbit farming community has been supported by cucumber studies demonstrating that drought stress causes increased bitterness in test plants (Haynes and Jones 1975, Gould 1978). Exposure to ambient levels of ultraviolet-B irradiance also induces cucurbitacin increases in cucurbit foliage (Tallamy and Krischik 1989). Responses to these stresses in addition to various mechanical injuries (Carroll and Hoffman 1980, Tallamy 1985) suggest that an understanding of the role of cucurbitacins in processes of general wound repair is essential to interpreting the evolutionary and ecological significance of these compounds. It is possible that cucurbitacins are bitter to herbivores serendipitously and that their induction is correlated with herbivory only as a means of repairing inflicted wounds, not as adaptations to discourage further wounding.

Evidence that cucurbitacins are active in gibberellin inhibition further supports the wound repair hypothesis (Guha 1974, Guha and Sen 1975, Mukherjee et al. 1986, Medeiros and Takahashi 1987). Cucurbitacin B inhibits growth in rice and cucumber seedlings, and antagonizes GA3 induction of amylase activity in cereal endosperm (Guha and Sen 1975). Similarly, mitosis in *Allium cepa* root tip cells is reduced by 38 and 62% 8 h following exposure to cucurbitacins at 1.2 and 1.8 g/L, respectively (Medeiros and Takahashi 1987). By slowing growth during periods of stress, be it from lack of water, UV irradiance, or herbivory, cucurbitacin induction might be a mechanism by which resources can be diverted toward speedy wound repair.

The defense hypothesis also becomes awkward when interspecific variation in cucurbitacin concentrations is considered (Tallamy and Krischik 1989). Although fruits and roots are significantly more bitter than leaves in *Cucurbita* species for which data are available, the interspecific range of cucurbitacin concentrations for each plant part is enormous (Fig. 10). If, for example, cucurbitacins accumulate in roots to discourage herbivory, why do some species have virtually no cucurbitacin in their roots while others produce as much as 4 mg/g fresh wt?

6 CONCLUSION

Our knowledge of the physiological mechanisms regulating the induction and relaxation of cucurbitacins remains in its infancy. We do know that two beetle herbivores of cucurbits respond physiologically and ecologically to these compounds, and that their behavior can be a powerful tool in the study of inducible cucurbit responses. Studies of diabroticite beetles have focused upon their compulsive feeding on cucurbitacins, but there are no data describing the interaction between diabroticites and the inducible properties of cucurbitacins. Ironically, the trenching behavior of *Epilachna* squash beetles that stimulated the discovery of mobile phytochemicals in *Cucurbita* may have only a coincidental relationship to the phenomenon. Recent evidence suggests that squash beetles snip feeding trenches around tissues to be consumed to minimize exposure to gelatinous proteins in phloem sap rather than to avoid the impending invasion of bitter cucurbitacins. In fact, evidence that cucurbitacins are induced at all by squash beetle herbivory is contradictory and demands further scrutiny. If trenching behavior evolved in response to the gelling properties of cucurbit sap, we predict that the absence of trenching in cucurbit feeding *Epilachna* such as *E. septima* reflects the absence of gelling proteins in the sap of their host plants. The data to test this hypothesis are not yet available.

Although the assumption that cucurbitacins are defensive adaptations is firmly entrenched in the literature, alternative explanations exist and they deserve critical attention. It is important to recognize that explanations for the evolution of constitutive and inducible cucurbitacins need not be mutually exclusive. Steroidal roles in wound repair, for example, could have served as a stage on which selection for antifeedant properties could act. Thus, defense and wound repair could be equally vital cofunctions of cucurbitacins that act independently or in concert during periods of plant stress. Evaluating the interaction between these compounds, specialist and generalist herbivores, and other environmental stresses will provide worthy challenges for scientists in the future.

ACKNOWLEDGMENTS

We are grateful for the comments and criticisms of M. J. Raupp, D. Dussourd, C. Plyler, and B. Binder; for cucumber beetle donations from K. Stoops, E. I. DuPont de Nemours & Co. and *Epilachna* donations from S. Nakano; for the statistical expertise of J. Pesek; and for helpful discussions with S. Nakano, H. Katakura, and K. Nakamura on the feeding behavior of Asian *Epilachna*. Portions of this work were supported by National Science Foundation Grant BSR-8506100 and a Sigma Xi Grant-in-Aid to E. McCloud. Published as Contribution No. 606 of the Department of Entomology and Applied Ecology and Contribution No. 131 of the Ecology Unit, School of Life and Health Sciences, University of Delaware, Newark, DE.

REFERENCES

Andersen, J. F., R. D. Plattner, and D. Weisleder. 1988. Metabolic transformations of cucurbitacins by *Diabrotica virgifera virgifera* Leconte and *D. undecimpunctata howardi* Baber. *Insect Biochem.* 18:71–78.

Baldwin, I. T. 1990. Damage-induced alkaloids in wild tobacco. 1991. In D. W. Tallamy and M. J. Raupp (eds.), *Phytochemical Induction by Herbivores.* New York: Wiley, pp. 47–70.

Bar-Nun, N., and A. M. Mayer. 1989. Cucurbitacins-responses of induction of laccase formation. *Phytochemistry* 28:1369–1371.

Brannon, L. W. 1937. Life-history studies of the squash beetle in Alabama. *Ann. Entomol. Soc. Am.* 30:43–51.

Carroll, D. R., and C. A. Hoffman. 1980. Chemical feeding deterrent mobilized in response to insect herbivory and counter adaptation by *Epilachina tredecimnotata. Science* 209:414–416.

Cheesman, O. D. 1989. An investigation of wound-induced chemical changes in the foliage of *Cucurbita pepo* cv. All Green Bush, and their effect on the behavior of 2 insect herbivores. Ph.D. Thesis. University of Southampton.

Craft, A. S., and C. E. Crisp. 1971. *Phloem Transport in Plants.* San Francisco: Freeman.

DaCosta, C. P., and C. M. Jones. 1971. Cucumber beetle resistance and mite susceptibility controlled by the bitter gene in *Cucumis sativa* L. *Science* 172:1145–1146.

David, A., and D. K. Vallance. 1955. Bitter principles of Cucurbitaceae. *J. Pharm. Pharmacol.* 7:295–296.

Davies, E. 1987. Plant responses to wounding. In D. D. Davies (ed.), *The Biochemistry of Plants,* Vol. 12. New York: Academic Press, pp. 243–264.

Dawson, R. F. 1941. The localization of the nicotine synthetic mechanism in the tobacco plant. *Science* 94:396–397.

Dixon, A. F. G. 1975. Aphids and translocation. In A. Pirson and M. H. Zimmerman (eds.), *Encyclopedia of Plant Physiology: Transport in Plants I; Phloem Transport.* New York: Springer, pp. 154–170.

Dussourd, D. E., and R. F. Denno. In press. Deactivation of plant defense: correspondence between insect behavior and secretory canal architecture. *Ecology.*

Dussourd, D. E., and T. Eisner. 1987. Vein-cutting behavior: insect counterploy to the latex defense of plants. *Science* 237:898–901.

Enslin, P. R. 1954. Bitter principles of Cucurbitaceae. I. Chemistry of cucurbitacin A. *J. Sci. Food Agric.* 5:410–416.

Enslin, P. R., F. J. Joubert, and S. Rehm. 1956. Bitter principles of the Cucurbitaceae III. Elaterase, an active enzyme for the hydrolysis of bitter principle glycosides. *J. Sci. Food Agric.* 7:646–655.

Esau, K. 1965. *Plant Anatomy*. New York: Wiley.

Esau, K., and V. I. Cheadle. 1959. Size of pores and their contents in sieve elements of Dicotyledons. *Proc. Nat. Acad. Sci.* 45:156–162.

Feeny, P. 1976. Plant apparency and chemical defense. *Rec. Adv. Phytochem.* 10:1–40.

Ferguson, J. E., and R. L. Metcalf. 1985. Cucurbitacins: plant derived defense compounds for diabroticites (Coleoptera: Chrysomelidae). *J. Chem. Ecol.* 11:311–318.

Ferguson, J. E., E. R. Metcalf, R. L. Metcalf, and A. M. Rhodes. 1983. Influence of cucurbitacin content in cotyledons of Cucurbitaceae cultivars upon feeding behavior of Diabroticina beetles (Coleoptera: Chrysomelidae). *J. Econ. Entomol.* 76:47–51.

Ferguson, J. E., R. L. Metcalf, and D. C. Fischer. 1985. Disposition and fate of cucurbitacin B in five species of diabroticites. *J. Chem. Ecol.* 11:1307–1321.

Fukuda, Y. 1967. Anatomical study of the internal phloem in the stems of dicotyledons, with special reference to its histogenesis. *J. Faculty Sci., Univ. Tokyo* 9:313–375.

Gordon, R. D. 1975. A revision of the Epilachninae of the Western hemisphere (Coleoptera: Coccinellidae). *ARS, USDA Tech. Bull. 1493.*

Gould, G. E. 1944. The biology and control of the striped cucumber beetle. *Ind. Purdue Agr. Exp. Station Bull. 490.*

Gould, F. 1978. Resistance of cucumber varieties to *Tetranychus urticae*: genetic and environmental determinants. *J. Econ. Entomol.* 71:680–683.

Green, T. R., and C. A. Ryan. 1972. Wound-induced proteinase inhibitors in plant leaves: a possible defense mechanism against insects. *Science* 175:776–777.

Guha, J. 1974. Studies on gibberellins and some gibberellin-antagonistic substances of the Cucurbitaceae. Ph.D. Thesis, Kalyani University.

Guha, J., and S. P. Sen. 1975. The cucurbitacins: a review. *Plant Biochem. J.* 2:12–28.

Hammerschmidt, R., and J. Kuc. 1980. Enhanced perloxidase activity and lignification in the induced systemic protection of cucumber. *Phytopathology* 70:689.

Hammerschmidt, R., and J. Kuc. 1982. Lignification as a mechanism for induced systemic resistance in cucumber. *Physiol. Plant Pathol.* 20:61–71.

Haynes, R. L., and C. M. Jones. 1975. Wilting and damage to cucumber by spotted and striped cucumber beetles. *Hort. Sci.* 10:256–266.

Iwamoto, M., H. Okabe, and T. Yamauchi. 1985a. Studies on the constituents of *Momordica cochinchinensis*. 2. Isolation and characterization of the root saponins, momordins I, II and III. *Chem. Pharm. Bull.* 33:1–7.

Iwamoto, M., H. Okabe, T. Yamauchi, M. Tanaka, Y. Rokutani, S. Hara, K. Mihashi, and R. Higuchi. 1985b. Studies on the constituents of *Momordica cochinchinensis*. 1. Isolation and characterization of the seed saponins, momordica saponins I and II. *Chem. Pharm. Bull.* 33:464–478.

Kleinig, H. 1975. Filament formation *in vitro* of a sieve tube protein from *Cucurbita maxima* and *Cucurbita pepo*. *Planta* 127:163–170.

Kuć, J. 1982. Induced immunity to plant diseases. *BioScience* 32:854–860.

Kupchan, S. M. 1970. Recent advances in the chemistry of terpenoid tumor inhibitors. *Pure Appl. Chem.* 21:227.

Lavie, D., and E. Glotter. 1971. The cucurbitacins, a group of tetracyclic triterpenes. *Forts. Chem. Organ. Naturstoffe* 29:306–362.

Lewis, P. A., R. L. Lampman, and R. L. Metcalf. 1990. Kairomonal attractants for *Acalymma vittatum* (Coleoptera: Chrysomelidae). *Environ. Entomol.* 19:8–14.

MacGibbon, D. B., and J. D. Mann. 1986. Inhibition of animal and pathogenic fungal proteases by phloem exudate from pumpkin fruits (Cucurbitaceae). *J. Sci. Food Agric.* 37:515–522.

McCloud, E. S. 1989. Induction of cucurbitacins: influence on squash beetle foraging behavior. M.S. Thesis, University of Delaware.

McCloud, E. S., and D. W. Tallamy. Unpublished. Squash beetle trenching behavior: Avoidance of cucurbitacin induction or mucilaginous plant sap. *Ecology*.

McEuen, A. R., and H. A. O. Hill. 1982. Superoxide, hydrogen peroxide, and the gelling of phloem sap from *Cucurbita pepo*. *Planta* 154:295–297.

Medeiros, M. D. G., and C. S. Takahashi. 1987. Effects of *Luffa operculata* on *Allium cepa* root tip cells. *Cytologia* 52:255–259.

Metcalf, R. L. 1985. Plant kairomones and insect pest control. *Illinois Natural History Survey Bulletin* 33:175–198.

Metcalf, R. L., and R. L. Lampman. 1989. The chemical ecology of Diabroticites and Cucurbitaceae. *Experientia* 45:240–247.

Metcalf, R. L., A. M. Rhodes, R. A. Metcalf, J. Ferguson, E. R. Metcalf, and Po-Yung Lu. 1982. Cucurbitacin contents and diabroticite (Coleoptera: Chrysomelidae) feeding upon *Cucurbita* spp. *Environ. Entomol.* 11:931–937.

Metraux, J. P., and T. H. Boller. 1986. Local and systemic induction of chitinase in cucumber plants in response to viral, bacterial and fungal infections. *Physiol. Mol. Plant Pathol.* 28:161–169.

Mukherjee, S., A. K. Shaw, and S. N. Ganguly. 1986. Amarinin: a new growth inhibitor from *Luffa amara*. *Plant Cell Physiol.* 27:935–937.

Nielson, J. K., M. Larsen, and H. J. Sorenson. 1977. Cucurbitacins E and I in *Iberis amara*: Feeding inhibitors for *Phyllotreta nemorum*. *Phytochemistry* 16:1519–1522.

Rehm, S., P. A. Enslin, A. D. J. Neeuse, and J. H. Wessels. 1957. Bitter principles of the Cucurbitaceae VII. The distribution of bitter principles in the plant family. *J. Sci. Food Agric.* 8:679–686.

Rhoades, D. F. 1985. Offensive–defensive interactions between herbivores and plants: their relevance in herbivore population dynamics and ecological theory. *Am. Natur.* 125:205–238.

Rhodes, A. M., R. L. Metcalf, and E. R. Metcalf. 1980. Diabroticite beetle responses to cucurbitacin kairomones in *Cucurbita* hybrids. *J. Am. Soc. Hort. Sci.* 105:838–842.

Richards, A. 1983. The *Epilachna vigintioctopunctata* complex (Coleoptera: Coccinellidae). *Int. J. Entomol.* 25:11–41.

Richardson, P. T., and D. A. Baker. 1982. The chemical composition of cucurbit vascular exudates. *J. Exp. Bot.* 33:1239–1247.

Ryan, C. A. 1983. Insect-induced chemical signals regulating natural plant protection responses. In R. F. Denno and M. S. McClure (eds.), *Variable Plants and Herbivores in Natural and Managed Systems*. New York: Academic, pp. 43–60.

Schaefer, P. W. 1983. Natural enemies and host plants of species in the Epilachninae (Coleoptera: Coccinellidae): a world list. *Univ. Delaware Agric. Exp. Sta. Bull. 445*.

Sharma, G. C., and C. V. Hall. 1973. Relative attractance of spotted cucumber beetles to fruits of fifteen species of Cucurbitaceae. *Environ. Entomol.* 2:154–156.

Stroesand, G. S., A. Jaworski, S. Shannon, and R. W. Robinson. 1985. Toxicologic response in mice fed *Cucurbita* fruit. *J. Food Protection* 48: 50–51.

Tallamy, D. W. 1985. Squash beetle feeding behavior: an adaptation against induced cucurbit defenses. *Ecology* 66:1574–1579.

Tallamy, D. W. 1986. Behavioral adaptations in insects to plant allelochemicals. In L. B. Brattsten and S. Ahmad (eds.), *Molecular Aspects of Insect–Plant Associations*. New York: Plenum, pp. 273–300.

Tallamy, D. W., and V. A. Krischik. 1989. Variation and function of cucurbitacins in *Cucurbita*: an examination of current hypotheses. *Am. Natur.* 133:766–786.

VanSambeek, J. W., and B. G. Pickard. 1976. Mediation of rapid electrical, metabolic, transpirational, and photosynthetic changes by factors released from wounds. I. Variation potentials and putative action potentials in intact plants. *Can. J. Bot.* 54:2642–2650.

Walker, T. S., and R. Thaine. 1971. Proteins and fine structural components in exudate from seive tubes in *Cucurbita pepo* stems. *Ann. Bot.* 35:773–790.

Whitaker, T. W., and W. P. Bemis. 1975. Origin and evolution of the cultivated *Cucurbita*. *Bull. Torrey Bot. Club* 102:362–368.

Whitaker, T. W., and G. N. Davis. 1962. *Cucurbits: Botany, Cultivation, and Utilization*. New York: Interscience.

CHAPTER 8

RESPONSES OF LEAF BEETLES TO INJURY-RELATED CHANGES IN THEIR SALICACEOUS HOSTS

MICHAEL J. RAUPP and CLIFFORD S. SADOF
Department of Entomology, University of Maryland, College Park, MD 20742

1   Introduction
2   Responses of willows and associated herbivores to injury
    2.1   Physical and chemical responses of willows to leaf injury
    2.2   Short-term responses of weeping willows to injury
    2.3   Behavioral responses of *Plagiodera versicolora* to injured willow leaves
    2.4   Physiological responses of *Plagiodera versicolora* to injured willow leaves
3   Injury-related changes in leaf suitability and spatial patterns of herbivory
4   Injury-related changes in leaf suitability and temporal patterns of herbivory
5   Impact of injury-related changes in leaves on higher trophic levels
6   Summary
    References

1   INTRODUCTION

There is an ever-increasing body of data indicating that damage alters the suitability of plant leaves as a food resource for herbivorous insects. In the majority of the systems investigated the alteration has been to the detriment of a herbivore associated with the plant. For example, in the well-studied lepidopteran fauna associated with the genus *Betula*, natural and artificial damage has been shown to reduce caterpillar feeding rates (Edwards and Wratten 1982, Wratten et al. 1984), alter selection of feeding sites (Bergelson et al. 1986), and reduce larval growth and development (Haukioja and Niemela 1977, 1979, Wallner and Walton 1979, Haukioja and Hahimaki 1985, Haukioja and Neuvonen 1985, Haukioja et al. 1985, Fowler and MacGarvin 1986, Neuvonen and Haukioja 1984). In relatively few cases, previous damage had no detectable effect on the performance of specific herbivores (Valentine et al. 1983, Hartley and Lawton, this volume, Chapter 5) or actually enhanced the suitability of the plant as a food source for herbivores (Neu-vonen and Danell 1987). The trend for injury to reduce the suitability of plant leaves is not unique to *Betula*. In their review of the literature Coleman and Jones (this volume, Chapter 1) found that in some thirty-nine accounts of induced changes in plants, more than two-thirds reported negative effects of induced changes on herbivore feeding behaviors or life history attributes such as survivorship, growth, and development.

This chapter examines the relationship between injury to willow leaves and the direct and indirect effects it has on herbivores. We attempt to elu-cidate phytochemical responses of willows to injury and to interpret spatial and temporal patterns of herbivory with respect to these responses. First, we examine the physical and chemical responses of willows to injury. Next, we focus on short-term injury-related changes in the foliage of two types of closely related weeping willows, *Salix babylonica* and *S. alba* "Tristis." Using these cultivars as a model system, we examine the direct effects of leaf injury on the survivorship, growth, and development of a leaf beetle, *Plagiodera versicolora*. Behavioral responses of this beetle to leaf injury are discussed and used to interpret patterns of herbivory and the distribution of beetles observed in the field. We suggest that injury-related changes in wil-low leaves may be an important key to resolving ambiguous patterns of herbivory associated with this species.

2   RESPONSES OF WILLOWS AND ASSOCIATED HERBIVORES TO INJURY

2.1   Physical and Chemical Responses of Willows to Injury

Responses of several willow species to herbivory and other types of leaf and stem injury have been well documented in a variety of studies. Bryant et al. (1985, this volume, Chapter 6) and Chapin et al. (1985) found the regrowth twigs of the feltleaf willow *S. alexensis* to be less palatable to snowshoe hares than mature stems. They attributed this effect to changes in the chemical composition of regrowth shoots, primarily increased lignin and phenolic content, rather than to alterations in morphology or accessi-bility of twigs (Bryant et al. 1985). The bebb willow, *S. bebbiana*, showed no induced response in several attributes such as resin, tannin, nutrient, and carbohydrate content two years after shoots were removed (Chapin et al. 1985).

Less is known about short-term changes in foliage quality of willows following leaf injury. There is little difference in total phenolic or proantho-cyanidin content in the leaves of damaged sitka willow, *S. sitchensis* relative to leaves of control trees (Rhoades 1983a) within several weeks following leaf damage. However, in the closely related genus, *Populus*, several changes in plant chemistry have been noted after defoliation. Mattson and

Palmer (1987) found increases in total phenolics in *P. tremuloides* within two weeks of defoliation. Specific phenolic compounds such as the phenolic glycosides salicin and tremuloiden increase in the leaves of *P. tremuloides* within hours of injury (Clausen et al., this volume, Chapter 3). Long-term changes in phenolic glycosides, tannins, and nitrogen levels have been demonstrated in *P. tremuloides* following defoliation (Clausen et al., this volume, Chapter 3). One year following defoliation nitrogen and tremulacin levels were higher and condensed tannins were lower in previously damaged plants (Clausen et al., this volume, Chapter 3).

2.2   Short-Term Responses of Weeping Willows to Leaf Injury

Recently, we have investigated the short-term responses of *S. alba* "Tristis" to natural and artificial defoliation. In this chapter we use short-term to mean time intervals of days rather than months or years.The willow species considered here are the ornamental weeping willows, *Salix babylonica* and *S. alba* "Tristis." *Salix babylonica* is probably a native of China and is now widely planted in North America where it often escapes from cultivation and grows in unmanaged sites (Santamour and McArdle 1988). *Salix alba* "Tristis" is a cultivar derived from *S. babylonica* and *S. alba* (Santamour and McArdle 1988). The trees used in the studies reported here are components of landscape plantings in and around the College Park campus of the University of Maryland.

To date, four attributes of leaf quality thought to be important to herbivore feeding ecology have been assayed: foliar nitrogen, moisture, and tannin content, and leaf toughness. Foliar nitrogen and water content play important roles in the performance of a wide variety of herbivores (McNeill and Southwood 1978, Mattson 1980, Scriber and Slanksy 1981, Slansky and Scriber 1985, Mattson and Scriber 1987). Furthermore, foliar nitrogen and water content have been shown to change following defoliation (Landsberg and Wylie 1983, Piene and Percy 1984, Ericsson et al. 1985, Wagner and Evans 1985, Svejcar and Christiansen 1987, Mattson and Palmer 1987, Wagner 1987). Although the role of tannins as allelochemicals active against herbivorous insects is still widely debated (Bernays 1981), several authors attribute an antiherbivore function to tannins, and many studies revealed increased tannin levels following leaf damage (Baldwin and Schultz 1985, Wratten et al. 1984, Faith 1985, 1986, Leather et al. 1987). The role of leaf toughness as a herbivore defense is well known (Williams 1954, Tanton 1962, Djamin and Pathak 1969, Rausher 1981, Coley 1983, Raupp 1985). Physical defenses that contribute to leaf toughness such as silica content are known to change following herbivory (McNaughton and Tarrants 1983, Myers and Bazely, this volume, Chapter 14).

Injured leaves were taken from shoots where each leaf had been torn transversely at intervals of 3–7 days prior to harvesting. These intervals were selected because they coincide with periods during which beetle (*P.

TABLE 1   Responses of *Salix alba* "Tristis" to Mechanical Damage

Days after Injury	Treatment	Attributes of Leaf Quality (means + standard error)			
		Nitrogen[a] (%)	Moisture[b] (%)	Toughness[c] (g)	Tannin[d] (mg)
Day 3	Uninjured	2.69 + 0.11	68.5 + 0.45	29.2 + 2.45	0.972 + 0.012
	Injured	2.89 + 0.15*	67.4 + 0.50**	36.9 + 3.15*	0.867 + 0.074
Day 7	Uninjured	2.29 + 0.12	66.7 + 0.59	39.3 + 4.35	0.666 + 0.007
	Injured	2.64 + 0.16*	66.5 + 0.59	41.2 + 4.39*	0.700 + 0.007

[a] Percentage of dry weight determined by macro-Kjeldahl as in Horwitz (1965).
[b] Percentage of total leaf weight.
[c] Grams necessary to pierce leaf as in Raupp (1985).
[d] Milligrams of tannin/100 mg leaf tissue as in Hagerman (1987) acetone extract.
* Comparison between treatments within dates significant at .05 (paired sample t test, $n = 15$).
** Comparison between treatments within dates significant at .01 (paired sample t test, $n = 15$).

versicolora) activity and defoliation were sampled in other studies (see sections 2.3 and 2.4). Control leaves were removed from shoots on the same branches. Only leaves of the preferred age class for adult beetles were analyzed (Raupp and Denno 1983).

Three measures of leaf quality differed between control and injured leaves following partial defoliation (Table 1). Nitrogen content was significantly higher in injured leaves at both time intervals. Moisture content was significantly higher in control leaves relative to injured leaves 3 days following injury. However, the difference in average moisture content was small and did not persist. Injured leaves were approximately 26% tougher than control leaves during the first time period. These differences did not persist for one week. Consistently elevated nitrogen levels suggest that injured leaves were somewhat more nutritious than nearby uninjured ones.

A similar comparison was made between uninjured leaves and nearby leaves injured naturally by the feeding activities of herbivorous insects, primarily adults of the leaf beetle *P. versicolora*. The amount of leaf area lost on naturally injured leaves amounted to about 17% and was typical for leaves attacked by *P. versicolora* (Raupp and Sadof 1989). Of nitrogen, moisture, and tannin content, and leaf toughness, the only attribute of plant quality that differed significantly between naturally injured leaves and nearby uninjured leaves was nitrogen content. As before, the percent dry weight of nitrogen was significantly greater in injured leaves (2.64 vs. 2.29, $p < .05$, paired t test). Caution must be exercised in interpreting the results of the comparison between naturally damaged leaves and controls. In this case the assignment of treatments was biased due to an implicit nonrandom selection of leaves by herbivores (Neuvonen and Haukioja 1985). Nevertheless, results of both analyses indicate that injured leaves may be a better quality food resource because of their higher nitrogen content. The ramifications of this observation on the behavior and performance of herbivores is investigated in the following sections.

2.3 Behavioral Responses of *Plagiodera versicolora* to Injured Willow Leaves

The short-term response of willow leaves to wounding has several impli-cations for mobile herbivores associated with the plant. Rapid increases in foliar nitrogen content should promote herbivory and a concentration of herbivores on nitrogen-rich leaves. Such responses to nitrogen are well doc-umented in many orders of herbivorous insects and mites (McNeill and Southwood 1978, Denno et al. 1980, Mattson 1980, Scriber and Slansky 1981, Denno 1983, Raupp and Denno 1983, Rodriguez and Rodriguez 1987, Slansky and Rodriguez 1987). In contrast, several recent investigations of herbivo-rous insects and mites demonstrate weak to very strong avoidance of injured leaves or plants (Parker 1984 , Wratten et al. 1984, Harrison and Karban 1986, Bergelson et al. 1986, Bergelson and Lawton 1988, Raupp and Sadof 1989).

A resolution of these contrasting results has been sought with the leaf beetle *P. versicolora*. This beetle was first reported in North America in 1911 (Johnson and Lyon 1988). In Maryland it feeds primarily on willows and its feeding ecology is relatively well known. Raupp and Denno (1983) found that adults and larvae of *P. versicolora* differed in their within-tree patterns of resource utilization on *S. babylonica*. Adults feed primarily on younger leaves and exhibit greater fecundity on these leaves. Larvae feed on somewhat more mature leaves. Other aspects of the bionomics of this beetle in natural settings have been described by Hood (1940) and Wade and Breden (1986).

Plagiodera versicolora is also known to respond to the secondary chem-istry of willows during host selection. Rowell-Rahier et al. (1987) reported a significant positive correlation between the phenolic glucoside content of individual willow trees and the abundance of beetles on those trees. Tah-vanainen et al. (1985) reported a preference in *P. versicolora* adults for wil-low species with moderate to low levels of phenolic glycosides. Recently, Coleman and Jones (1988) have found that exposure to an environmental stress (ozone) produces changes in poplar leaves that affect foraging deci-sions and performance of *P. versicolora* adults and larvae. The wealth of information on *P. versicolora* and its hosts makes it an ideal candidate for studying injury related phenomena.

We have investigated the behavioral response of *P. versicolora* to leaf injury under differing conditions of artificial and natural injury to leaves and at a variety of temporal scales ranging from less than 6 h to more than 1 week. Our first behavioral assay investigated the response of leaf beetles to artificially injured (torn) leaves at a very short time interval after injury. Leaves were removed from *S. alba* "Tristis" trees immediately after injury and offered to *P. versicolora* adults in binary choice tests for a period of 6 h. Under these conditions beetles did not discriminate between injured and uninjured leaves of the same age taken from nearby shoots (Raupp and Denno 1984).

A second series of laboratory bioassays investigated the response of beetles to naturally injured willow leaves. The willow leaves in these experiments were attacked by defoliators, primarily *P. versicolora* adults. The precise period of time between the induction event and behavioral assay was unknown. However, since the phyllotaxy of the leaves was controlled and the rate of leaf production for these trees was known, the induction event was estimated to have occurred from 0 to 12 days prior to the assay. Under these conditions there was a strong tendency for beetles to prefer uninjured leaves. During the course of the assay, beetles were observed almost twice as frequently on uninjured compared to injured leaves where they consumed more than twice as much leaf area (Raupp, unpublished results).

The previous experiments provided important insights into the responses of *P. versicolora* to leaf injury. However, they both had critical limitations. First, it was difficult to project the outcome of experiments conducted under laboratory conditions to responses of herbivores in the field (Risch 1985). Furthermore, in the case of the second bioassay, the generality of the conclusions is limited because of the initial bias introduced into the experimental design through the nonrandom selection of leaves (Neuvonen and Haukioja 1985).

To further clarify the response of leaf beetles to leaf injury, two sets of assays were conducted in the field. They utilized the same *S. alba* "Tristis" trees for which the attributes of leaf quality had been quantified (section 2.2). The first experiment determined the effect of artificial leaf injury on the spatial distribution of beetles placed on the trees at a controlled density from 1 to 5 days following artificial injury (Raupp and Sadof 1989). Results of the study indicated that for each of the 5 days beetles were much more abundant on uninjured compared to injured shoots (Raupp and Sadof 1989) (Fig. 1).

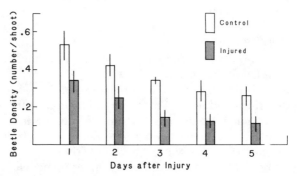

Figure 1   Abundance of *Plagiodera versicolora* beetles on uninjured (control) and injured willow shoots from 1 to 5 days following injury. Bars represent means and vertical lines represent standard errors. (From Raupp and Sadof 1989, p. 156; reprinted with permission.)

A second and similar experiment with the same trees determined the response of indigenous beetles to artificial leaf injury. As with the previous experiment, the location of beetles was observed relative to whether shoots were injured or uninjured. Observations were made 6–9 days following injury. During this time, beetles were significantly more likely to be found on uninjured compared to injured shoots (paired sample t test, $p < .037$, $n = 10$). The average total number of beetles observed on uninjured shoots was 13.9 (± 2.46, standard error) compared to 9.6 (± 2.58) observed on injured shoots. In total these studies demonstrated that elevated levels of nitrogen in injured leaves (section 2.2) did not predict the behavioral responses of beetles to injury. Furthermore, they indicate that natural and artificial defoliation produce changes in willows leaves that beetles respond to in similar ways.

2.4 Physiological Responses of *Plagiodera versicolora* to Injured Willow Leaves

Elevated nitrogen levels suggested an increase in suitability of injured willow leaves as a food source, yet the behavioral assays indicated a strong avoidance of injured leaves by adult beetles. To determine the physiological consequence of consuming previously injured leaves, we forced *P. versicolora* adults to eat two species of willow, *S. babylonica* and *S. alba* "Tristis" that were previously injured. *Salix babylonica* was used to test the effect of natural injury on the fecundity of *P. versicolora*. Leaves were injured through the feeding activities of several willow herbivores, including *P. versicolora*. The interval between initial leaf injury and the assay was not precisely known. However, because leaf phyllotaxy was controlled and leaf production rates were known, the initial injury occurred between 0 and 15 days prior to the assay. Under these conditions beetles forced to consume leaves from injured shoots experienced more than a 20% reduction in fecundity during a 1-week interval (Raupp and Denno 1984).

In a related experiment the effect of artificial injury on leaf suitability was evaluated with *P. versicolora* adults at a time interval of 4–7 days after the induction event. In this case, *S. alba* "Tristis" leaves were torn transversely. The effect of artificial injury was the same as that observed for natural injury. Beetles that consumed injured leaves experienced almost a 20% reduction in fecundity during a period of 1 week (Raupp and Denno 1984).

Thus far, discussion has centered on the effects of injury on the suitability of leaves as a food for adult leaf beetles. A question of equal interest is how injury affects the suitability of leaves as a food for the larvae of *P. versicolora*. Raupp and Denno (1984) investigated this question with leaves of *S. babylonica*. Larvae fed leaves adjacent to those injured by tearing developed slower and achieved a lower weight at pupation compared to larvae fed leaves of the same age from nearby untorn shoots. This observation led Raupp and Denno (1984) to conclude that injury to leaves reduced the suitability of nearby leaves born on the same shoot. Locally systemic responses such as this have been observed in other plants such as birches (Haukioja and Niemela 1979, Niemela et al. 1979).

These sets of studies lead to the following conclusions. Natural and artificial injuries bring about changes in the suitability of willow leaves as a food source for adults and larvae of *P. versicolora*. Despite short-term increases in foliar nitrogen levels, injured leaves are a poorer quality food than uninjured ones: Fecundity is substantially reduced when injured leaves are consumed. Adult beetles detect these changes and alter their feeding accordingly by avoiding injured leaves. The behavioral responses of larvae to leaf injury are, as yet, unknown.

3   INJURY-RELATED CHANGES IN LEAF SUITABILITY AND SPATIAL PATTERNS OF HERBIVORY

The system involving *P. versicolora* and willows provides an excellent opportunity to examine spatial aspects of herbivory because many willows exhibit indeterminant growth for extended periods. This growth habit provides an array of leaves produced on a single shoot that vary in age and suitability (Raupp and Denno 1983). A general pattern of feeding has emerged for adults and larvae of *P. versicolora*. Adults feed primarily on young leaves of willow borne near the terminals of actively growing shoots. Raupp and Denno (1983) and Raupp (1985) argued that young leaves were nutritionally superior to more mature leaves. Larvae were generally found on leaves significantly older than those consumed by adults. Raupp and Denno (1983) explained this distribution by the female's tendency to oviposit on leaves older than the ones they consumed. Possible reasons for this displacement include direct interactions between adult beetles and their young, nonrandom search by predators, and increased harshness of young leaves as ovipositional sites (Raupp and Denno 1983). These patterns of herbivory are now reexamined with respect to known responses of willows to injury.

How can injury affect the spatial pattern of herbivory caused by *P. versicolora* on willows? The most likely way is by increasing the dispersion of damage among shoots and age classes of leaves. Two lines of evidence support this contention. The first comes from an examination of willow leaves borne on shoots that were previously injured by tearing (Raupp and Sadof 1989, and section 2.3). When leaves of the preferred age class were harvested from previously injured shoots, two trends were clear. First, after leaves were injured, they were significantly less likely to be attacked by *P. versicolora* adults (Raupp and Sadof 1989). This is reflected by the greater proportion of leaves with no damage in Figure 2: Beetles avoid injured leaves. The second trend was for beetles to consume less of injured leaves compared to uninjured ones. The average amount of leaf area removed from injured leaves that were attacked was 0.34 cm^2. The average amount re-

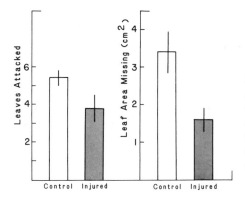

Figure 2   Numbers of leaves attacked and total leaf area missing in samples of leaves removed from uninjured (control) and injured willow shoots 6 days following injury. Bars represent means and vertical lines represent standard errors. (From Raupp and Sadof 1989, p. 156; reprinted with permission.)

moved from uninjured leaves was 0.59 cm². For *P. versicolora* there is a direct relationship between rates of leaf consumption and egg production (Raupp 1985). Therefore, to produce the same number of eggs per unit time, beetles must visit more previously injured than uninjured leaves. Thus, a consequence of this response should be that more leaves are attacked throughout the canopy of the tree. This suggestion has been made by Edwards and Wratten (1982), Edwards et al. (this volume, Chapter 9), and Silkstone (1987) and is supported by their work and that of others (Faith 1985, 1986, Raupp and Sadof 1989). However, the pattern of increased dispersion of herbivory is by no means universal. Bergelson and Lawton (1988) found no evidence that geometrid larvae avoided damaged birch leaves.

A second way in which injury-related changes in leaf suitability might affect the spatial distribution of *P. versicolora* on its hosts and resultant patterns of herbivory is through selective avoidance of leaves with a high risk of injury as oviposition sites. Injury has been shown to affect ovipositional choices of herbivorous insects in other systems. For example, Faith (1986) found the leaf miner, *Stilbosis juvantis*, selecting intact leaves in favor of damaged ones as oviposition sites. Leather et al. (1987) observed pine beauty moths, *Panolis flammea*, avoiding previously defoliated pines for oviposition. Raupp and Denno (1983) reported disjunct spatial distributions of adult and juvenile stages of *P. versicolora* with respect to leaf age class. The pattern results from adults consuming young leaves and ovipositing on more mature ones. A strong negative relationship exists between the amount of area an age class of leaves loses and the number of egg masses it receives. Because some age classes of leaves are more likely to be consumed by adult beetles, they are more likely to be a lower quality food due to induced changes for early instar larvae that are relatively immobile. By placing their eggs on leaves less likely to be injured, adult beetles reduce the risk that their relatively sessile progeny will be forced to contend with wound-induced deterioration of their host. It appears that adult beetles avoid such leaves as oviposition sites.

Injury could also affect the distribution of herbivores if leaves vary predictably in their inducibility and herbivores respond to this variation. For example, if the suitability of young willow leaves declines more so than older leaves following injury, adult leaf beetles may reduce exposure of their young to leaves of potentially poor quality by selecting more mature leaves as oviposition sites. Several studies have demonstrated greater changes in the suitability of young expanding leaves compared to older leaves following injury (Haukioja and Niemela 1979, Wratten et al. 1984, Hartley and Lawton, this volume, Chapter 5).

We examined the possibility of age-dependent responses of *S. babylonica* leaves by removing approximately 25% of the total leaf area of several willow shoots and rearing ten cohorts of five larvae on two age categories of leaves from injured shoots and nearby uninjured shoots. Leaves were removed 3–7 days following injury and larvae were raised from hatch to eclosion on young leaves (leaves 1–3 from the apical meristem) and old leaves (leaves 8–10) from both damage categories. Weights of adults produced from the four treatments did not differ. However, survivorship was significantly lower and developmental period was significantly longer for larvae on young injured leaves compared to any other treatment (Table 2). Analyses of variance revealed a significant interaction between leaf age and previous injury for both survivorship and developmental period. These results indicate a strong differential response of leaves to injury that is age dependent.

The ovipositional pattern of *P. versicolora* is consistent with the differential response of leaf age classes to injury. It appears likely that female beetles avoid young leaves as oviposition sites due to the radical decline in the suitability of young leaves following injury. Older leaves are less likely to be injured by the feeding of adult beetles and a variety of other willow herbivores (Raupp and Denno 1983). Also, older leaves appear to change less if they are injured. By ovipositing on more mature leaves, *P. versicolora* adults reduce the likelihood of direct interactions with their young which have secretions repellent to adults (Raupp et al. 1986). In addition, this

TABLE 2 Survivorship, Adult Weight, and Developmental Period of *Plagiodera versicolora* Larvae Fed Injured and Uninjured Willow Leaves from Two Age Categories (young and old)[a]

Treatment	Age	Survivorship	Weight (mg)	Developmental Period (days)
Injured	Young	0.5 + 0.26a	5.5 + 1.73a	16.8 + 1.24a
	Old	3.4 + 0.36b	5.4 + 0.70a	13.4 + 0.22b
Uninjured	Young	2.5 + 0.48b	5.8 + 0.71a	12.9 + 0.17b
	Old	3.5 + 0.53b	6.0 + 0.41a	12.9 + 0.126

[a] Means within a column that share the same letter did not differ by a Tukey's pairwise comparison test at the level of $p = .05$. Values are means and standard errors of 10 cohorts of 5 larvae per treatment.

behavior reduces the likelihood that larvae will injure young leaves which serve as a critical food resource for adults. Finally, by placing eggs on mature leaves adults may be insuring that relatively immobile larvae are not confronted with rapidly deteriorating leaves from which they must escape.

It is unknown what effect, if any, injury to mature leaves has on the suitability of young leaves as a food source for adults. Coleman and Jones (this volume, Chapter 1) argued that this effect is likely to be negligible. Nevertheless, it is plausible that injury-related changes in the suitability of young willow leaves are responsible for beetles laying eggs on older leaves and thereby spreading herbivory to age classes of leaves that would not be damaged by this herbivore otherwise.

4   INJURY-RELATED CHANGES IN LEAF SUITABILITY AND TEMPORAL PATTERNS OF HERBIVORY

In the previous section we suggest that induced changes in willow play an important role in determining spatial distributions of different life stages of *P. versicolora* and the damage associated with its feeding. We now examine the relationship between temporal patterns of leaf injury and patterns of beetle feeding. Others have suggested that herbivory at one time in the season can have important consequences for herbivores at a later time. For example, West (1985) suggested that the appearance of leaf miners late in the season was the result of indirect interactions with caterpillars whose feeding reduced foliage quality earlier in the season. Harrison and Karban (1986) reported a similar asymmetric interaction between two species of lepidoptera utilizing lupine.

The seasonal phenology of *P. versicolora* in North America was described by Hood (1940). He reported three complete generations and a partial fourth generation of *P. versicolora* on willows in Massachusetts. We studied the life history of *P. versicolora* on three *S. babylonica* trees in Maryland during the spring, summer, and fall of 1980 and 1981. In contrast to the earlier report of Hood (1940) we found the phenology of beetles to be characterized by two prominent peaks (Fig. 3). The first peak in April and May represents the return of overwintered adults to the canopies of the willows. Each year within two weeks of their appearance, adults began reproducing and eggs were observed on the leaves. These progeny produced a second cohort of adults similar in abundance to that produced by overwintered adults. Thereafter, both adults and juveniles were relatively rare, although not totally absent, for the remainder of the season.

During the late spring and early summer important changes in both the quantity and quality of willow foliage occur. Earlier studies demonstrated the importance of young willow leaves as a food for adult *P. versicolora* beetles (Raupp and Denno 1983, Raupp 1985). Mature leaves are a very poor resource for adult feeding and reproduction. Early in the season actively

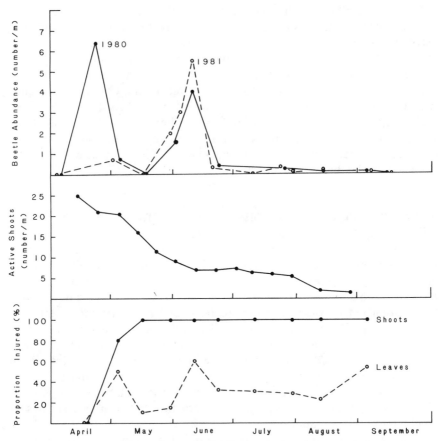

Figure 3   Abundance of leaf beetles, active willow shoots, and proportions of injured willow leaves of favored age on three willow trees. Plotted points are means of three trees. Leaf beetle abundance was observed in 1980 and 1981 and other parameters were observed in 1981.

growing shoots are extremely abundant on willow branches (Fig. 3). This creates a ubiquitous supply of young leaves for adults returning to the trees from overwintering sites. However, by late spring and early summer more than two-thirds of the shoots active at the season's beginning are no longer producing young leaves critical to the beetle's reproduction. Thus, the abundance of the beetle's resource appears to diminish with time.

The problem of diminishing resources may be further exacerbated by injury-related changes in the quality of willow leaves through time. Earlier in this chapter and elsewhere (Raupp and Denno 1984) we reviewed the evidence supporting a decline in suitability of willow leaves following injury. On a seasonal basis, willow foliage exhibits characteristic patterns of injury. We analyzed these patterns at two levels of plant organization. First, we observed patterns of injury to individual willow shoots. The existence of systemic responses of plants to injury among leaves borne on the same shoot has been observed in a variety of systems (Haukioja and Niemela 1979, Karban and Carey 1984, Karban 1985, 1986, Tallamy 1985) including willow (Rhoades 1983 a,b, Raupp and Denno 1984). We also studied patterns of injury at the individual leaf level where changes in leaf suitability following injury are well known.

An analysis of leaf injury revealed two distinct patterns (Fig. 3). First, prior to the arrival of beetles to the trees both shoots and leaves were free of defoliation. However, within one month of their arrival, adults injured one or more leaves on every shoot. Damage to individual leaves of the favored age class on actively growing shoots produced a different pattern of injury (Fig. 3). Two distinct peaks of injury to leaves were observed early in the season. These coincided with the periods of greatest abundance of adult beetles. The trough between these peaks coincided well with the period when juveniles were observed feeding on older foliage (Raupp and Denno 1983). Later in the season on the few shoots that produced leaves, the proportion of favored leaves receiving injury hovered between 20 and 40%. Only late in the season when active shoots were very rare did the proportion of injured leaves in the favored age category rise to early season levels.

What are the consequences of temporal changes of leaf injury on the suitability of willow leaves as a resource for *P. versicolora*? Clearly, by the time first-generation adults are mature they are confronted with a resource that is considerably more rare and less suitable than the resource available to their predecessors. If, on a single shoot, herbivory to a single leaf alters not only that leaf but others produced later, and if these changes persist throughout the season, then by May all food resources are lower in suitability than those present on the tree early in the season. A similar situation exists if induced responses are localized to single leaves or those in close spatial or physiological proximity. Early in the season virtually every vegetative shoot has an uninjured leaf of the favored age. In June more than 70% of these shoots are no longer producing such leaves and on the shoots still active more than 60% of the leaves have been previously injured. A seasonal decline in the suitability of willow leaves as a food source for adult beetles was previously reported (Hood 1940). Whether these changes were the result of injury-related changes in willows is unknown.

What are the possible responses of beetles to this changing spectrum of resources? One response is to consume injured leaves despite the decline in suitability. Evidence for this response comes from two sources. First, under laboratory conditions adult beetles will consume previously injured foliage (see section 2.3). Although injured leaves are less preferred than uninjured ones, they are consumed. Moreover, under laboratory conditions adults that consume injured leaves can produce offspring, although not as prolifically as those that consume uninjured ones (Raupp and Denno 1984). Second, under field conditions injured leaves will be attacked and consumed

TABLE 3 **Comparisons of Variation in *Salix babylonica* Leaves Consumed by *Plagiodera versicolora* Adults Early (April) and Later (June) in the Season**[a]

Tree	Variance		F	p
	April (*n*)	June (*n*)		
1	0.24 (20)	1.61 (17)	6.71	<.001
2	0.81 (53)	2.99 (49)	3.69	<.001
3	1.49 (80)	11.16 (90)	7.49	<.001

[a] Differences in the temporal variation in feeding site were evaluated with a variance ratio test of leaves eaten by beetles in April and June.

albeit not as frequently or completely as uninjured ones (Raupp and Sadof 1989).

Another possible strategy to cope with increasing levels of injury would be to avoid injured leaves in the most favored age categories and accept leaves in less favored age classes that either lack or may be less responsive to injury. This should be reflected in greater variation in leaf age classes consumed as the season progresses. A temporal comparison of variation in adult feeding site on three *S. babylonica* supports this contention (Table 3). Variance in leaf age classes consumed by adults was significantly greater later in the season compared to earlier when fewer favored leaves were injured.

It is tempting to speculate that injury-related declines in leaf suitability contribute to a lack of reproduction by first-generation adult beetles and the relative rarity of all life stages on *S. babylonica* after June. However, Hood (1940) reported no such trend but found, instead, multiple generations of *P. versicolora* on willows in New England. It appears that either the willows studied by previous authors differed from ours in seasonal patterns of resource quality or that other factors such as adverse temperature regimes limit the seasonal abundance of *P. versicolora* in Maryland.

5 IMPACT OF INJURY-RELATED CHANGES IN LEAVES ON HIGHER TROPHIC LEVELS

Relatively low levels of herbivory seen on individual leaves caused by the feeding of *P. versicolora* adults results from adult beetles abandoning or avoiding leaves that are deteriorating or have declined in suitability due to prior injury. An alternative hypothesis explaining the beetle's avoidance of injured leaves involves potential interactions with predators or parasitoids. Several authors have suggested or documented that vertebrate predators such as birds and parasitoids use visual and chemical cues associated with leaf damage to help them locate prey (Heinrich 1979, Heinrich and Collins 1983, Odell and Godwin 1984, Faith, this volume, Chapter 13). However, results of field experiments conducted with *P. versicolora* adults have thus far failed to reveal any significant effect of leaf damage on predator related mortality (Raupp, unpublished results). Due to their physical and chemical defenses, increased levels of movement and injury related reductions in development may have less severe consequences for the survival of *P. versicolora* than for less well-protected herbivores such as many lepidopteran species that lack physical or chemical defenses (Heinrich 1979, Lawton and McNeill 1979, Price et al. 1980, Heinrich and Collins 1983, Stamp and Bowers 1988, Raupp and Sadof 1989).

Unlike *P. versicolora*, other species of leaf beetles may benefit from injury-related changes in levels of phytochemicals found in their host. Clausen et al. (this volume, Chapter 3) report rapid short-term increases in the levels of several phenolic glycosides following injury to poplar leaves. One of these glycosides, salicortin, is believed to be converted to salicin and a more toxic by-product in the insect gut (Pasteels et al. 1989, Clausen et al. this volume, Chapter 3). For the larvae of many species of leaf beetles that feed on willows and poplars, salicin and to some extent salicortin are the precursors of a potent allelochemical, salicylaldehyde, which is secreted by the larvae to ward off antagonists (Pasteels et al. 1989 and references therein). Phenolic glycoside levels play an important role in host selection of several leaf beetles, one of which utilizes salicin as a precursor of its defensive secretion (Pasteels et al. 1989). If herbivory increases levels of phenolic glycosides that serve as precursors of defensive secretions and no other phytochemicals change to countermand this effect, then phytochemical induction could clearly benefit these adapted beetle species and ultimately increase levels of herbivory (Pasteels et al. 1989). Similarly, Smiley et al. (1985) found increased levels of herbivory associated with an adapted leaf beetle on willow clones containing high levels of salicin. Smiley et al. (1985) suggest that where adapted specialists such as the beetles utilizing salicin are abundant and inflict high levels of injury, selection will mitigate the presence of these defensive precursors in plants. The same argument holds for willows and poplars that utilize these compounds as inducible phytochemicals.

6 SUMMARY

We have reviewed the evidence for injury-related changes in the closely related weeping willows, *S. babylonica* and *S. alba* "Tristis," through phytochemical analyses and experiments with an associated leaf beetle, *P. versicolora*. Our studies demonstrate that willow leaves change phytochemically following both natural and artificial injury. However, phytochemical changes, particularly increases in foliar nitrogen levels, do not adequately predict behavioral and physiological responses of *P. versicolora* to injured leaves. Injured leaves are a less-preferred food resource for beetles and less suitable if consumed. The response of willow leaves to injury is locally systemic, at least at the level of adjacent leaves, occurs within one day of injury, and is leaf age dependent. Young, unexpanded leaves appear more responsive to injury than mature ones.

The responses of willows to injury help explain spatial and temporal patterns of *P. versicolora* and its associated herbivory that are observed in the field. Injury-related changes in willow leaves help to explain heretofore ambiguous patterns of oviposition by *P. versicolora*. The utilization of young leaves as feeding sites and mature leaves as oviposition sites insures that relatively sessile larvae encounter leaves less likely to be injured by adults and with lower levels of response to injury. Because adult beetles utilize different age classes of leaves for food and oviposition, herbivory is more dispersed than it would be if beetles consumed and oviposited on only young leaves. Another outcome of this resource partitioning is the reduction of antagonistic interspecific interactions between adults and larvae mediated by the host plant (Price 1980, Raupp and Denno 1983, Raupp et al. 1986, Pasteels et al. 1989). Temporal patterns of leaf injury and production indicate a food resource that declines dramatically in quality and quantity seasonally. At certain times uninjured leaves of the favored age classes may be difficult to find. One result of this relative scarcity of high-quality food appears to be an expansion of the age classes of leaves utilized by adult beetles as the season progresses, thereby increasing the dispersion of herbivory among leaves of different age. At present there is no indication that responses of *P. versicolora* to injured willow leaves are mediated through higher trophic levels. This may not be the case for other leaf beetles such as those utilizing phytochemicals as precursors for allelochemicals. It is unknown if the responses of *P. versicolora* to injury-related changes in *Salix* are idiosyncratic or if they represent a general pattern in the Chrysomelidae.

ACKNOWLEDGMENTS

We thank Robert F. Denno, Douglas W. Tallamy, and Pedro Barbosa for stimulating discussions and helpful comments on earlier drafts of this manuscript. This is Contribution No. 8094 of the Maryland Agricultural Experiment Station.

REFERENCES

Baldwin, I. T., and J. C. Schultz. 1985. Rapid changes in tree leaf chemistry induced by damage: evidence for communication between plants. *Science* 221:277–278.

Bergelson, J. S., and J. H. Lawton. 1988. Does foliage damage influence predation on the insect herbivores of birch? *Ecology* 69:434–445.

Bergelson, J., S. Fowler, and S. Hartley. 1986. The effects of foliage damage on casebearing moth larvae, *Coleophora seratella*, feeding on birch. *Ecol. Entomol.* 11:241–250.

Bernays, E. A. 1981. Plant tannins and insect herbivores: an appraisal. *Ecol. Entomol.* 6:353–360.

Bryant, J. P., G. D. Wieland, T. Clausen, and P. Kuropat. 1985. Interactions of snowshoe hares and feltleaf willow (*Salix alaxensis*) in Alaska. *Ecology* 66:1564–1573.

Bryant, J., K. Danell, F. Provenza, P. Reichardt, and T. Clausen. 1991. Effects of mammal browsing on the chemistry of deciduous woody plants. In D. W. Tallamy and M. J. Raupp (eds.), *Phytochemical Induction by Herbivores*. New York: Wiley, pp. 135–154.

Chapin, F. S., III, J. P. Bryant, and J. F. Fox. 1985. Lack of induced chemical defense in juvenile Alaskan woody plants in response to simulated browsing. *Oecologia (Berlin)* 67:457–459.

Clausen, T., J. P. Bryant, P. B. Reichardt, and R. A. Werner. 1991. Long-term and short-term induction in quaking aspen: related phenomena? In D. W. Tallamy and M. J. Raupp (eds.), *Phytochemical Induction by Herbivores*. New York: Wiley, pp. 71–84.

Coleman, J. S., and C. G. Jones. 1988. Plant stress and insect performance: cottonwood, ozone and a leaf beetle. *Oecologia* 76:57–61.

Coleman, J. S., and C. G. Jones. 1991. A phytochemical perspective of phytochemical induction by herbivores. In D. W. Tallamy and M. J. Raupp (eds.), *Phytochemical Induction by Herbivores*. New York: Wiley, pp. 3–46.

Coley, P. D. 1983. Herbivory and defensive characteristics of tree species in a lowland tropical forest. *Ecol. Monogr.* 53:209–233.

Denno, R. F. 1983. Tracking variable host plants in space and time. In R. F. Denno and M. S. McClure (eds.), *Variable Plants and Herbivores in Natural and Managed Systems*. New York: Academic, pp. 291–341.

Denno, R. F., M. J. Raupp, D. W. Tallamy, and C. F. Reichelderfer. 1980. Migration in heterogeneous environments: differences in habitat selection between wingforms of the dimorphic planthopper, *Proklesia marginata* (Homoptera:Delphacidae). *Ecology* 61:859–867.

Djamin, A., and M. D. Pathak. 1969. Role of silica in resistance to Asiatic rice borer *Chilo suppressalis* (Walker) in rice varieties. *J. Econ. Entomol.* 60:347–451.

Edwards, P. J., and S. D. Wratten. 1982. Wound-induced changes in palatability of birch (*Betula pubescens* Ehrh. ssp. *pubescens*). *Am. Natur.* 119:816–818.

Edwards, P. J., S. D. Wratten, and R. M. Gibberd. 1991. The impact of inducible phytochemicals on grazing behavior of insect herbivores. In D. W. Tallamy and M. J. Raupp (eds.), *Phytochemical Induction by Herbivores*. New York: Wiley, pp. 205–222.

Ericsson, A., C. Hellquist, B. Langtrom, S. Larsson, and O. Tenow. 1985. Effects on growth of simulated and induced shoot pruning by *Tomicus pinipedra* as related to carbohydrate and nitrogen dynamics in Scots pine. *J. Appl. Ecol.* 22:105–124.

Faith, S. H. 1985. Host selection by leaf miners: interactions among three trophic levels. *Ecology* 66:870–875.

Faith, S. H. 1986. Indirect interactions between temporally separated herbivores mediated by the host plant. *Ecology* 67:479–494.

Faith, S. H. 1991. Inducible oak responses: interactions among oak folivores. In D. W. Tallamy and M. J. Raupp (eds.), *Phytochemical Induction by Herbivores.* New York: Wiley, pp. 293–323.

Feeny, P. P. Plant apparency and chemical defense. 1976. *Rec. Adv. Phytochem.* 10:1–40.

Fowler, S. V., and M. MacGarvin. 1986. The effects of leaf damage on the performance of insect herbivores on birch, *Betula pubescens. J. Anim. Ecol.* 55:565–573.

Hagerman, A. E. 1987. Radial diffusion method for determining tannin in plant extracts. *J. Chem. Ecol.* 13:437–449.

Harrison, S., and R. Karban. 1986. Behavioral responses of spider mites (*Tetranychus urticae*) to induced resistance of cotton plants. *Ecol. Entomol.* 11:181–188.

Hartley, S. S., and J. H. Lawton. 1991. The biochemical basis and significance of rapidly induced changes in birch. In D. W. Tallamy and M. J. Raupp (eds.), *Phytochemical Induction by Herbivores.* New York: Wiley, pp. 105–132.

Haukioja, E., and S. Hahimaki. 1985. Rapid wound-induced resistance in white birch (*Betula pubescens*) foliage to the geometrid *Eppirita autumnata*: a comparison of trees within and outside the outbreak range of the moth. *Oecologia* 65:223–228.

Haukioja, E., and S. Neuvonen. 1985. Induced long-term resistance of birch foliage against defoliators: defense or incidental? *Ecology* 66:1303–1308.

Haukioja, E., and P. Niemela. 1977. Retarded growth of geometrid larva after mechanical damage to leaves of its host tree. *Ann. Zool. Fenn.* 14:48–52.

Haukioja, E., and P. Niemela. 1979. Birch leaves as a resource for herbivores: seasonal occurrence of increased resistance in foliage after mechanical damage of adjacent leaves. *Oecologia* 39:151–159.

Haukioja, E., J. Suomela, and S. Neuvonen. 1985. Long-term inducible resistance in birch foliage: triggering cues and efficacy on a defoliator. *Oecologia* 65:363–369.

Heinrich, B. 1979. Foraging strategies of caterpillars: leaf damage and possible predator avoidance strategies. *Oecologia (Berlin)* 42:325–337.

Heinrich, B., and S. L. Collins. 1983. Caterpillar leaf damage and the game of hide and seek with birds. *Ecology* 64:592–602.

Hood, C. E. 1940. Life history and control of the imported willow leaf beetle. *USDA Circular. No. 572.*

Horwitz, W. 1965. *Official Method of Analysis of the Association of Agricultural Chemists.* Washington, DC: Association of Official Agricultural Chemists.

Hough, J. A., and D. Pimentel. 1978. Influence of host foliage on development, survival, and fecundity of the gypsy moth. *Environ. Entomol.* 7:97–102.

Johnson, W. T., and H. H. Lyon. 1988. *Insects That Feed on Trees and Shrubs.* Ithaca, NY: Cornell University Press.

Karban, R. 1985. Resistance against spider mites in cotton induced by mechanical abrasions. *Entomol. Exp. Applic.* 37:137–141.

Karban, R. 1986. Induced resistance against spider mites in cotton: field verification. *Entomol. Exp. Applic.* 42:239–242.

Karban, R. 1987. Environmental conditions affecting the strength of induced resistance against mites in cotton. *Oecologia* 73:414–419.

Karban, R., and J. R. Carey. 1984. Induced resistance of cotton seedlings to mites. *Science* 223:53–54.

Landsberg, T., and F. R. Wylie. 1983. Water stress, leaf nutrients and defoliation: a model of dieback in rural eucalyptus. *Aust. J. Ecol.* 8:27–41.

Lawton, J. H., and S. McNeill. 1979. Between the devil and the deep blue sea: on the problems of being a herbivore. *Br. Ecol. Soc. Symp.* 20:223–244.

Leather, S. R., A. D. Watt, and G. I. Forrest. 1987. Insect-induced chemical changes in young lodgepole pine (*Pinus contorta*). The effect of previous defoliation on oviposition, growth and survival of the pine beauty moth, *Panolis flammea*. *Ecol. Entomol.* 12:275–281.

Mattson, W. J. 1980. Herbivory in relation to plant nitrogen content. *Annu. Rev. Ecol. Syst.* 11:119–162.

Mattson, W. J., and S. R. Palmer. 1987. Changes in levels of foliar minerals and phenolics in trembling aspen, *Populus tremuloides*, in response to artificial defoliation. In W. J. Mattson, J. Levieux, and C. Bernard-Dagan (eds.), *Mechanisms of Woody Plant Defenses against Insects: A Search for Pattern*. New York: Springer, pp. 157–169.

Mattson, W. J., and J. M. Scriber. 1987. Nutritional ecology of insect folivores of woody plants: nitrogen, water, fiber, and mineral considerations. In F. Slansky and J. G. Rodriguez (eds.), *Nutritional Ecology of Insects, Mites, Spiders, and Related Invertebrates*. New York: Wiley, pp. 105–146.

Mattson, W. J., R. K. Lawrence, R. A. Haack, D. A. Herms, and P. J. Charles. 1988. Defensive strategies of woody plants against different insect feeding guilds in relation to plant ecological strategies and intimacy of association with insects. In W. J. Mattson, J. Levieux, and C. Bernard-Dagan (eds.), *Mechanisms of Woody Plant Defenses against Insects: A Search for Pattern*. New York: Springer, pp. 3–38.

McNaughton, S. J., and J. L. Tarrants. 1983. Grass leaf silicification: natural selection for an inducible defense against herbivores. *Proc. Nat. Acad. Sci. U.S.A.* 80:790–791.

McNeill, S., and T. R. E. Southwood. 1978. Role of nitrogen in the development of insect-plant relationships. In J. B. Harborne (ed.), *Biochemical Aspects of Plant and Animal Coevolution*. New York: Academic, pp. 77–98.

Moran, N., and W. E. Hamilton. 1980. Low nutritive quality as defense against herbivores. *J. Theoret. Biol.* 86:247–254.

Myers, J. H., and Bazely, D. 1991. Thorns, spines, prickles, and hairs: are they stimulated by herbivory and do they deter herbivores? In D. W. Tallamy and M. J. Raupp (eds.), *Phytochemical Induction by Herbivores*. New York: Wiley, pp. 325–344.

Myers, J. H., and K. S. Williams. 1987. Lack of short or long term inducible defenses in the red alder-western tent caterpillar system. *Oikos* 48:73–78.

Neuvonen, S., and K. Danell. 1987. Does browsing modify the quality of birch foliage for *Epirrita autumnata* larvae? *Oikos* 49:156–160.

Neuvonen, S., and E. Haukioja. 1984. Low nutrient quality as defense against herbivores: induced responses in birch. *Oecologia* 63:71–74.

Neuvonen, S., and E. Haukioja. 1985. How to study induced plant resistance? *Oecologia* 66:456–457.

Niemela, P., E. Aro, and E. Haukioja. 1979. Birch leaves as a resource for herbivores: damage-induced increase in leaf phenols with trypsin-inhibiting effects. *Rep. Kevo Sub. Res. Sta.* 15:37–40.

Odell, T. M., and P. A. Godwin. 1984. Host selection by *Belpharipa pratensis*, a tachinid parasite of gypsy moth. *J. Chem. Ecol.* 10:311–320.

Parker, M. A. 1984. Local food depletion and the foraging behavior of a specialist grasshopper, *Hesperotettix viridis*. *Ecology* 65:824–835.

Pasteels, J. M., M. Rowell-Rahier, and M. J. Raupp. 1989. Plant derived defense in chrysomelid beetles. In P. Barbosa and D. K. Letourneau (eds.), *Novel Aspects of Insect–Plant Interactions*. New York: Wiley, pp. 231–272.

Piene, H., and K. Percy. 1984. Changes in needle morphology, anatomy, and mineral content during the recovery of protected balsam fir trees initially defoliated by spruce budworm. *Can. J. For.* 14:238–245.

Price, P. W., C. E. Bouton, P. Gross, B. A. McPheron, J. N. Thompson, and A. E. Weis. 1980. Interactions among three trophic levels: influence of plants on interactions between herbivores and natural enemies. *Annu. Rev. Ecol. Syst.* 11:41–65.

Raupp, M. 1985. Effects of leaf toughness on mandibular wear of the leaf beetle, *Plagiodera versicolora*. *Ecol. Entomol.* 10:73–79.

Raupp, M. J., and R. F. Denno. 1983. Leaf age as a predictor of herbivore distribution and abundance. In R. F. Denno and M. S. McClure (eds), *Variable Plants and Herbivores in Natural and Managed Systems*. New York: Academic, pp. 91–124.

Raupp, M. J., and R. F. Denno. 1984. The suitability of damaged willow leaves as food for the leaf beetle, *Plagiodera versicolora*. *Ecol. Entomol.* 9:443–448.

Raupp, M. J., and C. S. Sadof. 1989. Behavioral responses of a leaf beetle to injury-related changes in its salicaceous host. *Oecologia* 80:154–157.

Raupp, M. J., F. R. Milan, P. Barbosa, and B. Leonhardt. 1986. Methylcyclopentanoid monoterpenes mediate interactions among insect herbivores. *Science* 232:1408–1409.

Rausher, M. D. 1981. Host plant selection by Battus philenor butterflies: the roles of predation, nutrition and plant chemistry. *Ecol. Monogr.* 51:1–20.

Rhoades, D. F. 1983a. Responses of alder and willow to attack by tent caterpillars and fall webworms: evidence for pheremonal sensitivity of willows. In P. A. Hedin (ed.), *Plant Resistance to Insects*. Washington, DC: American Chemical Society, pp. 55–68.

Rhoades, D. F. 1983b. Herbivore population dynamics and plant chemistry. In R. F. Denno and M. S. McClure (eds.), *Variable Plants and Herbivores in Natural and Managed Systems*. New York: Academic, pp. 155–220.

Rhoades, D. F. 1985. Offensive–defensive interactions between herbivores and plants: their relevance in herbivore population dynamics and ecological theory. *Am. Natur.* 125:205–238.

Risch, S. J. 1985. Effects of induced chemical changes on feeding preference test. *Entomol. Exp. Applic.* 39:81–84.

Rodriguez, J. G., and L. D. Rodriguez. 1987. Nutritional ecology of phytophagous mites. In F. Slansky Jr. and J. G. Rodriquez (eds.), *Nutritional Ecology of Insects, Mites, Spiders and Related Invertebrates.* New York: Wiley, pp. 177–208.

Rowell-Rahier, M., P. H. Soetens, and J. M. Pasteels. 1987. Influence of phenol-glucoisides on the distribution of herbivores on willows. In V. Labeyrie, G. Fabres, and D. Lachaise (eds.), *Insects–Plants.* Dordrecht: Junk, pp. 91–95.

Santamour, F. S., Jr., and A. J. McArdle. 1988. Cultivars of *Salix babylonica* and other weeping willows. *J. Arboric.* 14:180–184.

Schultz, J. 1983. Impact of variable plant defensive chemistry on susceptibility of insects to natural enemies. In P. A. Hedin (ed.), *Plant Resistance to Insects. American Chemical Society Symposium Number 208.* Washington, DC: American Chemical Society, pp. 37–54.

Scriber, J. M., and F. Slansky. 1981. The nutritional ecology of immature insects. *Annu. Rev. Entomol.* 26:183–211.

Silkstone, R. E. 1987. The consequences of leaf damage for subsequent insect grazing on birch (*Betula* spp.). *Oecologia* 74:149–152.

Slansky, F., Jr., and J. G. Rodriguez. 1987. *Nutritional Ecology of Insects, Mites, Spiders, and Related Invertebrates.* New York: Wiley.

Slansky, F., Jr., and J. M. Scriber. 1985. Food consumption and utilization. G. A. Kerkut and L. I. Gilbert (eds.), *Comprehensive Insect Physiology, Biochemistry, and Pharmacology.* Oxford: Pergamon, pp. 87–163.

Smiley, J. T., J. M. Horn, and N. E. Rank. 1985. Ecological efffects of salicin at three trophic levels: new problems from an old adaptation. *Science* 229:649–650.

Stamp, N. E., and M. D. Bowers. 1988. Direct and indirect effects of predatory wasps (*Polistes* sp: Vespidae) on gregarious caterpillars (*Hemileuca lucina*:Saturniidae). *Oecologia* 75:619–624.

Svejcar, T., and S. Christiansen. 1987. Grazing effects on water relations of caucasian bluestem. *J. Range Manag.* 40:15–18.

Tahvanainen, J., R. Julkunenen-Tiitto, and R. Kettumen. 1985. Phenolic glycosides govern the food selection pattern of willow feeding leaf beetles. *Oecologia* 67:52–56.

Tallamy, D. 1985. Squash beetle feeding behavior: an adaptation against induced cucurbit defenses. *Ecology* 66:1574–1579.

Tanton, M. T. 1962. The effects of leaf 'toughness' on the feeding of larvae of the mustard beetle. *Entomol. Exp. Applic.* 5:74–78.

Valentine, H. T., W. E. Wallner, and P. M. Wargo. 1983. Nutritional changes in host foliage during and after defoliation, and the relation to the weight of gypsy moth pupae. *Oecologia* 57:298–302.

Wade, M. J., and F. Breden. 1986. Life history of natural populations of the Imported willow leaf beetle, *Plagiodera versicolora* (Coleoptera:Chrysomelidae). *Ann. Entomol. Soc. Am.* 79:73–79.

Wagner, M. R. 1987. In W. J. Mattson, J. Levieux, and C. Bernard-Dagan (eds.), Induced Defenses in Ponderosa Pine against Defoliating Insects. *Mechanisms of Woody Plant Defenses Against Insects: Search for Pattern.* New York: Springer, pp. 141–155.

Wagner, M. R., and P. D. Evans. 1985. Defoliation increases nutritional quality and allelochemics of pine seedlings. *Oecologia* 67:235–237.

Wallner, W. E., and G. S. Walton. 1979. Host defoliation: a possible determinant of gypsy moth population quality. *Ann. Ent. Soc. Am.* 72:62–67.

Wellings, P. W., and A. F. G. Dixon. 1987. Sycamore aphid numbers and population density. III. The role of aphid-induced changes in plant quality. *J. Anim. Ecol.* 56:161–170.

West, C. 1985. Factors underlying the late seasonal appearance of the lepidopteran leaf-mining guild on oak. *Ecol. Entomol.* 10:111–120.

Williams, L. H. 1954. The feeding habits and food preference of Acrididae and the factors which determine them. *Trans. R. Entomol. Soc. London* 105:423–454.

Wratten, S. D., P. J. Edwards, and I. Dunn. 1984. Wound-induced changes in the palatability of *Betula pubescens* and *B. pendula*. *Oecologia* 61:372–375.

CHAPTER 9

THE IMPACT OF INDUCIBLE PHYTOCHEMICALS ON FOOD SELECTION BY INSECT HERBIVORES AND ITS CONSEQUENCES FOR THE DISTRIBUTION OF GRAZING DAMAGE

P. J. EDWARDS, S. D. WRATTEN, and R. M. GIBBERD
Department of Biology, University of Southampton, Southampton S09 3TU, United Kingdom

1   Introduction
2   Effects of damage on plant palatability to insects
    2.1   Laboratory bioassays of leaf palatability
    2.2   Effects of damage on leaf palatability: between-leaf effects
    2.3   Effects of localized damage on within-leaf grazing patterns
3   Field bioassays of between-leaf effects of damage
4   Effects of leaf damage on insect foraging: an experiment
    4.1   Movement between leaves
    4.2   Feeding damage
5   Discussion
    References

1   INTRODUCTION

The topic of phytochemical induction embraces a wide range of biochemical and physiological effects as this volume so clearly illustrates. It encompasses phenomena that operate at a wide range of spatial and temporal scales, from local chemical changes, which appear within a few minutes following damage, to long-term changes in the chemistry of foliage, which become apparent in the years following an insect outbreak (Edwards and Wratten 1983). In understanding the ecological significance of phytochemical induction and, in particular, its possible defensive role, the diversity of plant responses should not be regarded as different examples of the same ecological phenomenon. In this chapter we are concerned with rapidly induced chemical changes that affect the food quality of foliage, and the way such changes affect the food selection and feeding behavior of insects. Our emphasis is upon deciduous trees, notably birch (*Betula pubescens* and *B. pendula*), alder (*Alnus glutinosa*), and hawthorn (*Crataegus monogyna*), for which we now have the most detailed information about the effects of damage on insect feeding. In the discussion we consider the possible defensive role of rapid wound-induced effects.

2   EFFECTS OF DAMAGE ON LEAF PALATABILITY TO INSECTS

2.1   Laboratory Bioassays of Leaf Palatability

Although workers such as Haukioja (1982) and Haukioja and Niemela (1976, 1977) have conducted laboratory experiments in which insect larvae have been reared on damaged leaf tissue, this section will be concerned only with choice experiments. The simplest of these is typified by the work of Edwards and Wratten (1982), in which disks of birch leaves (*Betula pubescens* ssp. *pubescens*) were cut from damaged leaves or those adjacent to them. These disks (cut 4–17 days after artificial damage to leaves in the field) were placed in choice experiments in humid polystyrene boxes, and the polyphagous snail, *Helix aspersa*, was added as the bioassay organism. Although significant reductions in palatability of leaf disks (proportion of disk eaten with respect to that of controls) were demonstrated following damage, the experiments were limited by the use of a noninsect herbivore and by the use of leaf disks; there is a possibility that chemical changes following the cutting of disks could influence any demonstrated differences between control and experimental leaves. These early experiments did, however, make a connection between herbivore choice and performance when confined on damaged and undamaged material (cf. Haukioja and Niemela 1977).

The temporal and spatial effects of leaf damage on larval feeding choice were investigated further in a bioassay using the armyworm, *Spodoptera littoralis*, as a test animal (Edwards and Wratten 1985) and tomato, a known host of *Spodoptera* (Bishara 1934). Strong effects of damage in reducing palatability could be detected within 8 h on damaged leaves, and within 24 h on adjacent but undamaged leaves. These effects persisted for at least 7 days, and leaves above and below those damaged were affected. There was up to a ninefold reduction in the leaf area consumed. Evidence for the possible behavioral basis of the effects came from the observation that the pattern of feeding on previously damaged leaves commonly revealed numerous small holes, indicating that feeding had been attempted but was soon aborted. These early observations led to more behavioral work described in section 3 of this chapter.

2.2   Effects of Damage on Leaf Palatability: Between-Leaf Effects

We have used the bioassay technique (Edwards et al. 1986) to investigate effects of damage upon leaf palatability for a wide range of trees and herbs. In these experiments relatively high levels of damage were inflicted with scissors (without removing any leaf area), and both the damaged leaves and undamaged leaves in the same shoot were tested against controls. In a study of 16 broad-leaved deciduous tree species, 10 species exhibited significantly reduced palatability of damaged leaves in bioassays using *Spodoptera littoralis*. In six species there was a significant reduction in palatability on undamaged adjacent leaves. In these experiments no induced reduction in palatability was found in *Betula pendula*, though such an effect was demonstrated in other experiments (Wratten et al. 1984) using different individual trees. Overall, the results suggest that wound-induced reduction in palatability is a relatively common, albeit variable, phenomenon. The bioassay technique has also been used to study in detail the temporal patterns of induced changes in palatability in four species, *B. pendula*, *B. pubescens*, *A. glutinosa*, and *C. monogyna*. The results are summarized in this section.

The work on birch (both *B. pendula* and *B. pubescens* (Wratten et al. 1984) used both larvae of *S. littoralis* (an exotic species) and *Orgyia antiqua* (a native polyphagous moth), which were offered choices between whole leaves in the laboratory. Leaves of birch were damaged artificially in April, June, and August. In the April series of experiments (which used *Spodoptera* as the test animal) palatability in both species was significantly lower than in controls in both damaged and adjacent leaves after 6 h, and the difference remained detectable for up to two months. Reductions in palatability were less clear in the June series of experiments (which used *Orgyia* as the test animal), and there was no detectable effect in the August series. Soluble tannin levels also changed markedly in the April series of experiments, but these changes did not closely parallel changes in palatability. These experiments thus reveal that (1) the induced response is rapidly transmitted to nearby leaves and (2) the strength of the induced response is greatest at the beginning of the season and thereafter declines.

Figure 1 summarizes the results of a similar series of bioassay experiments conducted in early May with alder and hawthorn in which the test animal was again *S. littoralis* (Gibberd 1987). On alder (Fig. 1*a*) both damaged leaves and adjacent undamaged leaves became significantly less palatable than controls within 24 h, though the effect was strongest in the damaged leaves. The strongest reduction in palatability in both classes of leaves was recorded after 2 days and there was no detectable effect after 14 days. Hawthorn (Fig. 1*b*) also showed significantly reduced palatability after 24 h with a very strong effect in the damaged leaves. The reduction in palatability of adjacent leaves increased gradually and was greatest at 7 days after damage. Figure 2 shows the relative palatability of leaves at 2 and 14 days after damage for a series of bioassays conducted with alder and hawthorn at different times

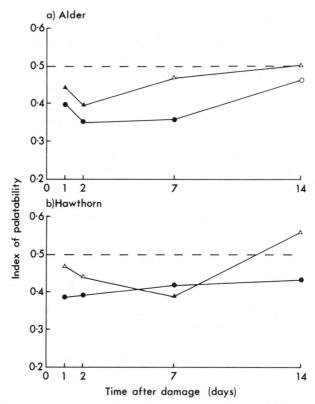

Figure 1 Changes in the palatability of damaged leaves (●) and adjacent undamaged leaves (▲) of (*a*) *Alnus glutinosa*, and (*b*) *Crataegus monogyna* following artificial damage using larvae of *Spodoptera littoralis* as test animals. The palatability index is the ratio of the area of the experimentally treated leaves consumed to the total area consumed of both treatment and control leaves (mean values for all replicates following arc sine transformation). A value of less than 0.5 indicates that the treatment leaves are eaten less than the control leaves. Each point is the mean value for 24–60 replicates. An open symbol indicates no significant difference from the control in the amount consumed; a solid symbol indicates a difference significant at $p < .05$ (*t* test on arc sine transformed data).

of year. Both species exhibited seasonal trends similar to those recorded for birch. In both there was a strong reduction in palatability after 48 h at the beginning of the growing season, but the response declined and was not detectable by late July. The strength of the induced response was always greater in the damaged leaves than in undamaged adjacent leaves, though both classes of leaves showed the same trends.

2.3   Effects of Localized Damage on within-Leaf Grazing Patterns

In the experiments described above a relatively high level of experimental damage was inflicted upon the foliage so that the treatment simulated the

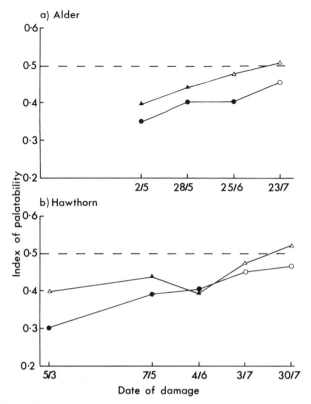

Figure 2   The palatability of damaged leaves (●) and adjacent undamaged leaves (▲) of (*a*) *Alnus glutinosa*, and (*b*) *Crataegus monogyna* at 2 days following damage, for leaves damaged at different times of year. See caption to Figure 1 for details of the palatability index and significance testing.

level of wounding that might occur during an outbreak of an insect population. However, it is of considerable interest to discover whether phytochemical induction is a response only to outbreak conditions, or whether very low levels of damage such as typically occur in deciduous woodland may also affect the quality of foliage for insects. To investigate this question, Gibberd et al. (1988) punched single small holes (4 or 5 mm diameter) on one side of the lamina of leaves of hawthorn (*C. monogyna*), alder (*A. glutinosa*), and birch (*B. pendula*) in the field. At a range of intervals following damage, single leaves were offered to *Spodoptera* larvae in bioassays in which the distribution of subsequent grazing was recorded by eye independently by two observers. The extent of grazing at the cut edge of the initial hole, within a "halo" a few millimeters around it, and on the side of the leaf bearing the initial hole was compared in each case with a matched site on the control (undamaged) side of the leaf (Fig. 3).

Bioassays conducted in May revealed that within 1 day, acceptability of

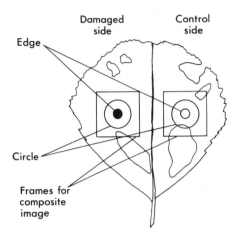

Figure 3 Diagram of an alder leaf to show the positions of the hole originally punched (black), and its control on the opposite side. Grazing was scored (1) at the cut edges of the 5-mm holes, (2) within the 14-mm circle, and (3) for each side of the leaf.

the damaged side of leaves of alder and birch was significantly reduced compared with the undamaged side. The most extreme reductions in acceptability occurred in the immediate vicinity of the damage in both species. These effects persisted for at least 7 days in birch and at least 14 days in alder. On hawthorn, grazing levels were significantly reduced on the damaged side of the leaf within 6 h following damage, and the effect persisted for at least 14 days. When the experiments were repeated in August, the effects were less pronounced but significant reductions in grazing were detected on the damaged side of alder leaves 14 days after damage and similarly on hawthorn leaves 2 days following damage.

These experiments suggest that even small amounts of damage may cause at least local changes in food quality. They provide some support for the earlier suggestion (Edwards and Wratten 1983) that induced changes caused by insect feeding affect the foraging behavior of insects and thus may affect the subsequent patterns of insect feeding damage within the canopy.

Although these experiments revealed how local leaf damage can affect larval food choice, they were limited by both the artificial imposition of spatial scales at which the distribution of grazing was scored, and by the limitations of a visual assessment of the extent of grazing damage. For these reasons Edwards and Wratten (1987) working with alder, and Croxford et al. (1989) using cotton and soybean quantified the within-leaf distribution of grazing using a computer-based image analysis system. An image was generated by superimposing the silhouettes of a large number of grazed leaves to provide what is in effect a contour map of levels of grazing in relation to the initial artificial damage (Plate 1). For the three species studied using this technique the images clearly revealed a zone surrounding the initial damage which had been avoided by insects. They thus lent support to the concept of a "halo" of reduced palatability which spreads out from the site of damage (Edwards and Wratten 1983). Much more work is needed to confirm whether

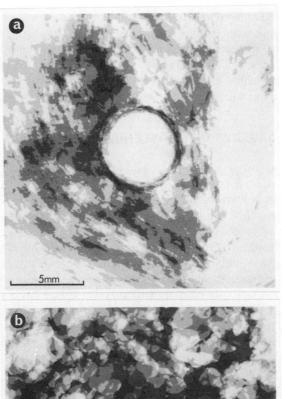

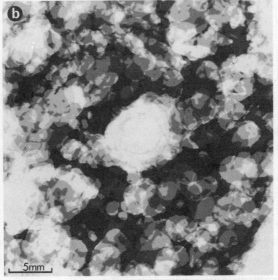

Plate 1 Effects of single holes (5 mm diameter) punched in leaves of birch (*Betula pendula*) and alder (*Alnus glutinosa*) upon subsequent feeding by larvae of *Spodoptera littoralis*. The images represent the palatability index (see caption to Fig. 1) at every picture point (pixel) on the damaged side, in each case treating the corresponding pixel on the undamaged side of the leaf as the control. The least palatable parts of the leaf following damage are shown by the darkest gray. (*a*) Image (shown as the LHS of a leaf) based on 40 birch leaves bioassayed 96 h after damage. Successively deeper shades of gray represent palatability indices of ≥ 0.5, $0.475 < 0.5$, $0.45 < 0.475$, $0.425 < 0.45$, $0.40 < 0.425$, respectively; (*b*) image (shown as RHS of a leaf) based on 36 alder leaves bioassayed 48 h after damage. Successively deeper shades of gray represent palatability indices of ≥ 0.5, $0.4 < 0.5$, $0.3 < 0.4$, $0.2 < 0.3$, $0.1 < 0.2$, respectively.

211

such effects are widespread or ecologically important. Overall the data from feeding trials present a confusing picture (Hartley and Lawton, this volume, Chapter 5), and some experiments have not shown any significant effects of wounding upon subsequent patterns of grazing damage.

3 FIELD BIOASSAYS OF BETWEEN-LEAF EFFECTS OF DAMAGE

Laboratory bioassays can be criticized because of the artificial nature of the choice with which a confined insect is presented, and because the plants tested are often not natural hosts of the experimental animal. Field bioassay experiments in which the influence of damage upon feeding by the natural community of phytophagous insects are therefore particularly useful. Fowler and Lawton (1985) described a field experiment with birch (*B. pubescens* ssp. *pubescens*) in which trees received different levels of artificial defoliation and were then left for 4 weeks while they were grazed by the natural insect fauna. In that experiment they found no evidence that damage affected the level of subsequent insect attack.

In the experiment of Fowler and Lawton (1985) the defoliation treatments were applied to entire trees, and the effects on insect grazing were scored for the trees as a whole. Edwards and Wratten (1985) described an experiment conducted at a smaller scale in which various levels of damage were inflicted on individual leaves, and subsequent grazing to those leaves was recorded. Holes of 1 mm diameter were punched in young but fully expanded leaves of birch (*B. pubescens*) and hazel (*Corylus avellana*). The leaves were collected in late summer and the amount of subsequent grazing was assessed by counting the number of holes made by insects. There was a highly significant negative correlation for both birch ($r^2 = .94$, $p < .001$) and hazel ($r^2 = .89$, $p < .001$) between the number of subsequent attacks and the extent of initial damage.

Silkstone (1987) performed a similar experiment in which one or three holes 8 mm in diameter (representing approximately 5 and 15%, respectively, of leaf area) were punched in dispersed leaves of *B. pubescens* and *B. pendula*. The leaves were left on the trees until the end of the season when the percentage of leaf area subsequently removed by insects was recorded. In both species significantly fewer of the damaged leaves received subsequent grazing than did undamaged controls. In addition, the mean amount of damage to those leaves that were subsequently grazed was lower on those that had been experimentally damaged.

4 EFFECTS OF LEAF DAMAGE ON INSECT FORAGING: AN EXPERIMENT

The patterns of insect grazing damage that we see in foliage at the end of the year reflect feeding choices made by thousands of insects throughout the growing season. The laboratory bioassays reviewed in the previous section clearly show how previous damage may influence these choices. Schultz (1983) has argued that the heterogeneity in food quality of leaves, due in part to induction effects, may force insects to forage widely, with consequent adverse effects upon larval development and survivorship. However, there has been very little work on the effect of leaf damage upon insect foraging. In this section we describe an experiment to investigate how artificial damage to birch foliage affects the foraging activity of larvae of the common quaker moth, *Orthosia stabilis* (Noctuidae).

The tree used in these experiments was a mature specimen of silver birch (*B. pendula*) in the Botanic Gardens at Southampton University. The experimental animals were larvae raised in the laboratory from eggs laid by moths collected in light traps in the Botanic Gardens. Prior to the experiment the larvae had been fed on a plentiful supply of birch (*B. pendula*) leaves which was renewed each day.

Birch shoots each with 10 leaves were selected in mid-June 1986. There were three treatments with 10 replicates of each:

1. Control shoots with no artificial damage
2. "Full damage," in which each leaf was experimentally damaged
3. "Partial damage," in which alternate leaves of the shoot were damaged.

Leaves were damaged in the field by making six 2-cm cuts in each leaf with a pair of scissors, but removing no leaf lamina. Four independent experiments were performed with shoots removed at intervals of 12, 24, 48, and 96 h between field damage and laboratory assessment.

The base of each twig was trimmed and pushed through a small hole in the lid of a 6 × 2.5-cm specimen tube filled with water. An inverted cone made of paper was fitted around the base of each shoot to catch falling larvae and provide access back to the shoot. At the start of the experiments two larvae were placed in a random position on each twig; one larva was in the 2nd-3rd instar, and the other in the 4th-5th instar, so that they could be easily distinguished by size.

The larvae were introduced onto the twigs at about 0900 h. Observations commenced 20 min later and continued at hourly intervals until 2100 h, with a final observation the next day at about 0900 h. At each observation the position of the larvae on the shoot and the amount of feeding since the previous observation were recorded. From the combined results the following classes of data were recorded for each shoot:

1. Total number of separate leaves visited
2. Total number of visits to leaves
3. Total number of individual leaves damaged

For example, if a larva moved onto leaf 6, and from this to leaf 7 and then back to leaf 6, the number of separate leaves visited would be 2, and the number of visits would be 3. In addition, the area removed by grazing was measured at the end of the experiment.

4.1   Movement between Leaves

The data (summarized in Table 1) reveal greater activity by larvae of *O. stabilis* on shoots that had been previously damaged. For each of the four time intervals following damage, a larger proportion of leaves were visited on shoots that had been damaged (both partial and full damage treatments) than on control shoots. These differences were statistically significant ($p <$ .05) only in the experiment that commenced 12 h after damage. In this experiment, 71 of a possible 100 leaves were visited on the control shoots (pooled data for 10 replicates), compared with 92 and 93 leaves visited on the full and partial treatments respectively. The combined data show significant differences ($p <$ .05) between the treatments and the control, but no significant difference between the treatments themselves.

The effect of damage on movement can be seen most clearly in the data for the total number of separate visits to leaves (Table 1a). For all time intervals there were more visits on shoots that had been damaged than on controls, though once again individual results were significant only for the 12-h interval. For the combined data the mean number of visits in both the partial and complete damage treatments was about 25% greater than in controls and the differences were highly significant ($p <$ .01). On the partially damaged shoots there were also differences in the numbers of visits to the damaged and undamaged leaves; in the experiments commencing 24 and 48 h after damage, significantly more visits were made to undamaged leaves. For the combined data the difference was highly significant ($p <$ .001), with visits to the undamaged leaves accounting for 74% of the total.

4.2   Feeding Damage

The data on feeding damage revealed significant effects of damage upon both the numbers of leaves eaten and the total amount of feeding, though the differences are in opposite directions. In both damage treatments, the numbers of leaves grazed was greater than in controls (Table 1b). Individual differences were significant in only three cases (partial treatment versus control after 24 h, full treatment versus control after 48 and 96 h), while the results for the combined data are highly significant ($p <$ .01). There were 35% more leaves fed upon in the full damage treatment compared with controls, and 42% more in the partial treatment; differences between the two treatments were not significant. A similar pattern is shown in the data for the number of feeding bouts, with significantly more bouts in the full damage treatment than in the control (combined data). However, in this case, the

TABLE 1   Effects of Artificial Damage to Leaves of Birch (*Betula pendula*) on Foraging and Feeding by Larvae of *Orthosia stabilis*

Interval (h)	U	D	P	PU	PD
		(a) Number of Separate Visits			
12	83^a	116^b	120^b	63	57
24	65	81	83	53^a	30^b
48	79	90	94	55	39
96	86	105	93	47	46
Mean	7.8^a	9.8^b	9.8^b	5.5^a	3.6^b
		(b) Number of Leaves Eaten			
12	51	62	68	35	33
24	33^a	43^{ab}	57^b	33	24
48	27^a	44^{ab}	35^b	18	17
96	47^a	65^b	64^{ab}	32	32
Mean	3.9^a	5.3^b	5.6^b	2.9	2.6
		(c) Total Leaf Area Grazed (cm²)			
12	133	118	139	88^a	51^b
24	148^a	78^b	115^b	81^a	34^b
48	126^a	78^b	131^a	99^a	32^b
96	163^a	100^b	92^b	59^a	33^b
Mean	142^a	93^b	119^b	82^a	37^b
		(d) Mean Area Grazed per Grazed Leaf (cm²)			
12	2.62	1.91	2.04	2.51^a	1.55^b
24	4.47^a	1.81^b	2.02^b	2.46^a	1.42^b
48	4.67^a	1.76^b	3.75^a	5.52^a	1.87^b
96	3.46^a	1.54^b	1.44^b	1.86^a	1.03^b
Mean	3.81^a	1.76^b	2.31^b	3.09^a	1.47^b

Note: Each experiment used 10 control twigs each with 10 undamaged leaves (U), 10 twigs each with 10 damaged leaves (D), and 10 partially damaged twigs (P) each with 5 damaged leaves (PD) and 5 undamaged leaves (PU). Treatments (U, D, and P) were compared using χ^2 for a and b. Within partially damaged shoots, differences between damaged leaves (PD) and undamaged leaves (PU) were compared in the same way. Means that share a common letter, or where there is no letter, are not significantly different ($p < .05$).

partial damage treatment does not give consistently higher results than the control and none of the differences are significant.

Whereas experimental damage led to an increase in the numbers of leaves grazed, the total leaf area removed was less (Table 1c). In each of the four experiments the total amount of grazing in the full damage treatment was less than that of the control. The difference was least in the 12-h experiment in which the fully damaged shoots received 89% of the grazing of the controls, and was greatest after 24 h with grazing levels of 53% of those in controls. The results of the partial damage treatments varied inconsistently between experiments; the combined data show that grazing was 84% of that in controls. However, in all experiments, there was substantially less grazing of the damaged leaves in the partial treatment than of the undamaged leaves. This difference was least after 12 h when the damaged leaves received 37% of total grazing in the partial damage treatment, and greatest after 48 h when they received only 24% of total grazing.

The contrary effects seen in the data for numbers of leaves grazed and levels of grazing mean that the influence of damage is seen most clearly when we consider mean area grazed *per grazed leaf* (Table 1d). In the experiment 12 h after damage, the mean area removed per grazed leaf in the full damaged treatment was 73% of that of the control, but by 48 h was only 40% of that of the control. The equivalent values for the partial damage treatment were 78 and 49%, respectively. There were also highly significant differences between damaged and undamaged leaves in the partial damage treatment; thus, in the 12h experiment the mean area grazed on previously undamaged leaves was 96% of that on control shoots, but was 59% on damaged leaves. In the 24-h experiment, the undamaged leaves were also grazed substantially less than the controls (55% of control values), but substantially more than damaged leaves on the same shoots which received 34% of the grazing of control shoots.

5 DISCUSSION

A reduction in the palatability of foliage following wounding is a widespread phenomenon that has been reported in many plant species. In most cases there is little or no evidence of the chemical basis of the effect. Notable exceptions are the wound-induced production of proteinase inhibitors in the foliage of tomato (Green and Ryan 1972) and the production of cucurbitacins in *Cucurbita* spp. (Tallamy 1985). Although there is accumulating evidence for the role of phenolic metabolism in the wound-induced responses of birch, it is not clear that phenolics are responsible for any reduced palatability to insects (Hartley and Lawton, this volume, Chapter 5). Indeed, it has been suggested that in some cases apparent changes in palatability may have no chemical basis but are simply visual or tactile responses by a herbivorous insect to a cut edge. Although such explanations may occasionally apply, they do not account for reduced palatability in undamaged leaves, or for the characteristic sequence of changes in palatability following damage. It may be that leaf damage leads to increased water loss and that insects avoid tissues with reduced hydration, though we have been unable to detect reduced water content in several of our experiments, and this idea would not explain the between-leaf effect described earlier. (See Wratten et al. 1990.)

Among those plant species that exhibit a rapidly induced reduction in palatability there are some notable similarities in the rates of induction and relaxation of the responses. In the five species we have studied in most detail (tomato, alder, hawthorn, and two birch species), reduced palatability develops rapidly and is detectable in both damaged and undamaged adjacent leaves within 24 h. In tomato we have detected a significant reduction in palatability within 8 h, and in birch and hawthorn within 6 h. However, the bioassay we use is not sufficiently sensitive to detect precisely the onset of reduced palatability since a significant period of feeding by the test animal is necessary to produce a result. Even more rapid responses have been demonstrated in other species; for example, reduced palatability following damage occurs within 3 h in *Cucurbita pepo* (Tallamy 1985). The subsequent development and relaxation of the induced response varies, but a typical pattern is for the intensity to increase to maximum after 2–4 days and then to decline so that it is scarcely detectable after 14 days.

There are also similarities between many plant species in the spatial scales over which the damage-induced effects can be detected. In several broad-leaved trees a high level of damage imposed on a proportion of leaves on a shoot leads to reduction in palatability in undamaged leaves on the same shoot, though usually the intensity of the effects is less than in the damaged leaves (Edwards et al. 1986). However, it is clear that high levels of damage are not necessary to produce a reduction in leaf acceptability to insects, since a single small hole may induce at least local effects (Gibberd et al. 1988). The use of an image-analyzing computer has allowed us to see more clearly how a halo of reduced palatability spreads out from the original hole. The concept of an induced change in leaf chemistry spreading out from the site of damage is supported by the work of Bergelson et al. (1986) who found that pinpricks in leaf tissue surrounding the feeding sites of the leaf miner *Coleophora serratella* on birch led to a local increase in phenolic levels which gradually spread outward.

At first sight there appears to be an inconsistency between the evidence that wounding leads to a rapidly dispersed response that affects other leaves on the same shoot, and the evidence for a reduction in palatability of tissues in the immediate vicinity of damage. For example, in alder we find clear evidence of reduced palatability in adjacent leaves (Figs. 1 and 2), and also of a halo of reduced feeding surrounding a small hole (Plate 1). We are not yet in a position to resolve this apparent inconsistency. Possibly the two observations simply reflect a declining gradient in the intensity of the induced effect with distance from the site of damage. Alternatively, it may be that they represent two distinct phenomena, and that high levels of damage in-duce a more widely dispersed signal which causes reduced palatability in distant leaves. Whatever the answer, it seems that insects feeding and for-aging on foliage may induce complex patterns of variation in the quality of their food resource. Although it may be difficult to demonstrate statistically, the pattern of grazing damage is likely to be more dispersed than it otherwise would be in the absence of induced effects. In particular, young leaves are likely to receive less grazing than they would in the absence of an induced response since induced changes in palatability are commonly strongest in these leaves.

In the last few years there has been considerable debate concerning the behavioral and ecological consequences of wound-induced changes in leaves and, in particular, their possible defensive significance. A frequent criticism of laboratory-based bioassay experiments in which confined polyphagous lepidopteran larvae are presented with a choice of food materials is that they have little relevance to ecological processes in the field. Although these bioassays may suggest significant changes in palatability following damage, it is argued that insects in a natural canopy are not presented with such simple choices from which to select their food. The results of field experiments such as that of Silkstone (1987) are therefore particularly valuable since they demonstrate that damage to foliage does affect food choices made by the natural community of phytophagous insects. A particularly interesting feature of that experiment was the finding that not only do insects feed less on leaves that have been previously damaged, but they tend to avoid a significantly higher proportion of them (than of controls) without feeding at all. The results suggest that damage-induced changes may affect leaf chemistry in such a way that damage can be detected without first tasting the leaf. The study described here of effects of damage on larval foraging behavior is similarly useful because it shows that damage does significantly increase the amount of foraging activity and affects larval food choices within a shoot.

Another argument against claims for a defensive role of wound-induced responses is that they are unlikely to have any significant effect upon herbivore populations. In temperate deciduous woodlands, the proportion of leaf production consumed by insects is usually not more than about 5%, suggesting that neither intra- nor interspecific competition among herbivores can be an important ecological process. In fact, intraspecific competition among insect herbivores has rarely been demonstrated (Strong et al. 1984), and it has been suggested that it is difficult to see how wound-induced changes in plants could have a major role in influencing single-species herbivore populations, or could be a factor in interspecific competition among herbivores (Fowler and Lawton 1985).

It has been argued that plant effects upon insect populations are likely to be so trivial that they are swamped by other processes such as natural enemies affecting the populations. However, Dempster (1983) analyzed a large number of age-specific life-table studies of Lepidoptera and suggested that host–plant resources were a major factor in the insects' population dynamics, with natural enemies constituting a minor role in most cases. Although age-specific life tables may be flawed with respect to their ability to detect spatial density dependence (Hassell et al. 1987), the role of inducible phytochemicals in herbivore population dynamics remains largely undemonstrated.

Clearly, any wound-induced change in the host plant that adversely affects food quality will contribute to any role the plant may have in reducing herbivore numbers, and could reduce the habitat's carrying capacity in a subtle way to a level well below that derived from a crude assessment of gross foliage quantity. This idea would link wound-induced changes in leaves to the *ceiling hypothesis* of herbivore population regulation (Dempster and Pollard 1981) rather than to the *equilibrium hypothesis* of Hassell (1985). It can be seen that the debate over the defensive significance of induced responses has focused upon the capacity of such responses to reduce insect populations. However, we argue that this is only one way in which wound-induced responses might serve a defensive role.

We recognize two models of how rapid wound-induced changes in leaf chemistry might protect a plant (Edwards and Wratten 1987). According to the first model, changes in secondary chemistry are induced in *all* foliage above a certain threshold level of damage, and the plant is thereafter more resistant to insect herbivores. As a result, the insect population is adversely affected and the plant suffers less subsequent damage from insects. Because there is a threshold of damage below which no chemical changes occur, we would expect such a defense to become effective only following an outbreak of an insect population, while for much of the time it would not operate. This is essentially the model tested by Fowler and Lawton (1985) in their experiment in which they damaged the foliage of *B. pubescens* and then studied the accumulation of natural grazing damage. In that experiment there was no evidence that even high levels of damage (25% defoliation) affected the subsequent level of insect grazing. However, we do not dismiss this model, which may apply in certain circumstances, particularly in the case of delayed induced resistance (Haukioja and Hakala 1975).

In the second model there is no threshold, and any damage to leaves, however small, causes at least local changes in leaf chemistry. An insect responds to the induced change by moving away from the vicinity of damage. The result for the plant is a distribution of grazing damage, both within and between leaves, that is more dispersed than it otherwise would be. This facet of the model is supported by evidence from the experimentation described above, showing that insects tend to feed some distance away from previous damage. In general, our experiments demonstrate that this deflection away from damage is strongest in young leaves.

How could alteration of the distribution of grazing damage act as a defense? The plant may benefit because insects spend more time foraging for suitable food. In addition, mortality risks through predation or falling off the plant may be greater for an insect that must continually move (Schultz 1983, Wratten et al. 1988). However, we suggest that the most important benefit of overdispersed grazing damage is its influence on plant competition. For plants growing in highly productive habitats the most important biotic factors are neighboring plants with which they must compete for resources. If water and nutrient conditions are adequate, then light is the principal limiting factor, and competition takes the form of a continuous struggle to occupy the outermost canopy. In competitive conditions the outermost leaves, which are usually the youngest leaves, are the most important to the plant for successful competition, and plants characteristic of such environments tend to produce new leaves throughout much of the growing season as they "forage" for light (Grime 1979). However, young leaves may be most vulnerable to insects since they are less tough and have a greater nitrogen concentration than older leaves. For insects, young leaves on a plant with an indeterminate growth pattern are a predictable resource of high-quality food. We suggest that the significance of wound-induced responses is that they minimize damage to young leaves by deflecting grazers away from them, which may or may not be associated with a detectable reduction in the numbers of the herbivore.

REFERENCES

Bergelson, J., S. Fowler, and S. Hartley. 1986. The effects of foliage damage on casebearing moth larvae, *Coleophora serratella*, feeding on birch. *Ecol. Entomol.* 11:241–250.

Bishara, I. 1934. The cotton worm, *Prodenia litura* F., in Egypt. *Bull. Soc. R. Entomol. d'Egypte* 18:223–404.

Croxford, A., P. J. Edwards, and S. D. Wratten. 1989. Temporal and spatial variation in palatability of leaves of cotton and soybean following wounding. *Oecologia* 79:520–525.

Dempster, J. P. 1983. The natural control of populations of butterflies and moths. *Biol. Rev.* 58:461–481.

Dempster, J. P., and E. Pollard. 1981. Fluctuations in resource availability and insect populations. *Oecologia* 50:412–416.

Edwards, P. J., and S. D. Wratten. 1982. Wound-induced changes in palatability of birch (*Betula pubescens* Ehrh. ssp. *pubescens*). *Am. Natur.* 120:816–818.

Edwards, P. J., and S. D. Wratten. 1983. Wound-induced defenses in plants and their consequences for patterns of insect grazing. *Oecologia* 59:88–93.

Edwards, P. J., and S. D. Wratten. 1985. Induced plant defenses against insect grazing: fact or artefact? *Oikos* 44:70–74.

Edwards, P. J., and S. D. Wratten. 1987. Ecological significance of wound-induced changes in plant chemistry. In V. Labeyrie, G. Fabres, and D. Lachaise (eds.), *Insects–Plants*. Dordrecht: Junk, pp. 213–218.

Edwards, P. J., S. D. Wratten, and H. Cox. 1985. Wound-induced changes in the acceptability of tomato to larvae of *Spodoptera littoralis*: a laboratory bioassay. *Ecol. Entomol.* 10:155–158.

Edwards, P. J., S. D. Wratten, and S. Greenwood. 1986. Constitutive and induced defenses against insect grazing in British trees. *Oecologia (Berlin)* 69:316–319.

Fowler, S. V., and J. H. Lawton. 1985. Rapidly induced defenses and talking trees: the devil's advocate position. *Am. Natur.* 126:181–195.

Gibberd, R. M. 1987. Wound-induced plant responses and their consequences for insect grazing. Ph.D. thesis, University of Southampton.

Gibberd, R., P. J. Edwards, and S. D. Wratten. 1988. Wound-induced changes in the acceptability of tree foliage to Lepidoptera: within-leaf effects. *Oikos* 51:43–47.

Green, T. R., and C. A. Ryan. 1972. Wound-induced proteinase inhibitor in plant leaves: a possible defense mechanism against insects. *Science* 175:776–777.

Grime, J. P. 1979. *Plant Strategies and Vegetation Processes.* New York: Wiley.

Hartley, S., and J. H. Lawton. 1991. The biochemical basis and significance of rapidly induced changes in birch foliage. In D. W. Tallamy and M. J. Raupp (eds.), *Phytochemical Induction by Herbivores.* New York: Wiley, pp. 105–132.

Hassell, M. P. 1985. Insect natural enemies as regulating factors. *J. Anim. Ecol.* 54:323–334.

Hassell, M. P., T. R. E. Southwood, and P. M. Reader. 1987. Detecting regulation in patchily distributed animal populations. *J. Anim. Ecol.* 56:705–713.

Haukioja, E. 1982. Inducible defenses of white birch to a geometrid defoliator *Epirrita autumnata. Proceedings of the 5th International Symposium on Insect–Plant Relationships.* Pudoc: Waageningen, pp. 200–230.

Haukioja, E., and T. Hakala. 1975. Herbivore cycles and periodic outbreaks: formulation of a general hypothesis. *Rep. Kevo Subarctic Res. Sta.* 12:109.

Haukioja, E., and P. Niemela. 1976. Does birch defend itself actively against herbivores? *Rep. Kevo Subarctic Res. Station* 13:44–49.

Haukioja, E., and P. Niemela. 1977. Retarded growth of a geometrid larva after mechanical damage to leaves of its host tree. *Ann. Zool. Fennici* 14:48–52.

Schultz, J. C. 1983. Impact of variable plant defensive chemistry on susceptibility of insects to natural enemies. In A. Hedin (ed.), *Plant Resistance to Insects. American Chemical Society Symposium No. 208.* Washington, DC: American Chemical Society.

Silkstone, B. E. 1987. The consequences of leaf damage for subsequent insect grazing on birch (*Betula* spp.): a field experiment. *Oecologia (Berlin)* 74:149–152.

Strong, D. R., J. H. Lawton, and T. R. E. Southwood. 1984. *Insects on Plants.* Oxford: Blackwell.

Tallamy, D. 1985. Squash beetle trenching behavior: an adaptation against induced cucurbit defenses. *Ecology* 66:1574–1579.

Wratten, S. D., P. J. Edwards, and A. M. Barker. 1990. Consequences of rapid feeding-induced changes in trees for the plant and the insect: individuals and populations in A. D. Watt, S. R. Leather, M. D. Hunter and N. A. C. Kidd (eds.), *Population Dynamics of Forest Insects.* Andover, Hampshire, U.K.: Intercept, pp. 137–145.

Wratten, S. D., P. J. Edwards, and I. Dunn. 1984. Wound-induced changes in the palatability of *Betula pubescens* and *B. pendula. Oecologia (Berlin)* 61:372–375.

Wratten, S. D., P. J. Edwards, and L. Winder. 1988. Insect herbivory in relation to dynamic changes in host plant quality. *Biological Journal of the Linnean Society* 35:339–350.

CHAPTER 10

THE EFFECTS OF INDUCED PLANT PROTEINASE INHIBITORS ON HERBIVOROUS INSECTS

JANE L. WOLFSON
Department of Entomology, Purdue University, West Lafayette, IN 47907

1   Introduction
2   Insect digestive proteinases
3   Proteinase inhibitors
    3.1   Plant proteinase inhibitors
    3.2   Impact of proteinase inhibitors on insect growth and development
4   Induction of proteinase inhibitors in plants
    4.1   Tomatoes: the model system
    4.2   Effects of plant age and environmental conditions on leaf quality of wounded tomato plants
    4.3   The impact of tomato proteinase inhibitor on larval growth
    4.4   Effect of plant age on proteinase inhibitor induction
5   Conclusion
    References

1   INTRODUCTION

When Green and Ryan (1972) described the induction of proteinase inhibitors in tomato and potato leaves in response to herbivory, they captured the attention of ecologists, entomologists, and plant physiologists. The possible defensive role of plant proteinase inhibitors against herbivorous insects had been suggested earlier (Lipke et al. 1954, Applebaum 1964), but the demonstration that plants could be induced to produce high levels of these inhibitors in response to herbivory made these compounds even more interesting.

Plant proteinase inhibitors traditionally are identified and characterized by their ability to inhibit certain mammalian digestive enzymes (Laskowski and Kato 1980). Since insects utilize digestive proteinases to obtain their

223

essential amino acids from dietary protein (House 1974), and many insect digestive proteinases have been described as resembling mammalian digestive enzymes (Applebaum 1985), it has been assumed that proteinase inhibitors should affect insect development by reducing rates of nitrogen assimilation. There are several ways in which the proteinase inhibitors could affect the rate of nitrogen and/or protein assimilation in insects. They could bind irreversibly with the digestive proteinases and thereby directly depress the level of available amino acids, or they could bind with the proteinases and, via feedback mechanisms, initiate increased production of digestive proteinases resulting in wasted energy and depletion of a limited amino acid pool. Regardless of the mechanism, since insects are known to be sensitive to changes in levels of dietary nitrogen and/or protein (Slansky and Feeny 1977, Mattson 1980, Wint 1981), there is clearly a potential for proteinase inhibitors to be active against insects. These facts, along with some circumstantial evidence associating the presence of proteinase inhibitors with poor rates of insect development (Lipke et al. 1954, Gatehouse et al. 1979), have supported the hypothesis that proteinase inhibitors, both induced and constitutive, affect herbivorous insects by their impact on insect digestive proteinases. While this theory is attractive, there has been little research directed toward testing it.

Proteinase inhibitors are generally active against only one class of proteinases; therefore, the potential of an inhibitor to be biologically active against an herbivore depends on the proteinase type(s) in the insect. Even though a proteinase inhibitor is active against a certain class of proteinase, it does not necessarily inhibit all proteinases within that class (i.e., all serine proteinase inhibitors do not inhibit all serine proteinases), and, of equal importance, an inhibitor that is active against an enzyme from one source is not necessarily equally potent against different sources of that same enzyme type (i.e., a trypsin inhibitor will not always inhibit different sources of trypsin equally (Rascon et al. 1985)).

The potential importance of a proteinase inhibitor for herbivorous insects in general or for a specific herbivore depends on the characteristics of the insect's digestive proteinase(s) and their susceptibility to different proteinase inhibitors. To assess this importance, we must first answer several questions. What kinds of digestive proteinases do insects utilize? Does a specific proteinase inhibitor inhibit the digestive proteinases found in the target insect? Can this inhibition (if it exists) affect larval growth and development? Once we know that proteinase inhibitors can inhibit insect digestive proteinases, then there are additional questions that should be answered. Do those proteinase inhibitors, when induced, exist at a titer sufficiently high to affect insect herbivores? Does the induction occur under physical/environmental conditions that make the phenomenon ecologically important (e.g., does it occur under heat, light, nutrient, etc., conditions that exist outside of environmental chambers and greenhouses)? However, to begin our examination of these questions we first must look at the targets of the induced proteinase inhibitors, the digestive proteinases of insects.

2 INSECT DIGESTIVE PROTEINASES

There are four classes of proteinases: serine, cysteine (also called thiol), aspartic (also called carboxylic acid), and metallo-proteinases (Barrett 1986). The names reflect the key catalytic moiety (an amino acid or cation) participating in the cleavage of the peptide bonds. Insect endopeptidases come from only the first three classes (Applebaum 1985). The mechanistic class of a proteinase is inferred from its in vitro characteristics, including (1) the pH range over which it is maximally active, (2) its ability to hydrolyze specific peptide bonds or specific substrates, (3) its similarity to well-characterized proteinases, and (4) its sensitivity to various inhibitors (North 1982, Barrett 1977, Wagner 1986). In brief, serine proteinases (e.g., trypsin and chymotrypsin) are generally active in the pH 7.0–10.0 range and are inhibited by Bowman-Birk, Kunitz, and lima bean inhibitors; cysteine proteinases (e.g., papain) are generally most active in the mildly acid range, pH 5.0–7.0 and are inhibited by heavy metals, cystatin, and E-64, and enhanced by reducing agents; aspartic proteinases (e.g., pepsin) are active in the acid pH range, generally below pH 4.5, and are inhibited by pepstatin (Barrett 1977, North 1982).

There have been relatively few studies on insect digestive proteinases. By far, the majority of insects studied are reported to utilize proteinases with alkaline pH optima which is characteristic of serine-type proteinases (Applebaum 1985). Well-characterized, serine-type enzymes have been described in many insects, among them the tobacco hornworm (*Manduca sexta*) and all other Lepidoptera studied to date, stable fly (*Stomoxys calcitrans*), Hessian fly (*Mayetiola destructor*), carabid beetles (*Carabus* spp. and *Calosoma calidum*), cockroach (*Leucophaea maderae*), sweetpotato weevil (*Cylas formicarius elegantulus*), and mosquitoes (*Culex nigripalpus* and *Aedes aegypti*) (Miller et al. 1974, Kunz 1978, Engelmann and Geraerts 1980, Baker et al. 1984, Vaje et al. 1984, Shukle et al. 1985, Borovsky 1986, Cheeseman and Gillott 1987, Houseman et al. 1987). In contrast to serine-type digestive proteinases, cysteine and acid proteinases have been described, until recently, in only a very few insects, mostly blood feeders (*Rhodnius prolixus* and *Cimex* spp.) (Garcia et al. 1978, Houseman 1978, Houseman and Downe 1982a,b, 1983). The reported abundance of serine-type digestive proteinases must be viewed with an understanding that the methods used to assess insect digestive proteolytic activity were heavily influenced by protocols developed to assess mammalian digestive proteinases of the serine type. These methods were often restricted to the alkaline pH range and therefore the only digestive proteinases detected would be serine-type enzymes. These methodological problems led some investigators looking for serine-type proteinases to conclude that certain insects may not possess digestive proteinases at all (Applebaum 1964).

Cysteine proteolytic activity has recently been discovered to be an important component of the midgut proteolytic activity in several economically important herbivorous and stored-product beetles, including the Colorado potato beetle (*Leptinotarsa decemlineata*), the boll weevil (*Anthonomus grandis*), red flour beetle (*Tribolium castaneum*), Mexican bean beetle (*Epilachna varivestis*), cowpea weevil (*Callosobruchus maculatus*), and bean weevil (*Acanthoscelides obtectus*) (Gatehouse et al. 1985, Kitch and Murdock 1986, Murdock et al. 1987, Weiman and Nielsen 1988). The presence of nonserine digestive proteinases in insects comes as no surprise. A brief look at the published reports on insect midgut pH values leads one to anticipate that many insects utilize digestive proteinase with nonalkaline pH optima (Swingle 1931, Grayson 1951, 1958). A survey of pH optima of the proteolytic activity in insect midgut homogenates indicates that there is a high diversity in digestive enzyme activity (Wolfson and Murdock 1990a). Using a proteinase assay based on release of radioactive fragments from tritium-labeled methemoglobin, we have detected, in insect midgut homogenates, peaks in proteinase activity ranging from pH 3.0 to 12.0. In some insect species we observe multiple peaks which could indicate a complex of proteinases of different classes. As examples, the pH profiles of Mexican bean beetle (*E. varivestis*), European pine sawfly (*Neodiprion sertifer*), and margined blister beetle (*Epicauta pestifera*) are presented in Figure 1.

3 PROTEINASE INHIBITORS

3.1 Plant Proteinase Inhibitors

Most of the early research on plant proteinase inhibitors came from the field of animal and human nutrition, so it is not at all surprising that the most thoroughly studied plant groups are important food sources. The most well-known inhibitors are from plants within the Leguminoseae (e.g., soybean, lima bean, cowpea, adzuki beans, marama bean), Gramineae (e.g., wheat, rye, triticale, barley, sorghum, millet), and Solanaceae (e.g., potato and tomato) (Ryan 1966, 1979, Ryan and Huisman 1970, Richardson 1977, Gatehouse et al. 1979, Kumar et al. 1979, Chang and Tsen 1981, Weder 1981, Manjunath et al. 1983, Elfant et al. 1985, Ishikawa et al. 1985, Mosolov and Shul'gin 1986). As might be expected, most of the described inhibitors inhibit serine proteinases, although other types of inhibitors are known. Inhibitors of cysteine proteinases have been described in potatoes and tomatoes (Akers and Hoff 1980, Rodis and Hoff 1984), in rice and corn (Abe et al. 1980, Abe and Arai 1985), and in cowpeas (Rele et al. 1980), but they appear to be less abundantly distributed among plant species than are serine proteinase inhibitors. Most inhibitors occur in seeds or storage organs with two important

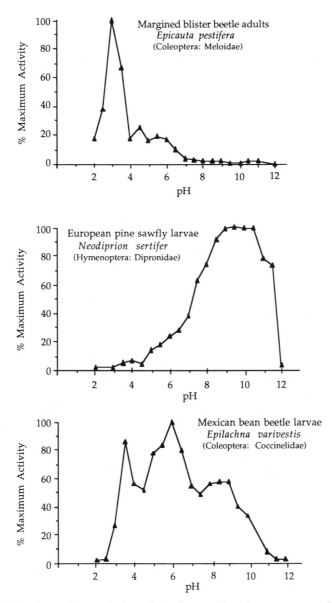

Figure 1   pH optima of proteolytic activity from midgut homogenates of selected insects. Substrate was [³H]-methemeglobin. (From Wolfson and Murdock 1990a.)

exceptions: Both the serine and the cysteine proteinase inhibitors in tomato and potato occur in the leaves and are inducible. The concentrations of induced inhibitor depend on the environmental conditions under which the plants are grown and in which they are placed during the period of induction (Green and Ryan 1972, 1973, Akers and Hoff 1980).

3.2   Impact of Proteinases Inhibitors on Insect Growth and Development

Considering that insect digestive proteinases are described in terms that reflect similarity to mammalian proteinases (e.g., trypsin-like, chymotrypsin-like), one might assume that trypsin inhibitors would be effective inhibitors of insect proteinase both in vitro and in vivo. In fact, no generalization can be made because too few experiments have been done. With the exception of studies characterizing an insect's digestive proteinase, there have been few in vitro inhibition studies. The relevance of many of the reported in vitro inhibition studies is potentially limited since many of the assays measure esterolytic activity of the proteinase and assume that the proteolytic activity shows similar activity and inhibitability. Inhibition of in vitro esterolytic or proteolytic activity of a proteinase might or might not be reflected in similar amounts of in vivo proteinase inhibition (Borovsky 1985), but it can potentially identify good candidate inhibitors. Using the radiometric assay described earlier we have evaluated the impact of selected inhibitors on insect midgut homogenates. Figure 2 presents data for *L. decemlineata* larvae indicating the impact of pepstatin, an aspartic proteinase inhibitor, and E-64, a cysteine proteinase inhibitor, on midgut proteolytic activity. E-64 inhibited proteolytic activity over the pH range where most of the activity was found, a range that included the gut pH (Wolfson and Murdock 1987a). These data suggested that E-64 could have an important impact on an insect's growth and development, a suggestion validated by bioassay experiments (see below).

There have been several studies in which proteinase inhibitors have been incorporated into an insect's food source to assess the potential impact of the ingested inhibitors. Until our recent work, only serine proteinase inhibitors had been evaluated. Soybean trypsin inhibitor (Kunitz inhibitor) incorporated into artificial diets has been shown to retard larval growth of *M. sexta* at 5% (wt/vol), and with *Ostrinia nubilalis* to retard larval growth at 2% and pupal development at 3% (Steffens et al. 1978, Shukle and Murdock 1983). When added to a casein-based artificial diet at 0.045% (wet wt), potato chymotrypsin inhibitor II retarded larval development of both *Heliothis zea* and *Spodoptera exiqua*, while soybean trypsin inhibitor inhibited larval growth of *H. zea* at 0.045% (wet wt) and *S. exiqua* at 0.09% (wet wt) (Broadway and Duffy 1986). When adult female *S. calcitrans* were fed erythrocytes in which between 1 and 3% soybean trypsin inhibitor was encapsulated, there was 50% mortality and egg production was eliminated in the survivors

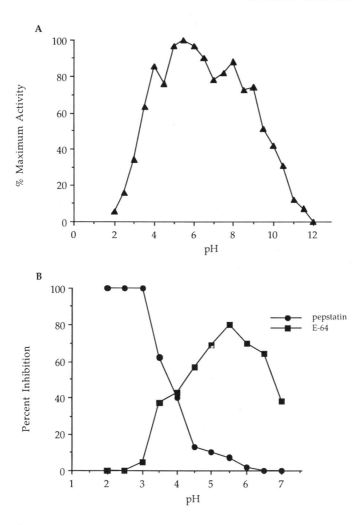

Figure 2   The impact of two proteinase inhibitors with different specificities on proteolytic activity in larval *Leptinotarsa decemlineata*. (*A*) pH optimum curve without inhibitors. (*B*) Percent inhibition over a pH range when the midgut homogenate is preinculated with pepstatin (1 μg/mL) or E-64 (100 μg/mL). (From Wolfson and Murdock 1987a.)

(Deloach and Spates 1980). Larval growth and development of *Tribolium confusum* was not affected by ingestion of soybean serine proteinase inhibitors, although a preparation of the Tribolium-specific inhibitor from soybean was toxic (Lipke et al. 1954). *Callosobruchus chinensis* and *A. obtectus* were not affected by soybean trypsin inhibitor in their diet at 0.2% wt/wt (Applebaum 1964), nor was *C. chinensis* affected by Bowman-Birk inhibitor at

0.7% in the diet (Applebaum et al. 1965). Larval development of *C. maculatus* on a diet containing 5% Kunitz inhibitor was delayed 21% whereas potato chymotrypsin inhibitor either had no effect on development (at 1%) or was lethal (at 5%) (Janzen et al. 1977). Incorporation of cowpea trypsin inhibitor at 2–20% in a diet for *C. maculatus* is reported to delay larval development, but the effect of the inhibitor on larval mortality cannot be evaluated since initial infestation levels are not given (Gatehouse and Boulter 1983). Shade et al. (1986) incorporated various inhibitors into artificial seeds and found that larval development in *C. maculatus* was significantly delayed by Bowman-Birk inhibitor at 1% and Kunitz inhibitor at 5% (lower concentrations had no effect). They also reported that 5% cowpea trypsin inhibitor had no effect on larval developmental time but that 5% Bowman-Birk and 5% cowpea trypsin inhibitor increased the larval mortality rate.

Recent evaluation of the effect of ingestion of cysteine proteinase inhibitors on insects known to demonstrate substantial amounts of cysteine proteolytic activity in their midgut homogenates have demonstrated that these inhibitors can severely impact growth and development of both stored product pests and leaf-feeding beetles. Growth and development of *C. maculatus* was examined in artificial seeds that contained concentrations of 0.0–0.25% E-64, a highly specific cysteine proteinase inhibitor isolated from *Aspergillus japonicus* (Hanada et al. 1978a, b, Murdock et al. 1988). Developmental time (egg hatch to adult emergence) was delayed with the addition of E-64 in a dose-dependent manner; for each 0.01% increase in E-64, developmental time increased about 4.5 days. Fecundity in these beetles was also affected but only at concentrations over 0.06% (Murdock et al. 1988).

The impact of E-64 on larval development has also been tested on leaf-feeding beetles. Different concentrations of E-64 were added to gelatin solutions which were then painted onto leaves from the natural host plants (e.g., potato for the Colorado potato beetle (*L. decemlineata*), jimsonweed

TABLE 1   Mean Weights in mg ± s.e. (number of observations in parentheses) of *Leptinotarsa decemlineata* Larvae Grown on Potato Leaves Treated with E-64, a Cysteine Proteinase Inhibitor. Treatment Values Indicate μg of Inhibitor per ml of Gelatin Solution Applied to the Leaves

| | *Weight of Larvae** | |
Treatment	After 6 Days	After 8 Days
Control	16.7 ± 1.36[a] (32)	42.8 ± 3.59[a] (27)
8 μg E-64/ml	15.9 ± 1.30[a] (35)	41.4 ± 2.53[a] (27)
40 μg E-64/ml	9.8 ± 0.64[b] (24)	20.1 ± 2.09[b] (20)
200 μg E-64/ml	6.5 ± 0.61[c] (26)	9.7 ± 1.13[c] (22)
1000 μg E-64/ml	4.4 ± 0.40[d] (26)	6.5 ± 0.72[d] (26)

* Means within columns followed by the same letter not significantly different, $p > 0.05$; Duncan's multiple range test. Data analysis performed on log transformed data. (Data from Wolfson and Murdock, 1987a.)

TABLE 2   Mean Weights in mg ± s.e. (number of observations in parentheses) of *Lema trilineata* Larvae Grown on Jimsonweed Leaves Treated with (A) E-64, a Cysteine Proteinase Inhibitor and (B) Kunitz Inhibitor, a Serine Proteinase Inhibitor. Treatment Values Indicate µg of Inhibitor per ml of Gelatin Solution Applied to the Leaves

	*Weight of Larvae**	
Treatment	After 4 Days	After 6 Days
A Control	10.7 ± 0.70[a] (20)	58.6 ± 3.52[a] (20)
40 µg E-64/ml	8.4 ± 0.31[b] (21)	47.7 ± 2.91[b] (21)
200 µg E-64/ml	3.6 ± 0.47[c] (16)	9.2 ± 1.20[c] (16)
B Control	9.1 ± 0.62[a] (25)	43.8 ± 2.69[a] (24)
10 mg Kunitz/ml	9.8 ± 0.65[a] (28)	47.2 ± 3.69[a] (28)

* Means within columns followed by the same letter not significantly different, $p > 0.05$; Duncan's multiple range test.

for the three lined potato beetle (*Lema trilineata*)) (see Wolfson and Murdock 1987b for bioassay method). After 6 days Colorado potato beetle larvae consuming potato leaves treated with a gelatin solution containing 40 µgE-64/ml gelatin solution (which is equivalent to 0.8 µg E-64 cm^2 leaf tissue) were significantly smaller than control insects (Table 1). *Lema trilineata* and *E. varivestis* feed on plants belonging to families known for their serine proteinase inhibitors. Therefore, in addition to testing the effect of E-64 we also tested the effect of Kunitz inhibitor on these insects (Wolfson and Murdock, unpublished results). Results of the experiments on these two beetle species are presented in Tables 2 and 3. Larval growth of *E. varivestis* and

TABLE 3   Mean Weights in mg ± s.e. (number of observations in parentheses) of *Epilachna varivestis* Grown on Lima Bean Leaves Treated with (A) E-64, a Cysteine Proteinase Inhibitor and (B) Kunitz Inhibitor, a Serine Proteinase Inhibitor. Treatment Values Indicate µg of Inhibitor per ml of Gelatin Solution Applied to the Leaves

	*Weight of Larvae**		
Treatment	After 4 Days	After 7 Days	After 10 Days
A Control	1.2 ± 0.06[a] (32)	3.4 ± 0.18[a] (29)	11.4 ± 0.24[a] (29)
50 µg E-64/ml	0.9 ± 0.04[b] (31)	2.9 ± 0.11[b] (29)	5.9 ± 0.39[b] (27)
100 µg E-64/ml	0.7 ± 0.04[c] (34)	1.5 ± 0.13[c] (31)	3.4 ± 0.27[c] (27)
200 µg E-64/ml	0.5 ± 0.04[d] (31)	0.9 ± 0.15[d] (22)	1.7 ± 0.36[d] (16)
400 µg E-64/ml	0.4 ± 0.02[d] (27)	0.5 ± 0.04[d] (22)	0.6 ± 0.06[e] (13)
B Control	1.2 ± 0.07[a] (27)	3.2 ± 0.18[a] (25)	9.2 ± 0.45[a] (24)
10 mg Kunitz/ml	1.1 ± 0.06[a] (25)	3.2 ± 0.12[a] (23)	10.1 ± 0.39[a] (23)

* Means within columns followed by the same letter not significantly different, $p > 0.05$; Duncan's multiple range test.

L. trilineata was delayed by consumption of E-64, a cysteine proteinase inhibitor, but not affected by Kunitz inhibitor, a serine proteinase inhibitor.

To return to the questions asked earlier, evidence has been presented that

1. Insect digestive proteinases can demonstrate activity typical of either serine, cysteine, or aspartic proteinase types.
2. Proteinase inhibitors can inhibit insect digestive proteinases.
3. Consumption by an insect of a potent inhibitor of its digestive proteinase can affect that insect's growth and development.

We now turn to induced proteinase inhibitors.

4   INDUCTION OF PROTEINASE INHIBITORS IN PLANTS

4.1   Tomatoes: The Model System

For over 20 years, Ryan and his colleagues have been studying the induction of serine proteinase inhibitors in potato and tomato leaves in response to leaf injury. The phenomenon was originally noted in excised leaves that were exposed to constant light (Ryan 1968). The most extensive studies have been on tomato, *Lycopersicum esculentum* (var. Bonny Best), and can be briefly described as follows: Under conditions of high light intensity (800–1000 foot candles), high heat (31–37 °C), and good fertility, wounding of a tomato leaf releases PIIF (proteinase inhibitor inducing factor) which causes a systemic induction and subsequent accumulation of proteinase inhibitor at sites distant from the site of injury. Both mechanical wounding and insect feeding can induce the response, but the response is light and temperature sensitive. Transport of PIIF from the site of injury is rapid. Within 1 h of injury sufficient signal has been transported from the injury site to insure induction in noninjured neighboring leaves. Other plant species also will respond to PIIF and have been identified as sources of PIIF-like materials, but the spontaneous accumulation of high levels of proteinase inhibitors in response to injury in intact plants has not been observed in other species (Ryan 1974, Ryan and Green 1974, Walker-Simmons and Ryan 1977).

The induction of proteinase inhibitors in wounded tomato plants has been the model for a rapidly induced response of plants to injury and reveals the potential of proteinase inhibitors as allelochemicals directed toward leaf-feeding insects (Rhoades 1979, Ryan 1979). These well-studied induced proteinase inhibitors are of the serine type and might be expected to inhibit the serine proteinases which are common in lepidopteran larvae. Wounding tomato leaves has been demonstrated to affect the feeding behavior of lepidopteran herbivores. Edwards et al. (1985) working with *Spodoptera littoralis* found that larvae consumed lower amounts of unwounded leaves from tomato plants on which certain leaves had been wounded, than of leaves from undamaged control plants. On the wounded plants, the wounded leaves themselves were less preferred than the neighboring unwounded leaves and the onset of nonpreference appeared more rapidly in the wounded leaves (8 h) than in the neighboring leaves (24 h). The pattern of change in acceptability was similar to the pattern of PIIF-induced responses in proteinase inhibitor level, but there was no indication that proteinase inhibitors were important for the feeding preferences.

Evidence that induced proteinase inhibitors are effective against herbivorous insects is scarce. In a recent study, Broadway et al. (1986) reported that *Spodoptera exiqua* grew more slowly when fed agar based diets containing tomato leaf tissue from wounded tomato plants than when fed on control diets that contained tomato foliage from similar, but nonwounded, plants. These researchers measured proteinase inhibitor level, protein, and phenol level in leaves from wounded and control tomato plants and found the lower quality of the leaves from the wounded plants was associated only with an increase in proteinase inhibitor level of those leaves. The possible effects of changes in larval feeding behavior associated with leaf damage on larval growth (described above) was not discussed.

Additional information on environmental conditions associated with changes in leaf quality is necessary to accurately evaluate the ecological importance of induction of proteinase inhibitors in tomatoes for herbivorous insects. To this end, I will report on some recent experiments on growth of *M. sexta* larvae consuming leaves from wounded tomato plants (Wolfson and Murdock 1990b). These experiments focused on plant age and environmental conditions at wounding. The questions asked were

1. Does wounding tomato plants at different ages and under different environmental conditions alter the impact of wounding on leaf quality as assessed by larval growth of *M. sexta*?
2. Does the plant's age at wounding change its ability to produce proteinase inhibitors in response to wounding?

4.2   Effects of Plant Age and Environmental Conditions on Leaf Quality of Wounded Tomato Plants

To assess the effect of age at wounding on changes in plant quality, I grew tomatoes under well-defined environmental conditions. The environmental conditions and the wounding protocol were intended to maximize induction of proteinase inhibitors in the wounded plants (Green and Ryan 1973) while minimizing possible induced responses in the control plants. All plants were injured similarly but the age at injury was different between experiments. Plants were injured at 14 days or at 1 month after transplanting. Tomato seeds (var. Bonny Best) were germinated in vermiculite and transplanted into 8-cm pots with commercial potting mix when the first leaf began to show between the two cotyledons. Two weeks before injuring the one month-at-injury plants, the tomatoes were transplanted into 11-cm pots. All plants were grown in a greenhouse in the early fall under a 14-h day length provided by supplemental mercury vapor lights until the start of wounding. In these experiments wounded plants were placed in an environmental chamber (24 h light, sodium and mercury vapor lamps, 29 °C) after the first wounding episode. Control plants remained in the greenhouse throughout their growth period in both experiments. In this study, the wounding (leaf crushing) occurred progressively over a 3-day period: day 1, cotyledons injured; day 2, leaf blade of lowest leaf injured; day 3, vein of lowest leaf and leaf blade of next higher leaf injured and 2 h later the vein on the second leaf was injured. On the fifth day after the first wounding episode, both wounded and control plants were fed to *M. sexta* larvae.

An additional set of experiments was run to evaluate the impact of the conditions in the environmental chamber (high light intensity and high temperatures) on plant quality. In this set of experiments, plant growing conditions and wounding protocol were similar to those described above except that both wounded and control plants were left in the greenhouse until they were used to feed larvae. Wounding began 17 days after transplanting rather than 14 days. In both sets of experiments transplanting and injuring were staggered within each treatment so that similarly aged and treated plants were available for insect growth experiments on any specific day.

The experiments utilized neonate larvae. After the first day, all larvae that had fed were transferred onto a new group of plants. Thereafter, larvae were transferred onto fresh plants every 2 days. Insects were maintained on the plants in an environmental chamber under low light intensity, 14 h light, and at mild temperatures (21 °C) so as to minimize induction of proteinase inhibitors in control plants. Larvae were reared as a group on 3–6 plants per 2-day period. Larvae never consumed all of the available foliage. All plants were sampled before the larvae started to feed and after the larvae had been transferred to fresh plants. Leaf tissue samples were frozen immediately after sampling and later evaluated for proteinase inhibitor level. The amount of proteinase inhibitor in the plant sample was calculated by comparing the amount of trypsin inhibitory activity in the leaf extract to a standard curve of trypsin inhibitory activity in a stock preparation of freeze-dried partially purified tomato proteinase inhibitor.

The results of these experiments are presented in Table 4A and Figure 3. The weights of *M. sexta* larvae after 7 days of feeding on the wounded tomato plants were significantly lower than those of insects fed control plants in all three experiments. As assessed by *M. sexta*, the leaf quality was decreased with wounding whether the plants were wounded at 14 days or 1 month after transplant, or whether they were placed in the environmental chamber or left in the greenhouse. There was some increase in proteinase inhibitor levels in the control plants during the feeding period. Therefore, I calculated the proteinase inhibitor concentration during the feeding period as the mean of pre- and postfeeding inhibitor levels. This method of cal-

TABLE 4   The Mean Weights in mg $\pm$ s.e. (number of observations in parentheses) of *Manduca sexta* Larvae Consuming (A-1) Wounded Tomato Plants of Two Different Ages Placed into Environmental Chamber at Wounding, (A-2) Wounded Tomato Plants Maintained in the Greenhouse after Wounding, and (B) Mature Tomato Leaves Coated with Gelatin Containing a Crude Tomato Proteinase Inhibitor Preparation. The Proteinase Inhibitor Concentration (mean μg of proteinase inhibitor equivalent per mg of leaf tissue) is also Presented

Treatment			7 Day Larval Weight	Proteinase Inhibitor Concentration
A-1	2 week old plants	Wounded	52.3 $\pm$ 3.2 (25)	2.51 $\pm$ 0.74
		Control	112.8 $\pm$ 8.7 (26)***	0.46 $\pm$ 0.05**
	Month old plants	Wounded	28.7 $\pm$ 2.3 (16)	0.35 $\pm$ 0.10
		Control	37.4 $\pm$ 2.6 (18)*	0.11 $\pm$ 0.02*
A-2	2.5 week old plants	Wounded	36.8 $\pm$ 1.16 (20)	1.03 $\pm$ 0.58
		Control	52.2 $\pm$ 2.4 (20)**	0.45 $\pm$ 0.03 ns
B	Proteinase inhibitor	Treated	37.0 $\pm$ 2.1 (19)	6.01 $\pm$ 0.96
	treated leaves	Control	42.6 $\pm$ 2.2 (18) ns	0.34 $\pm$ 0.11***

*, **, *** Means between treatments significantly different at $p <$ .05, .01 or .001 according to a t-test on log transformed data. (Data from Wolfson and Murdock, 1990b.)

culation provides a conservative estimate of the difference in proteinase inhibitor levels between wounded and control plants. Higher concentrations of proteinase inhibitor were always observed in wounded plants compared to control plants but the differences were statistically significant only for the plants that were placed into the environmental chamber.

Histograms in Figure 3 present larval growth patterns for experiments in which the environmental chamber was used. As assessed by larval weight gain, plant age has a much greater impact on leaf quality than does induction of proteinase inhibitors. After 7 days of growth, larvae feeding on the control month old plants were smaller than the larvae on the induced 2 week old plants, in spite of the higher concentrations of proteinase inhibitor in the younger, wounded plants. Larvae feeding on wounded plants held in the greenhouse during wounding also grew more slowly than those on the control plants even though there was no significant difference in plant proteinase inhibitor level between wounded and control plants.

Thus, the results of these experiments reveal the following:

1. When placed in an environmental chamber under conditions that maximize proteinase inhibitor production, younger plants produce higher levels of proteinase inhibitor than older plants.

2. According to *M. sexta*, plant age has a much greater impact on plant quality than does wounding.

3. Wounded tomato plants depress larval growth rates of *M. sexta* even without a concomitant increase in proteinase inhibitor levels.

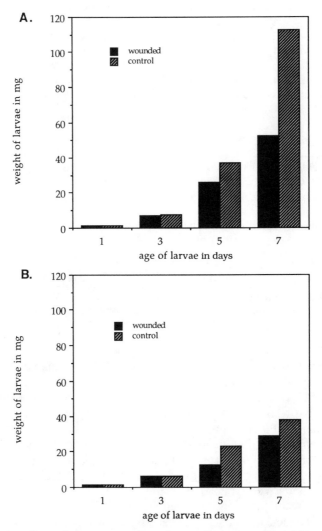

Figure 3   The effect of plant age at time of wounding on growth of *Manduca sexta* larvae. Plants were 2 weeks (**A**) or 1 month (**B**) post transplant at wounding. (From Wolfson and Murdock 1990b.)

4.3   The Impact of Tomato Proteinase Inhibitor on Larval Growth

These studies were designed to assess the impact of tomato proteinase inhibitor alone (in the absence of wounding) on larval *M. sexta* growth. A crude proteinase inhibitor preparation obtained following the procedures of Gustafson and Ryan (1976) was dissolved in water and incorporated into a 6% gelatin solution which was applied to the upper surfaces of excised, fully expanded tomato leaves from mature tomato plants according to the methods of Wolfson and Murdock (1987b). The level of proteinase inhibitor on the leaves was sampled after the larvae were transferred onto fresh leaves and is probably a conservative measure of the actual proteinase inhibitor concentration. In general we only recover 70% of gelatin applied material from tomato leaves. The results of these experiments are presented in Table 4B.

These data suggest that consumption of high concentrations of proteinase inhibitor in the absence of wounded leaves has no statistically significant effect on larval growth in *M. sexta*. They offer an interesting contrast to some of the data presented earlier. In this instance, there is no significant difference in larval weights even though there was a large difference in concentration of proteinase inhibitor consumed in the diets. Perhaps for some insects it has been another plant response to wounding rather than induced proteinase inhibitor levels, or some synergistic interaction between them that has been responsible for the lower growth rates observed with wounded tomato plants.

4.4 Effect of Plant Age on Proteinase Inhibitor Induction

Ryan and his colleagues have studied the proteinase induction system only in young tomato plants. To evaluate the ecological impact of the induction of proteinase inhibitors on herbivores, it was necessary to know whether the induction phenomenon was restricted to young plants or whether it continued to occur as plants matured. In contrast to the previous experiments reported in section 4.2, these experiments were intended to mimic field conditions. I grew plants in an environmental chamber under realistic growing conditions: 14 h light, 27 °C days, 20 °C nights. Plants were germinated in vermiculite and transplanted as described earlier. Samples of 10 plants were then injured at 10, 13, 16, 19, 22, 25, 28, and 31 days after transplanting. Prior to injury one leaflet from the leaf above the youngest injured leaf was sampled per plant; 2 days after injury the matching, remaining leaflet was sampled. The leaflets were frozen after sampling and later evaluated for proteinase inhibitor concentration. The wounding was in three discrete phases over 8 h. The amount of leaf tissue wounded was increased as the plants aged, although only the older, lower leaves of the plants were wounded. The results of these experiments are presented in Figure 4.

These results demonstrate that induction of proteinase inhibitors under the conditions described is an age-dependent phenomenon. It is possible that the decrease in proteinase inhibitor induction is related to root binding as discussed by Baldwin (this volume, Chapter 2). The plants used to assess induction for ages 10–22 days were in smaller pots than those used for days 25–31. If root binding were important one might expect to see a change in the downward trend between days 22 and 25, but that is not apparent.

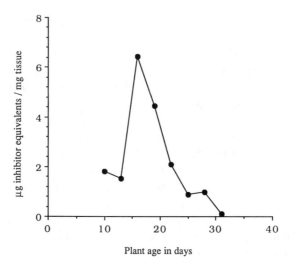

Figure 4   The effect of plant age on production of proteinase inhibitor in response to wounding. (From Wolfson and Murdock, 1990b.)

5   CONCLUSION

Research reported to date indicates that proteinase inhibitors can have a major impact on insect growth and development but not all proteinase inhibitors will have similar potency vis-à-vis an insect. The impact of a proteinase inhibitor on a specific insect will be determined by the class of proteinase(s) the insect utilizes, the class of proteinase inhibitor ingested, and the specificity of that inhibitor for the insect's proteinase. Although in vitro inhibition is not a perfect predictor of in vivo inhibition, it can point to potentially important classes of proteinase inhibitors.

Proteinase inhibitors can be induced in tomato leaves in response to wounding, but the level of induced inhibitor depends on the environmental conditions that the plants are exposed to at the time of wounding. Plant quality, as assessed by growth rates of larval *M. sexta*, was always negatively affected by wounding; that is, wounded plants fostered slower growth rates than the control plants. But lower plant quality of wounded plants was not necessarily associated with high levels of induced proteinase inhibitor in the consumed foliage. The independence between proteinase inhibitor consumption and growth rate was confirmed in studies in which *M. sexta* consumed unwounded tomato leaves with high concentrations of proteinase inhibitors; proteinase inhibitor in the absence of wounded leaves had no adverse affect on growth. In addition, the data indicate that very young tomato plants are more susceptible to induction than older plants. These results clearly indicate that the simple scenario in which induced proteinase inhibitors always have a great impact on important insect herbivores is inappropriate.

A substantial amount of work needs to be done on the potential impact of induced proteinase inhibitors as defensive compounds directed toward insect herbivores. Information on the responses of other tomato herbivores is required before any generalization about the ecological importance of induced proteinase inhibitors in tomatoes should be made. It is certainly possible that current tomato herbivores have adapted to tomato inhibitors but that these inhibitors could be potent inhibitors among nonadapted insects.

ACKNOWLEDGMENTS

I am grateful to Larry L. Murdock for discussions and comments on the content of this chapter. The author's research described in this chapter was done in collaboration with LLM and supported in part by National Science Foundation Grant BSR-8415661.

REFERENCES

Abe, K., and S. Arai. 1985. Purification of a cysteine proteinase inhibitor from rice, *Oryza sativa* L *japonica*. *Agric. Biol. Chem.* 49:3349–3350.

Abe, M., S. Arai, H. Kato, and M. Fujimaki. 1980. Thiol-protease inhibitors occurring in endosperm of corn. *Agric. Biol. Chem.* 44:685–686.

Akers, C. P., and J. E. Hoff. 1980. Simultaneous formation of chymopapain inhibitor activity and cubical crystals in tomato leaves. *Can. J. Bot.* 58:1000–1003.

Applebaum, S. W. 1964. Physiological aspects of host specificity in Bruchidae. I. General considerations of developmental compatibility. *J. Insect Physiol.* 10:783–788.

Applebaum, S. W. 1985. Biochemistry of digestion. In G. A. Kerkut and L. I. Gilbert (eds.), Vol. 4. *Comprehensive Insect Physiology, Biochemistry and Pharmacology*. Elmsford, NY: Pergamon, pp. 279–311.

Applebaum, S. W., B. Gestetner, and Y. Birk. 1965. Physiological aspects of host specificity in the Bruchidae. IV. Developmental incompatability for *Callosobruchus*. *J. Insect Physiol.* 11:611–616.

Baker, J. E., S. M. Woo, and M. A. Mullen. 1984. Distribution of proteinases and carbohydrases in the midgut of larvae of the sweetpotato weevil *Sylas formicarius elegantulus* and response of proteinases to inhibitors from sweetpotato. *Entomol. Exp. Applic.* 36:97–105.

Baldwin, I. 1990. Damaged-induced alkaloids in wild tobacco. 1991. In D. W. Tallamy and M. J. Raupp (eds.), *Phytochemical Induction by Herbivores*. New York: Wiley, pp. 47–69.

Barrett, A. J. 1977. Introduction to the history and classification of tissue proteinases. In A. J. Barrett (ed.), *Proteinases in Mammalian Cells and Tissues*. Amsterdam: Elsevier/North-Holland Biomedial, pp. 1–21.

Barrett, A. J. 1986. The classes of proteolytic enzymes. In M. J. Dalling (ed.), *Plant Proteolytic Enzymes*, Vol. 1. Boca Raton, FL: CRC, pp. 1–16.

Borovsky, D. 1985. Characterization of proteolytic enzymes of the midgut and excreta of the biting fly, *Stomoxys calcitrans* L. *Arch. Insect Biochem. Phys.* 2:145–159.

Borovsky, D. 1986. Proteolytic enzymes and blood digestion in the mosquito, *Culex nigripalpus*. *Arch. Insect Biochem. Physiol.* 3:147–160.

Broadway, R. M., and S. S. Duffey. 1986. Plant proteinase inhibitors: mechanism of action and effect on the growth and digestive physiology of larval *Heliothis zea* and *Spodoptera exiqua*. *J. Insect Physiol.* 32:827–833.

Broadway, R. M., S. S. Duffey, G. Pearce, and C. A. Ryan. 1986. Plant proteinase inhibitors: a defense against herbivorous insects? *Entomol. Exp. Applic.* 41:33–38.

Chang, C. R., and C. C. Tsen. 1981. Isolation of trypsin inhibitors from rye, triticale, and wheat samples. *Cereal Chem.* 58:207–210.

Cheeseman, M. T., and C. Gillott. 1987. Organization of protein digestion in adult *Calosoma calidum* (Coleoptera: Carabidae). *J. Insect Physiol.* 33:1–8.

Deloach, J. R., and G. E. Spates. 1980. Effect of soybean trypsin inhibitor-loaded erthrocytes on fecundity and midgut protease and hemolysis activity in stable flies. *J. Econ. Entomol.* 73:590–594.

Edwards, P. J., S. D. Wratten, and H. Cox. 1985. Wound-induced changes in the acceptability of tomato to larvae of *Spodoptera littoralis*: a laboratory bioassay. *Ecol. Entomol.* 10:155–158.

Elfant, M., L. Bryant, and B. Starcher. 1985. Isolation and characterization of a proteinase inhibitor from marama beans. *Proc. Soc. Exp. Biol. Med.* 180:329–333.

Engelmann, F., and W. P. M. Geraerts. 1980. The proteases and the protease inhibitor in the midgut of *Leucophaea maderae*. *J. Insect Physiol.* 26:703–710.

Garcia, E. S., J. A. Guimaraes, and J. L. Prado. 1978. Purification and characterization of a sulfhydryl-dependent protease from *Rhodnius prolixus* midgut. *Arch. Biochem. Biophys.* 188:315–322.

Gatehouse, A. M. R., and D. Boulter. 1983. Assessment of the antimetabolic effects of trypsin inhibitors from cowpea (*Vigna unguiculata*) and other legumes on development of the bruchid beetle *Callosobruchus maculatus*. *J. Sci. Food Agric.* 34:345–350.

Gatehouse, A. M. R., J. A. Gatehouse, P. Dobie, A. M. Kilminster, and D. Boulter. 1979. Biochemical basis of insect resistance in *Vigna unguiculata*. *J. Sci. Food Agric.* 30:948–958.

Gatehouse, A. M. R., K. J. Butler, K. A. Fenton, and J. A. Gatehouse. 1985. Presence and partial characterisation of a major proteolytic enzyme in the larval gut of *Callobruchus maculatus*. *Entomol. Exp. Applic.* 39:279–286.

Grayson, J. M. 1951. Acidity–alkalinity in the alimentary canal of twenty insect species. *Va. J. Sci.* 2:46–59.

Grayson, J. M. 1958. Digestive tract pH of six species of Coleoptera. *Ann. Entomol. Soc. Am.* 51:403–405.

Green, T. R. and C. A. Ryan. 1972. Wound-induced proteinase inhibitor in plant leaves: a possible defense mechanism against insects. *Science* 175:776–777.

Green, T. R., and C. A. Ryan. 1973. Wound-induced proteinase inhibitor in tomato leaves. *Plant Physiol.* 51:19–21.

Gustafson, G., and C. A. Ryan. 1976. Specificity of protein turnover in tomato leaves. *J. Biol. Chem.* 251:7004–7010.

Hanada, K., M. Tamai, S. Morimoto, T. Adachi, S. Ohmlura, J. Sawada, and I. Tanaka. 1978a. Inhibitory activities of E-64 derivatives on papain. *Agric. Biol. Chem.* 42:537–541.

Hanada, K., M. Tamai, S. Morimoto, T. Adachi, S. Ohmlura, J. Sawada, and I. Tanaka. 1978b. Isolation and characterization of E-64, a new thiol protease inhibitor. *Agric. Biol. Chem.* 42:523–528.

House, H. L. 1974. Digestion. In M. Rockstein (ed.), *The Physiology of Insecta*, Vol. 5. New York: Academic, pp. 63–117.

Houseman, J. G. 1978. A thiol-activated digestive proteinase from adults of *Rhodnius prolixus*. *Can. J. Zool.* 56:1140–1143.

Houseman, J. G., and A. E. R. Downe. 1982a. Identification and partial characterization of digestive proteinases from two species of bedbug (Hemiptera:Cimicidae). *Can. J. Zool.* 60:1837–1840.

Houseman, J. G., and A. E. R. Downe. 1982b. Characterization of an acidic proteinase from the posterior midgut of *Rhodnius prolixus* Stäl (Hemiptera:Reduviidae). *Insect Biochem.* 12:651–655.

Houseman, J. G., and A. E. R. Downe. 1983. Cathepsin D-like activity in the posterior midgut of hemipteran insects. *Comp. Biochem. Physiol.* 75B:509–512.

Houseman, J. G., F. C. Campbell, and P. E. Morrison. 1987. A preliminary characterization of digestive proteases in the posterior midgut of the stable fly *Stomoxys calcitrans* (L.) (Diptera: Muscidae). *Insect Biochem.* 17:213–218.

Ishikawa, C., K. Watanabe, N. Sakata, C. Nakagaki, S. Nakamura, and K. Takahashi. 1985. Adzuki bean (*Vigna angularis*) protease inhibitors: isolation and amino acid sequences. *J. Biochem.* 97:55–70.

Janzen, D. H., H. B. Juster, and E. A. Bell. 1977. Toxicity of secondary compounds to the seed-eating larvae of the bruchid beetle *Callosobruchus maculatus*. *Phytochemistry* 16:223.

Kitch, L. W., and L. L. Murdock. 1986. Partial characterization of a major gut thiol proteinase from larvae of *Callosobruchus maculatus* F. *Arch. Insect Biochem. Physiol.* 3:561–575.

Kumar, P. M. H., T. K. Virupaksha, and P. Vithayathil. 1979. Sorghum proteinase inhibitors. *Int. J. Peptide Protein Res.* 13:153–160.

Kunz, P. A. 1978. Resolution and properties of the proteinases in the larva of the mosquito, *Aedes aegypti*. *Insect Biochem.* 8:43–51.

Laskowski, M., Jr., and I. Kato. 1980. Protein inhibitors of proteinases. *Annu. Rev. Biochem.* 49:593–626.

Lipke, H., G. S. Fraenkel, and I. E. Liener. 1954. Effect of soybean inhibitors on growth of *Tribolium confusum*. *Agric. Food Chem.* 2:410–414.

Manjunath, N., P. S. Veerabhadrappa, and T. Virupaksha. 1983. Isolation and characterization of a trypsin inhibitor from finger millet. *Phytochemistry* 22:2349–2357.

Mattson, W. J., Jr. 1980. Herbivory in relation to plant nitrogen content. *Annu. Rev. Ecol. Syst.* 11:119–161.

Miller, J. W., K. J. Kramer, and J. H. Law. 1974. Isolation and partial characterization of the larval midgut trypsin from the tobacco hornworm, *Manduca sexta* Johannson (Lepidoptera: Sphingidae). *Comp. Biochem. Physiol.* 48B:117–129.

Mosolov, V. V., and M. N. Shul'gin. 1986. Protein inhibitors of microbial proteinases from wheat, rye and triticale. *Planta* 167:595–600.

Murdock, L. L., G. Brookhart, P. E. Dunn, D. E. Foard, S. Kelley, L. Kitch, R. E. Shade, R. H. Shukle, and J. L. Wolfson. 1987. Cysteine digestive proteinases in Coleoptera. *Comp. Biochem. Physiol.* 87B:783–787.

Murdock, L. L., R. E. Shade, and M. Pomeroy. 1988. Effects of E-64, a cysteine proteinase inhibitor, on cowpea weevil growth, development and fecundity. *Environ. Entomol.* 17:467–469.

North, M. J. 1982. Comparative biochemistry of the proteinases of eucaryotic microorganisms. *Microbiol. Rev.* 46:308–340.

Rascon, A., D. Seidl, W. G. Jaffe, and A. Aizman. 1985. Inhibition of trypsins and chymotrypsins from different animal species: a comparative study. *Comp. Biochem. Physiol.* 82B:375–378.

Rele, M. V., H. G. Vartak, and V. Jagannathan. 1980. Proteinase inhibitors from *Vigna unguiculata* subsp. *cylindrica. Arch. Biochem. Biophys.* 204:117–128.

Rhoades, D. F. 1979. Evolution of plant chemical defense against herbivores. In G. A. Rosenthal and D. H. Janzen (eds.), *Herbivores: Their Interaction with Secondary Plant Metabolites*. New York: Academic, pp. 4–54.

Richardson, M. 1977. The proteinase inhibitors of plants and micro-organisms. *Phytochemistry* 16:159–169.

Rodis, P., and J. E. Hoff. 1984. Naturally occurring protein crystals in the potato. *Plant Physiol.* 74:907–911.

Ryan, C. A. 1966. Chymotrypsin inhibitor I from potatoes: reactivity with mammalian, plant, bacterial, and fungal proteinases. *Biochemisry* 5:1592–1596.

Ryan, C. A. 1968. An inducible protein in potato and tomato leaflets. *Plant Physiol.* 43:1880–1881.

Ryan, C. A. 1974. Assay and biochemical properties of the proteinase inhibitor-inducing factor, a wound hormone. *Plant Physiol.* 54:328–332.

Ryan, C. A. 1979. Proteinase inhibitors. In G. A. Rosenthal and D. H. Janzen (eds.). *Herbivores: Their Interaction with Secondary Plant Metabolites*. New York: Academic, pp. 599–618.

Ryan, C. A., and T. R. Green. 1974. Proteinase inhibitors in natural plant protection. *Recent Adv. Phytochem.* 8:123–140.

Ryan, C. A., and W. Huisman. 1970. The regulation of synthesis and storage of chymotrypsin inhibitor I in leaves of potato and tomato plants. *Plant Physiol.* 45:484–489.

Shade, R. E., L. L. Murdock, D. E. Foard, and M. A. Pomeroy. 1986. Artificial seed system for bioassay of cowpea weevil (Coleoptera:Bruchidae) growth and development. *Environ. Entomol.* 15:1286–1291.

Shukle, R. H., and L. L. Murdock. 1983. Lipoxygenase, trypsin inhibitor, and lectin from soybeans: effects on larval growth of *Manduca sexta* (Lepidoptera:Sphingidae). *Environ. Entomol.* 12:787–791.

Shukle, R. H., L. L. Murdock, and R. L. Gallun. 1985. Identification and partial characterization of a major gut proteinase from larvae of the Hessian fly, *Mayetiola destructor* (Say) (Diptera:Cecidomyiidae). *Insect Biochem.* 15:93–101.

Slansky, F., Jr., and P. Feeny. 1977. Stabilization of the rate of nitrogen accumulation by larvae of the cabbage butterfly on wild and cultivated food plants. *Ecol. Monogr.* 47:209–228.

Steffens, R., F. R. Fox, and B. Kassell. 1978. Effect of trypsin inhibitors on growth and metamorphosis of corn borer larvae *Ostrinia nubilalis* (Hubner). *J. Agric. Food Chem.* 26:170–174.

Swingle, M. C. 1931. Hydrogen ion concentration within the digestive tract of certain insects. *Ann. Entomol. Soc. Am.* 24:489–495.

Vaje, S., D. Mossakowski, and D. Gabel. 1984. Temporal, intra- and interspecific variation of proteolytic enzymes in carabid-beetles. *Insect Biochem.* 14:313–320.

Wagner, F. W. 1986. Assessment of methodology for the purification, characterization, and measurement of proteases. In M. J. Dalling (ed.), *Plant Proteolytic Enzymes*, Vol. 1. Baton Rouge, FL: CRC.

Walker-Simmons, M., and C. A. Ryan. 1977. Wound-induced accumulation of trypsin inhibitor activities in plant leaves. *Plant Physiol.* 59:437–439.

Weder, J. K. P. 1981. Protease inhibitors in the Leguminosae. In R. M. Polhill and P. H. Raven (eds.), *Advances in Legume Systematics*, Part 2. Kew, UK: Royal Botanic Gardens.

Wieman, K. F., and S. S. Nielsen. 1988. Isolation and partial characterization of a major gut proteinase from larval *Acanthoscelides obtectus* Say (Coleoptera:Bruchidae). *Comp. Biochem. Physiol.* 89B:419–426.

Wint, G. R. W. 1981. The effect of foliar nutrients upon the growth and feeding of a lepidopteran larva. In *Nitrogen as an Ecological Factor. 22nd Symposia of the British Ecological Society*. Oxford: Blackwell Scientific, pp. 301–320.

Wolfson, J. L., and L. L. Murdock. 1987a. Suppression of larval Colorado potato beetle growth and development by digestive proteinase inhibitors. *Entomol. Exp. Applic.* 44:235–240.

Wolfson, J. L., and L. L. Murdock. 1987b. Method for applying chemicals to leaf surfaces for bioassay with herbivorous insects. *J. Econ. Entomol.* 80:1334–1336.

Wolfson, J. L., and L. L. Murdock. 1990a. Diversity in digestive proteinase activity among insects. *J. Chem. Ecol.* 16:1089–1102.

Wolfson, J. L., and L. L. Murdock. 1990b. Growth of *Manduca sexta* on wounded tomato plants: role of induced proteinase inhibitors. *Entomol. Exp. Applic.* 54:257–264.

CHAPTER 11

INDUCED DEFENSIVE REACTIONS IN CONIFER–BARK BEETLE SYSTEMS

KENNETH F. RAFFA

Department of Entomology, University of Wisconsin, Madison, WI 53706

1 Introduction
2 Life history and biology of bark beetles
3 Role of induced reactions in conifer resistance to bark beetle–microbial complexes
4 Induced conifer responses to bark beetle attack
   4.1 Histological changes
   4.2 Chemical changes
   4.3 Biological effects
5 Elicitation of induced responses
   5.1 Biotic and purified elicitors
   5.2 Sequential stages of induction
6 Beetle counteradaptations to induced defenses
   6.1 Exhaustion of host defenses
   6.2 Tolerance and avoidance of host toxins
   6.3 Integration of colonization strategies
   6.4 Conflicting rate reactions: accumulation versus depletion of allelochemicals
   6.5 Reproductive costs of mass attack
7 Factors affecting the extent of induced responses
   7.1 Host stress
   7.2 Effects of prior exposure, inoculum load, and site of attack
   7.3 Seasonal and age-related variation
   7.4 Genetic and environmental components of constitutive and induced defenses
8 Role of host resistance in the population dynamics and management of bark beetles
   8.1 Population dynamics
   8.2 Management considerations
   References

1 INTRODUCTION

Bark beetles (Coleoptera:Scolytidae) have been the object of intense study for several decades, primarily because of the severe economic problems that they pose. However, they also possess a variety of traits that make them extremely useful organisms for studying coevolution between plants and insects (Mitton and Sturgeon 1982, Raffa and Berryman 1987). Among the most important of these are beetle associations with symbiotic fungi, the ability to metabolically convert host allelochemicals into aggregation and antiaggregation pheromones, and kairomonal relationships with natural enemies and competitors (Waters et al. 1985). Bark beetles can impose strong selective pressures on host populations by killing large numbers of mature healthy trees. Induced defenses play a major role in tree survival, but counteradaptations by beetles can interfere with normal functioning of the tree's physiological responses. Interactions between tree metabolic processes and beetle behavior are shaped by environmental and genetic factors. The outcome of each colonization attempt is usually discrete: Either the cohort of attacking beetles kills the tree and reproduces successfully, or the tree resists attack and remains an unsuitable breeding substrate.

2 LIFE HISTORY AND BIOLOGY OF BARK BEETLES

Bark beetle development occurs entirely within the subcortical region of trees. Different species vary in the substrate they typically colonize. Species colonizing relatively healthy trees are called primary species because they are the first organisms to invade otherwise uninfested tissue, or aggressive species because they kill individuals capable of mounting some resistance (Rudinsky 1962). These terms can be used interchangeably, although there are some exceptions in which primary beetles are not particularly aggressive. Conversely, species that attack previously infested hosts or weakened trees are called secondary or nonaggressive species, respectively. Many other species are saprophytic, colonizing only dead trees or dead tree parts. All points along this continuum are represented among the Scolytidae.

Here I will concentrate on primary species that colonize main stems. The rate and extent of active defensive responses in these systems often determine whether the host lives or dies, and whether beetles successfully reproduce or are repelled. The principle genera in this group, *Dendroctonus*, *Ips*, and *Scolytus*, are primarily associated with conifers (Wood 1982).

The subcortical substrate can usually support only one generation of bark beetles. As a resource it is exhausted by the high beetle densities that typically kill the tree, and by numerous secondary insects and microorganisms that follow. Therefore, each adult must locate a new tree suitable for oviposition (Wood 1972).

As beetles bore into selected trees, they oxidize and/or synergize host monoterpenes into aggregation pheromones that elicit landing by conspecifics of both genders (Renwick and Vite 1970, Borden 1984). These conversions are performed both by the beetles themselves, and by specialized bacteria and fungi present in the gut (Brand et al. 1975, 1976, Borden et al. 1986). Beetles also transport a species specific flora of phytopathogenic fungi that disrupt host translocation and assist in killing the tree (Mathre 1964, Whitney 1982). Mating, gallery construction, and oviposition occur in the phloem. Depending on host responses, beetle attacks may be discontinued before these activities are completed, lethally confined by induced reactions, or successful in killing the host. If the tree can be converted into a passive substrate by aggregating adults, eggs hatch and developing larvae feed on host phloem and bark. Nutrient uptake is facilitated by additional fungal, yeast, and bacterial symbionts. Pupation occurs in the bark through which emerging adults drill exit holes and resume dispersal. Recontamination of callow adults with microbial symbionts is an elaborate process, involving spinning and biting by the last-instar larvae, development of specialized fungal-containing structures termed mycangiae during pupation, and sometimes transport of phloem-limited phytopathogenic fungi by phoretic mites. Bearing fungal spores on specialized sporothecae, these mites and attach to emerging beetles (Moser 1985). Because of these symbiotic relationships, the invading agent must be considered a complex of interacting organisms.

3   ROLE OF INDUCED REACTIONS IN CONIFER RESISTANCE

Conifer genera vary in the relative degree to which constitutive versus induced defensive properties confer resistance against bark beetle attack. Members of the genus *Pinus* contain an elaborate system of resin canals throughout the stem (Bannan 1936). Once beetles bore through the outer bark, they sever these canals and resin flows into the wound, thereby impeding their progress (Keen 1938). Duct resin contains monoterpenes that are toxic to a wide variety of insects and microorganisms (Brattsten 1983). These chemicals are also toxic to bark beetles, but apparently not at concentrations actually encountered in the constitutive resin of healthy hosts (Smith 1963, 1965). Once mass attack has commenced, beetles may shovel resin for several days or weeks before finally ovipositing safely (Raffa and Berryman 1983a).

Attempts to correlate aspects of either the constitutive or induced resin defenses of pines with survival during outbreaks have yielded mixed results. For example, physical properties of constitutive resin agree with interspecific rankings in susceptibility to the southern pine beetle, *Dendroctonus frontalis* (Hodges et al. 1979). Conversely, the likelihood of lodgepole pines, *Pinus contorta* var *latifolia*, surviving a severe mountain pine beetle, *D. ponderosae*, outbreak is better correlated with chemical aspects of the induced response than with any constitutive factor (Raffa and Berryman

1982a). Other studies have shown constitutive and/or induced properties to affect the likelihood of host survival; most evidence suggests that both are important (Vite and Wood 1961, Mason 1966, Mahoney 1978, Stark 1965, Hain et al. 1983, Gambliel et al. 1985, Hodges et al. 1985, Paine et al. 1985). The two systems probably function in an integrated fashion that minimizes the chances of beetle colonization success (Berryman 1972, Raffa and Berryman 1983a).

Other members of the Pinaceae have less pronounced resin systems (Bannan 1936). For example, *Abies*, *Tsuga*, and *Cedrus* species do not have a constitutive network of resin canals, but store pockets of resin in subcortical glands. Beetles avoid these glands when entering the tree (Ferrell 1983) and thus can tolerate the host environment. These tree species rely almost exclusively on induced reactions for defense (Berryman 1972, Wong and Berryman 1977, Russell and Berryman 1976, Wright et al. 1979, 1984, Raffa and Berryman 1982b). Some genera, such as *Picea*, and *Larix*, have intermediate constitutive resin systems, with lower duct densities, shorter duct lengths, and lower proportions of functional resin secretory cells within ducts than that which occurs in pines (Bannan 1936).

4 INDUCED CONIFER RESPONSES TO BARK BEETLE ATTACK

4.1 Histological Changes

Rapid cellular reactions are initiated within the host as soon as beetles enter the living tissue beneath the bark. Endogeneous and exogeneous elicitors diffuse from the wounded cells, resulting in a zone of killed tissue (Bernard-Dagan 1988, Berryman 1988, Cheniclet et al. 1988). An elliptical necrotic lesion forms in advance of the insect–fungal complex (Reid et al. 1967). The center of the lesion is filled with resin, which floods the beetle galleries (Fig. 1). The beetle–fungal complex may become confined within this reaction zone and cease development; in such cases the attack is contained.

Cells continue to die in advance of the beetle as long as the insect–fungal complex continues to progress, so lesion length varies (Wong and Berryman 1977, Raffa and Berryman 1983a, Lieutier and Berryman 1988). Under controlled-inoculation conditions, lesion expansion is sigmoidal with time. Subsequent wound periderm formation protects the living tissue from the killed cells. A layer of nonsuberized impervious tissue prevents the movement of potentially diffusive materials into healthy plant tissue (Hain et al. 1983, Lieutier and Berryman 1988). The wounded area then heals over as new tissue is laid down during the next several years. These scars remain in the growth rings for the rest of the tree's life (Ferrell 1973).

4.2 Chemical Changes

A major feature of host response to bark beetle–fungal attack is the broad array of chemical groups involved in the defense. Table 1 summarizes con-

Figure 1 Induced response of lodgepole pine to mountain pine beetle attack. Note the necrotic lesion and pronounced resinosis about beetle gallery. The attacking beetle is killed within the reaction tissue, and adjacent tissue remains undamaged.

ifer–scolytid–fungal systems for which such changes have been quantified. Although some chemical moieties have been more intensely studied than others, it is clear that no one group is totally responsible for halting the successful development of the invading organisms.

The most thoroughly studied chemical responses to bark beetle attack occur in the monoterpene fraction. Total monoterpene content within the reaction zone increases by about 100-fold within a few weeks. Monoterpenes accumulate exponentially with time during the early stages of response (Raffa and Berryman 1982b).

In addition to the rapid increase in total monoterpene content, changes in the relative proportions of monoterpenes may also occur (Russell and Berryman 1976, Raffa and Berryman 1982a,b, Ferrell 1988). That is, each monoterpene increases in an absolute sense, but some are synthesized and transported more extensively than others. The degree to which these qual-

TABLE 1   Summary of Chemical Changes Occurring in Conifer Subcortical Tissue Following Natural and/or Simulated Attack by Primary Bark Beetles[a]

System			Monoterpenes		Resin Acid Content	Increased and de novo Phenol Accumulation	Sugar Content	References
Host	Beetle	Fungus	Content	Altered Ratios				
Pc	Dp	natr. atk.	4, 3.8	+	1.3	+	0.3, 0.8	1, 2
Pc	Dp	Cc	48.2, 2.6, 27	+	3.0	+	0.8	2–4
Pt	Df	Cm	>500, 94.5	+	58.8		0.26	5–8
Ppa	Df	Cm	533.3	+			0.17	6, 7
Pr	Ip	Ci	+					9, 10
Pr	Dv	Lt	+					11, 12
Pb	Ip	Ci	+					9, 10
Pb	Dv	Lt	+					11, 12
Ag	Sv	Ts	25.6	+		+		13–17
Ag	Sv	natr. atk.	+	+				15
Pa	It	Cp	15					18
Ppi	Tp	V	60[b]		60[b]			19

[a] Contents of monoterpenes, resin acids, and sugars are expressed as ratios of reaction to constitutive concentrations. Where an increase or decrease was determined but an absolute value is not available, a plus or minus is indicated, respectively. *Note:* Because of different sampling and analytical methods used by different researchers, comparisons of absolute values cannot be made between species. If several values are available for one system, each is included.
[b] Includes all terpenes.

Abbreviations. Host: Pc, *Pinus contorta*; Pt, *P. taeda*; Ppa, *P. palustris*; Pr, *P. resinosa*; Pb, *P. banksiana*; Ppi, *P. pinaster*; Ag, *Abies grandis*; Pa, *Picea abies*. Beetle: Dp, *Dendroctonus ponderosae*; Df, *D. frontalis*; Dv, *D. valens*; Ip, *Ips pini*; It, *I. typographus*; Sv, *Scolytus ventralis*; Tp, *Tomicinus piniperda*. Fungi: natr. atk., natural attack; Cc, *Ceratocystis clavigera*; Cm, *C. minor*; Ci, *C. ips*; Cp, *C. polonica*; Lt, *Leptographium terebrantis*; Ts, Trichosporium symbioticum; V, *Verticicladiella* spp.

References. 1, Shrimpton 1973a; 2, Miller et al. 1986; 3, Raffa and Berryman 1982a; Raffa and Berryman 1983b; 4, Shrimpton and Watson 1971; 5, Hain et al. 1983; 6, Cook and Hain 1985; 7, Cook and Hain 1986; 8, Gambliel et al. 1985; 9, Raffa and Smalley, 1988a; 10, Raffa and Smalley 1988c; 11, Raffa and Smalley 1988b; 12, Raffa and Smalley, unpublished data; 13, Russell and Berryman 1976; 14, Wright et al. 1979; 15, Raffa and Berryman 1982b; 16, Wong and Berryman 1977; 17, Ferrell et al. 1988; 18, Christiansen and Horntvedt 1983; 19, Cheniclet et al. 1988.

NOTE: The designation of most *Ceratocystis* and *Trichosporium* associated with Scolytidae has been changed to *Ophiostoma*. The older designations are retained here to facilitate literature review.

itative changes occur varies between tree species. In general, qualitative changes are more pronounced in *Abies* than in *Pinus* (Raffa and Berryman 1987). Interestingly, aggressive colonization behavior has not evolved among scolytids that colonize *Abies*. Such extensive changes in host chemistry during induction may lower the relative advantage to beetles that attack healthy trees (Raffa and Berryman 1987).

A complex of other chemical conversions occurs during induction, most of which are associated with the mevalonic acid or phenylalanine pathways. Resin acids and other oxygenated terpenes, sesquiterpenes, and phenolics undergo both qualitative and quantitative changes (Shrimpton 1973a,b, Wong and Berryman 1977, Raffa and Berryman 1983b, Gambliel et al. 1985).

4.3   Biological Effects

In cases where induced defensive responses are successful, the beetles and their symbiotic fungi are unable to develop. The beetles usually leave the tree, and the only evidence of their colonization attempt is the necrotic lesions described previously. If beetles are not soon joined by a sufficient number of other beetles to overwhelm the tree's defenses, the galleries become flooded, few eggs hatch, and those larvae that do eclose construct very short mines before dying. The adults are also eventually trapped and killed in a thick mass of resin (Fig. 1).

Chemicals that accumulate during defensive reactions are largely responsible for the failure of beetles and fungi to become established. Extractives from induced but not constitutive lodgepole pine phloem inhibit the growth of *Ceratocystis clavigera* and *C. montia*, the two major fungal symbionts of *D. ponderosae* (Shrimpton 1973a,b). Likewise, resin collected from induced reaction tissue is more repellent to the fir engraver, *Scolytus ventralis*, than resin from constitutive glands (Bordasch and Berryman 1977). Paine et al. (1985) simulated the effects of reaction tissue on *D. frontalis* by inducing responses on living trees, allowing defensive reactions to proceed for 6 weeks, and then forcing beetles to colonize these trees after felling. They found higher adult mortality in induced than control tissue and reproduction was reduced by 56%.

Purified components of reaction tissue have been shown to affect the invading complex in several ways. For example, synthetic monoterpenes

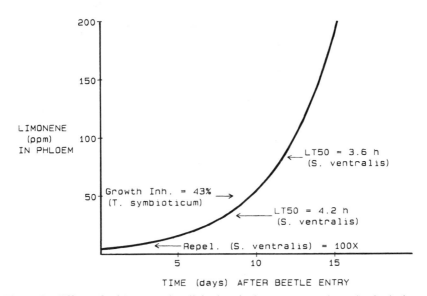

Figure 2 Effect of *Abies grandis* allelochemical transport and synthesis during an induced response to beetle attack on beetle behavior, survival, and symbionts. The accumulation of limonene through time is shown by the solid line fit to data from Raffa and Berryman (1982b). At various concentrations, results of beetle and fungal assays with synthetic limonene are shown (Bordasch and Berryman 1977, Raffa et al. 1985).

are toxic (Smith 1963, 1965, Coyne and Lott 1976, Raffa et al. 1985) and repellent (Bordasch and Berryman 1977) to adult beetles, ovicidal (Raffa and Berryman 1983b), and inhibitory to associated fungi (Cobb et al. 1968, Raffa et al. 1985) at concentrations present in reaction tissue. The significance of chemical changes induced by beetle attack in reducing the suitability of selected tissue is illustrated in Figure 2. The accumulation of limonene in *A. grandis* phloem following a controlled inoculation is shown through time, and a sampling of results from various bioassays are indicated along the curve. For example, an exposure of only 4.2 h to the concentration of limonene that is present 7 days after inoculation killed half of *S. ventralis* adults. All died within 22 h. The concentration of limonene present at this stage is also high enough to greatly reduce the growth of the fir engraver's most virulent symbiotic fungus, *Trichosporium symbioticum*. A single attack is

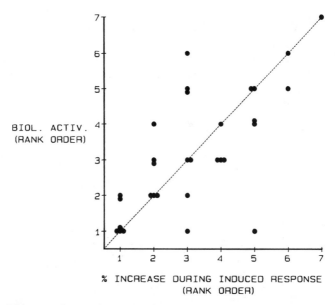

Figure 3 Disproportionate increase of monoterpenes most deleterious to bark beetles during conifer subcortical induced responses. For all systems where both compositional changes during induction and biological effects of synthetic materials are known, the relative increases are ranked along the abscissa and the relative biological activities are ranked along the ordinate. The dashed line indicates the hypothetical perfect correspondence: Spearman's coefficient of rank = 0.82, $t = 7.57, p < .001$. Chemical data are from *Pinus taeda* (Hain et al. 1983), *Abies grandis* (Raffa and Berryman 1982b), and *P. contorta* (Raffa and Berryman 1982b). Biological data include toxicity to *Dendroctonus frontalis* (Coyne and Lott 1976) and *Scolytus ventralis* (Bordasch and Berryman 1977), ovicidal effects against *D. ponderosae* (Raffa and Berryman 1983b), and fungistatic effects against *Trichosporium symbioticum* (Raffa et al. 1985).

unlikely to persist even this long, however, because the concentration of limonene in the phloem 3 days after attack is sufficient to repel most adults. Although Figure 2 shows the effects of only one monoterpene for simplicity, the total defensive profile of the tree is further augmented by corresponding increases in seven additional major monoterpenes, phenolics, and possible synergistic interactions. In summary, the invasion of the *S. ventralis–T. symbioticum* complex causes changes in a healthy host that eventually render the substrate intolerable to bark beetles.

This relationship between defensive efficacy and conformational changes during induction applies to several compounds and several conifer–scolytid systems. Raffa and Berryman (1987) summarized all of the published data where both proportionate changes in host monoterpenes during induction and biological effects of the purified compounds were known. In general, those monoterpenes most deleterious to beetles and their symbionts undergo the largest proportionate increases during induced reactions (Fig. 3). This conclusion is strengthened by the large number of independent studies from which these data were generated, and the variety of conifer–scolytid systems and biological modes of action represented. Interestingly, Blanche et al. (1985) found that the monoterpene content of lightning-struck trees differs from constitutive tissue in a pattern almost exactly opposite of that produced during induced reactions. Their results are ecologically important because lightning-struck trees are highly susceptible to bark beetle attack (Coulson et al. 1983).

Very few bioassays have been conducted with purified oxygenated or higher terpenes, phenolics, and combinations. Such studies are needed before we can fully characterize the effects of induced chemical conversions, because these groups are known to possess insecticidal and fungistatic properties.

5 ELICITATION OF INDUCED RESPONSES

5.1 Biotic and Purified Elicitors

The question of what actually initiates host defensive responses has been addressed by a progression of studies from observations with intact beetle–fungal complexes to purified biochemical probes. Reid et al. (1967) reported necrotic lesions and resinosis associated with naturally resisted mountain pine beetle attacks. Similar observations have been made with every host–beetle system examined to date. Subsequent studies demonstrated that inoculation with fungi vectored by bark beetles can induce responses that are morphologically and chemically similar to naturally occurring defensive reactions (Shrimpton 1973a,b, 1978, Raffa and Berryman 1982b, Hain et al. 1983, Cook and Hain 1985, Miller et al. 1986). The similarities between host responses to controlled fungal inoculations and natural attack are illustrated for two systems in Table 2.

TABLE 2   Comparison of Host Responses to Naturally Resisted Beetle Attacks and Simulated Attack by Inoculation with Fungal Symbiont[a]

System	Parameter	Natural Attack	Fungal Inoculation	Constitutive
Ag–Sv–Ts[b]	% Tricyclene	0.54	0.38	1.4
	% α-Pinene	46.39	33.85	48.1
	% Camphene	0	2.26	1.2
	% β-Pinene	30.46	24.35	42.1
	% Myrcene	14.71	14.54	4.5
	% Sabinene	0	8.72	2.2
	% δ-3-Carene	2.83	2.86	0.3
	% Limonene	2.33	1.32	0.23
	% β-Phellandrene	0.59	3.28	0.01
	% Terpinolene	0.98	2.61	0.17
Pc–Dp–Cc[c]	Monoterpenes (mg/g)	9.2	7.4	2.4
	Soluble sugars (mg/g)	53.9	58.9	67.8
	Starch (mg/g)	15.2	15.5	12.5

[a] Abbreviations as in Table 1. No comparisons between columns 3 and 4 are significant at $p < .05$.
[b] Raffa and Berryman 1982.
[c] Miller et al. 1986.

Comparisons between fungal-inoculated and aseptically injured tissue show that simple physical damage to host cells causes little monoterpene accumulation and no proportionate changes in chemical constituents (Table 3). Therefore, the induced reaction is not a mere wound response. Likewise, inoculations with killed fungi yield less pronounced host responses than with living inoculum (Stephen and Paine 1985, Gamliel et al. 1985, Raffa and Smalley 1988a,b). Depending on the fungus and tree involved, these reactions may be equivalent to aseptic-wound reactions, or intermediate between mechanical wounds and living inoculum. Thus, fungal cell wall constituents are not sufficient to induce a full reaction, and products associated with active fungal metabolism are probably also involved.

A beetle-vectored fungus can induce different levels of reaction in different host species. Conversely, a host species can react differently to various fungal associates of the major bark beetle species that colonize it. It is difficult to generalize at this time, but certain trends are apparent. First, trees respond more extensively to fungal symbionts of those bark beetle species that normally colonize them (Lieutier and Berryman 1988). For example, grand fir, *Abies grandis*, responds actively to fungal symbionts of the fir engraver, but not to symbionts of the mountain pine beetle. Mountain pine beetle symbionts induce active responses in lodgepole pine, but not in the nonhosts *Thuja plicata*, *Larix occidentalis*, and *Pseudotsuga menziesii*. Lieutier and Berryman (1988) proposed that constitutive chemicals or other forms of incompatibility prevent fungal establishment in nonhosts. A second

TABLE 3 Comparisons of Conifer Responses to Aseptic Wounds and Fungi Vectored by Bark Beetles[a]

System	Monoterpenes			Resin Acid Content		Lesion Formation	References
	Content		Altered Ratios				
	Mech/Const	Mech/Inoc[b]	Mech.[b]	Mech/Const[b]	Mech/Inoc[b]	Mech/Inoc	
Pc–Dp–Cc	2	0.16					1
Pt–Df–Cm	1.7	0.017, 0.002		1.72	0.0098	0.18, 0.29	2–4
Pr–Ip–Ci	>1	<1	—			0.27	5, 6
Pr–Dv–Lt	>1	<1	—			0.35	6, 7
Pb–Ip–Ci	>1	<1	—			0.52	5, 6
Pb–Dv–Lt	>1	<1	—			0.65	6, 7
Ppa–Dp–Cm						0.56	8
Ag–Sv–Ts	1.42	0.28	—			0.22	9
Ppi–Tp–V		0.08			0.08		10

[a] Contents of monoterpenes and resin acids are expressed as ratios of mechanical to inoculum treatment concentrations. *Note:* Because of different sampling methodologies, time intervals, and analytical methods, cross-comparisons are not intended.

[b] Negative signs indicate no difference between monoterpene ratios present in constitutive and mechanically wounded tissue.

Abbreviations. As in Table 1. Ppo, *Pinus ponderosa*; Db, *Dendroctonus brevicomis*; Mech, mechanical wound; Const, constitutive tissue; Inoc, inoculation.

References. 1, Miller et al. 1986; 2, Gambliel et al. 1985; 3, Paine et al. 1985; 4, Hain et al. 1983; 5, Raffa and Smalley 1988a; 6, Raffa and Smalley, unpublished data; 7, Raffa and Smalley 1988b; 8, Paine 1984; 9, Raffa and Berryman 1982b; 10, Cheniclet et al. 1988.

trend is that mycangial fungi induce less extensive responses than external symbionts (Paine and Stephen 1987a).

Although fungi alone have been used to induce responses similar to fungal–beetle complexes, I am not aware of any studies using aseptic beetles. Whether beetles alone can induce these responses is academic, however, as all beetles in nature are contaminated.

A number of chemicals have been identified that can elicit responses the same as or very similar to the defensive reactions described previously. Shrimpton (1978) found that cellulolytic enzymes such as proteinases, pectinases, and cellulases cause hypersensitive lesions in lodgepole pine. Phytopathogenic fungi produce such materials during cellular invasion, and presumably resistant hosts respond to their presence by initiating a series of biochemical events that culminate in successful defense.

Miller et al. (1986) found that chitosan, a common fungal cell wall fragment, and proteinase inhibiting inducing factor (PIIF), a pectic fragment released by tomato leaves during wounding that elicits systemic induced reactions (Green and Ryan 1972), induce monoterpene accumulation, soluble sugar reduction, elevated starch titres, and lesion formation in lodgepole pine. Moreover, the changes induced by inoculation with living mountain pine beetles, their symbiont *C. clavigera*, and PIIF were qualitatively identical and quantitatively very similar. Chitosan treatments yielded higher monoterpene accumulation than the other inocula. These results suggest that components of fungal cell walls are at least partially involved in the process of induction, and that diffusion of molecules released at the invasion site initiates cellular reactions in advance of the pathogen.

Not all products of fungi vectored by bark beetles elicit defensive reactions. For example, Hemingway et al. (1977) isolated three isocoumarins from *Ceratocystis minor* that are toxic to loblolly pine, *Pinus taeda*. These materials cause abnormal transpiration, membrane disruption, reduced waterflow, and reduced oleoresin flow, which render trees more favorable to southern pine beetle brood (Hodges et al. 1985). None of the morphological or chemical changes associated with induced defenses have been reported in response to these fungal metabolites.

5.2   Sequential Stages of Induction

Induced responses to bark beetle–fungal invasion can be metabolically costly, involving 90 molecules of ATP for each molecule of monoterpene synthesized (Croteau et al. 1972, Wright et al. 1979). Carbohydrates probably provide the substrate for terpene biosynthesis during induction; a 30% decrease in soluble sugars and a 15% decrease in reducing sugars correspond with this response (Miller and Berryman 1985). Reduced sugar content has been observed within 24–36 h of fungal inoculation. Berryman et al. (1988) proposed that two phases of elicitation are involved that together provide a flexible conservation of energy. First, plant cell wall fragments (such as PIIF) released during beetle feeding activate a controlled metabolic response to mechanical wounding. Neighboring cells undergo multiplication of ribosomes, dictyosomes, and mitochondria, development of endoplastic reticula, decreased lipid content, starch accumulation, and increased number and size of plastids (Cheniclet et al. 1988). This is followed by enlargement of the leucoplasts and differentiation of new secondary resin ducts. Thereafter, fungal products such as chitosan may stimulate hyperactivity of the parenchyma cells, which then produce large quantities of defensive chemicals causing cell wall ruptures (Cheniclet et al. 1988, Lieutier and Berryman 1988). Water-soluble enzymes such as prenyl transferase are activated. Endoelicitors diffuse from cells destroyed during hyperactivity and activate cells in advance of the invading complex (Berryman 1988, Cheniclet et al. 1988, Lieutier and Berryman 1988). Ethylene appears to contribute to this process by acting as a wound hormone (Raffa and Berryman 1982c, Popp et al. 1988). Thus, elicitor release continues until the beetle–fungal complex is contained.

6   BEETLE COUNTERADAPTATIONS TO INDUCED DEFENSES

A variety of mechanisms may allow insects and microorganisms to reproduce despite the presence of potential plant defensive traits (Rhoades 1985). For example, some insects are able to tolerate, sequester, or excrete host defensive chemicals. Others can avoid noxious chemicals through temporal asynchrony with allelochemical abundance (Feeny 1970) or specialization on plant parts low in allelochemicals (Jones 1972). All of these processes occur to varying degrees among bark beetles, but the most significant adaptation is direct interference with the tree's ability to intoxicate the brood.

6.1 Exhaustion of Host Defenses

Bark beetles overcome conifer resistance by mass attack and mechanical disruption of the vascular system (Thalenhorst 1958, Rudinsky 1962, Raffa and Berryman 1983a,b, Cook and Hain 1986). Each beetle cuts across resin ducts, causing the flow of oleoresins to drain out of the entrance tunnel. Beetles physically carry resin away from their galleries until resin flow ceases. This behavior is analogous to the trenching and vein-cutting actions that have been described among certain folivores (Carroll and Hoffman 1980, Dussourd and Eisner 1987), but on a more massive scale. Because host allelochemicals are converted by bark beetles into aggregation pheromones, recruits continue to arrive as long as the tree's defenses are viable.

Once a tree's defensive capacity is physically depleted, its previous ability to respond is never manifested. This can be demonstrated in three ways. First, if trees are artificially screened from further arrivals during natural mass attack, there is a direct relationship between the chances of a tree being killed and the density of entered beetles at the time of caging. This relationship is not linear, but instead reveals a rather clearly defined *threshold of resistance* (Thalenhorst 1958), above which trees are killed and below which attacks are contained (Table 4A). Although beetles enter an environment that is or will become lethal to them, it can be converted into a suitable breeding substrate by the combined effects of the original colonizers and a critical number of subsequent recruits.

Second, each phytochemical response depletes the tree's ability to resist further attacks (Horntvedt et al. 1983, Raffa and Berryman 1983b, Christiansen and Ericsson 1986). Multiple-inoculation experiments generate classic dose–response curves, and eventually the concentrations of allelochemicals fall within the beetles' tolerance limit (Table 4B). Moreover, a tree's response to a single inoculation is related to its inflection point along this curve, suggesting that each tree is characterized by a unique threshold of resistance.

Third, there is a marked decline in both the constitutive and induced defensive properties of a tree during mass attack, even though aggregation lasts only a few days. Unattacked trees do not show equivalent declines during the same period. Thus, beetles entering a tree during the later phase of mass attack encounter an environment with lower resin flow and lesion formation rates than those encountered by earlier arrivals (Table 4C).

6.2 Tolerance and Avoidance of Host Toxins

Primary bark beetles can survive the concentrations of allelochemicals present in the constitutive tissue of most trees. However, the continual syn-

TABLE 4   Evidence for Direct Exhaustion of *Pinus contorta* Defenses by *Dendroctonus ponderosae/Ceratocystis clavigera* Mass Attack

A. *Interruption of Natural Mass Attack with Screen Cages*
(Raffa and Berryman 1983a)

	Tree Condition (number of trees/category)		
	Living		
Attack Density (females/m^2)	Lightly Damaged	Strip- or Patch-Killed	Killed
Below 40	6	3	0
Above 40	0	2	21

B. *Monterpene Accumulation Following Artificial Inoculation*
(Raffa and Berryman 1983b)

	Monoterpene Accumulation (digitizer units)		
Inoculation/m^2	All Trees	Trees with Low Response to Single Inoculation	Trees with High Response to Single Inoculation
3.3	42.1	20.8	74.1
6.2	30.3	25.9	40.6
13.2	46.2	40.4	54.8
25.6	48.0	39.0	86.6
52.8	29.6	18.2	39.1
105.6	11.0	7.3	16.6

C. *Comparison of Pre- and Postattack Measures of Host Resistance*
(Raffa and Berryman 1983a)

Resistance Parameter	Preattack	Postattack
Constitutive: Oleoresin flow (mL/day)	1.10	0.39
Indeed: Necrotic lesion formation (cm)		
3 days postinoculation	24.56	12.33
7 days postinoculation	24.5	15.06

thesis and transport of a broad array of chemicals to the attack site results in higher concentrations than beetles and their progeny can endure. Because trees continue to produce resin as long as they and the beetles remain alive, there is little opportunity for the evolution of beetle tolerance. To be genetically transferred, tolerance would have to be conferred by a gene or genes against high concentrations of all chemical groups simultaneously, and among all life stages of an individual. Total reliance on tolerance does not appear to be common in conifer–scolytid systems.

Likewise, subcortical reactions make avoidance of host allelochemicals in space or time difficult to achieve, at least in the manner achieved by folivores (e.g., Feeny 1970). For example, fir engravers bore between pre-formed resin pockets (Ferrell 1983), but tissue disruption during induced reactions soaks the entire gallery with secondary resin. There is circum-stantial evidence, however, that some bark beetles may attack trees during predictable periods of relative host susceptibility. Trees show seasonal varia-tion in both their preformed (Lorio 1986) and induced (Paine 1984) reactions, but the ecological significance of this variation is still unclear; there are differences in the extent of the response, not in the frequency of its occur-rence. Mountain pine beetles commonly initiate attacks during the late af-ternoon when diurnal tree water deficits peak. Similarly, beetles are sea-sonally active only during mid summer, also a period of maximum water stress. However, these relationships have not been thoroughly studied and there are other plausible explanations of beetle activity rhythms, as well as exceptions to the patterns.

A common means by which plant pathogens avoid defensive reactions is suppression of host recognition mechanisms (Albersheim and Valent 1974, Sequeira 1980), but this option also is not available to bark beetle–fungal complexes. Shredding live tissue with sclerotized mandibles and legs seems too crude a method of attack to disguise. Moreover, all beetles are covered with an array of microbe species, some of which are simply random con-taminants. The likelihood of all components of this flora simultaneously evolving sophisticated nonrecognition properties is remote.

Although primary bark beetles have little opportunity to avoid induced defenses once they have entered healthy trees, they can greatly increase their chances of success by orienting solely to weakened hosts. Beetles undergo a sequence of visual, tactile, olfactory, and gustatory responses that govern host acceptance and discriminate against healthy trees (Elkinton et al. 1981, Hynum and Berryman 1980, Moeck et al. 1981, Raffa and Berryman 1980, 1982c, Raffa 1988). There are also some disadvantages to this behavior, however; stressed trees are normally rather rare, such breeding substrates are usually of lower quality and quantity, and interspecific competition for such limited resources can be severe. The relative advantage of avoiding healthy trees depends on the physiological allocation to defense that typifies most members of a particular host species. This, in turn, depends on the host's general life history parameters (Raffa and Berryman 1987).

Certain bark beetle species that breed just above the soil line seem to provide exceptions to the above generalizations. *Dendroctonus micans*, *D. valens*, and *D. terebrans* can reproduce in living trees without killing them. These species do not engage in mass attack. Rather, the brood feed together, mediated by larval aggregation pheromones (Gregoire et al. 1982). It is not entirely clear how these species tolerate and/or avoid host responses, es-pecially considering the consistent association of *D. valens* and *D. terebrans* with *Leptographium terebrantis* (Barras and Perry 1971), which induces strong defensive responses (Raffa and Smalley 1988b). Maternal care, in the form of extensive and sophisticated gallery architecture that facilitates resin canal drainage, probably contributes to brood survival.

6.3 Integration of Colonization Strategies

The relative importance of avoiding resistant trees, tolerating and/or avoiding resistant responses within trees, and overwhelming the defenses of healthy trees varies among different bark beetle species. The importance of concerted mass attack is related to the preattack vigor of selected trees. This parameter can be estimated by plotting the beetle replacement rate (brood/parent) achieved against the attack density of the parental generation. This curve is typically an inverted parabola with the rising portion attributable to the beetle's cooperative effect and the falling portion due to intraspecific competition (Berryman 1974, Raffa and Berryman 1983a, 1987). Aggressive species such as *D. ponderosae* achieve their maximum replacement rates at higher attack densities than do less aggressive species such as *D. rufipennis*, *S. ventralis*, or *D. pseudotsugae* (Table 5, column 1). Yet, when *D. ponderosae* colonize dead trees, the curve closely resembles that of less aggressive species.

The degree to which avoidance of healthy trees (Table 5, column 2) is critical for reproduction can be estimated by caging beetles onto trees. For example, *D. ponderosae* show higher entry rates and more frequent cases of fatal persistence than *S. ventralis*. Life table data can also reflect these differences, since losses during adult dispersal are presumably related to an

TABLE 5 **Proposed Mechanisms by Which Primary Bark Beetles in the Genera** *Dendroctonus*, *Ips*, **and** *Scolytus* **Contend with Host Defenses**[a]

Species	Exhaust Host Defenses	Avoid Host Defenses by Selecting Weakened Trees	Tolerate Host Defenses
D. frontalis	H	M	M
D. brevicomis	H	M	M
D. ponderosae	H	M	M
I. typographus	MH?	MH	M?
D. rufipennis	MH?	MH	M?
D. pseudotsugae	M	H	M?
I. pini	M?	H	L?
S. ventralis	ML?	H	L
I. grandicollis	L?	H	L?
D. terebrans	L	M	H
D. valens	L	M	H
D. micans	L	M	H

[a] The terms low (L), moderate (M), and high (H) beetle reliance on a particular strategy are intended to depict points along a continuum, not absolute categories.

inability to locate a suitable host. In general, less aggressive species experience higher relative losses during dispersal, and more aggressive species experience higher mortality due to resinosis (Raffa and Berryman 1987).

Estimates of tolerance are based on published laboratory studies with purified chemicals and field observations (Table 5, column 3). Species that can colonize trees without killing them and those colonizing pines with developed constitutive resin duct systems generally exhibit the greatest tolerance. Based on available data, it appears that most bark beetles fall into one of several clearly defined groups, although each species relies on a variety of adaptations to some extent (Table 5).

6.4 Conflicting Rate Reactions: Accumulation versus Depletion of Allelochemicals

The ability of bark beetles and/or their microbial symbionts to convert host compounds into aggregation pheromones and thereby attract conspecifics as long as host resistance persists may seem to render all trees susceptible. Yet under most conditions most trees are not attacked (Hodges et al. 1985). This is not because beetles fail to locate trees. Flight-trap data show that in some systems almost all (if not all) trees in a stand are located by bark beetles each growing season (Raffa and Berryman 1980). When beetles are caged onto trees, only a minority bore in, and many of these subsequently exit (Raffa and Berryman 1982c, 1983b). Beetles orient away from many trees of the preferred host species, despite the beetle population's apparent ability to kill any tree if enough join in the attack.

A complex system of host defense that integrates constitutive and induced mechanisms appears to explain why healthy trees can survive periods of moderate beetle abundance (Berryman 1972). A rapid flow of resin at the attack site may interfere with beetle communication, thereby reducing the interactions among beetles that exhaust host defense. Although monoterpenes from host resin are readily converted to oxidized attractants in the laboratory (Hughes 1973), a beetle tunneling in a pitch tube does not necessarily attract other beetles, even when populations are high. Only about half of the natural attacks that are initiated by single mountain pine beetles elicit aggregation (Raffa and Berryman 1983a). Likewise, when beetles are caged onto lodgepole pines, only a third of the beetles that enter are joined by conspecifics. These observations cannot be attributed to an absence of pheromone precursors in the host or the unavailability of recruits, because phytochemical analyses, concurrent aggregation on neighboring trees, and trapping data discount these possibilities. Although some of these beetles may simply be poor pheromone producers, there appears to be a host role as well. If only attacks that begin in true pioneer fashion are considered, trees on which natural entries do not elicit mass attack have constitutive resin flow rates that are nearly 6 times as high as those on which natural entries lead to mass attack (Raffa and Berryman 1983a).

Several mechanisms involving pheromone biosynthesis, volatile emission, and/or chemoreception could contribute to this interference (Raffa and Berryman 1983a). However, the simplest theory is that volatiles produced by beetles in resinous galleries cannot permeate the tree's gummy secretions, at least in a form that elicits beetle landing. Following initial beetle entry, induced responses begin, and the survival time of the insect is limited (Fig. 2). If additional colonizers do not arrive quickly, secondary resins augment the primary resin flow, and the colonization sequence is interrupted.

Attacks that begin through switching (Gara and Coster 1968, Geizler et al. 1980) from trees in the terminal phase of attack onto healthier neighbors are more likely to attract flying beetles than attacks initiated by solitary beetles (Raffa and Berryman 1983a). In the same stands described above, 90% of the lodgepole pines on which attacks were started by switching underwent mass attack. Likewise, if all trees are included in the analysis, there is no difference in constitutive resin flow rates between naturally entered trees that were or were not subsequently mass attacked. These results suggest that the tree's potential ability to limit pheromone emission is depleted by the simultaneous initiation of several attacks. The interactions between the rate of beetle arrival, determined by the number of entered beetles, local density of flying beetles, and constitutive resin content, and the intensity of the induced host response, determined by tree age, vigor, and heredity, largely determine the outcome of colonization attempts.

The above considerations refer to outbreaks in which there are sufficient beetles to respond to volatile attractants, a condition that may not be fulfilled when populations are low. However, even the most aggressive beetle species exist for decades in stands without undergoing an outbreak; beetles must survive and reproduce during lengthy periods when conspecifics are rare. Because the outcome of colonization attempts is strongly influenced by beetle population density, there may be behavioral differences between outbreak and nonoutbreak populations. That is, selection may favor relatively discriminating host entry behavior when the arrival of many additional recruits is unlikely, but favor more indiscriminant behavior at high beetle densities (Raffa and Berryman 1983a, 1987, Raffa 1988). This may partially explain repeated observations of foresters who have long described beetles as being more aggressive in their attacks during outbreaks (Keen 1938).

6.5 Reproduction Costs of Mass Attack

The ability of beetles to overcome relatively resistant trees incurs the cost of distributing the host among a large number of colonizers. Although each beetle increases the likelihood of successful colonization, it also decreases the available substrate per individual (Rudinsky 1969). This may provide an explanation as to why some trees are consistently avoided; there is little advantage to colonizing trees that require more beetles to overcome tree defenses than the tree can support in reproduction.

Selecting optimal hosts can be paradoxical, however, because trees that provide the most substrate are usually the most resistant (Raffa 1988). For example, very young trees often have less pronounced induced reactions than healthy mature trees (Safranyik et al. 1975, Raffa and Berryman 1982b), but they also tend to have thinner phloem, bark, and stem diameters, and can therefore support fewer larvae (Amman 1969, 1972, Cole 1962). Conversely, healthy mature trees provide a good food base, but resistance capacity is highest during this stage of development. Older trees that are beginning to decline and mature trees that are suddenly stressed have both sufficient phloem for development and a relatively low threshold of resistance. This trade-off between a tree's threshold for attack and its carrying capacity may have favored bark beetle behaviors that facilitate orientation to old or highly stressed trees (Raffa and Berryman 1987). Tree mortality pattens appear to reflect such beetle preferences (Keen 1938, Rudinsky 1962, Struble 1957, Amman and Cole 1983). While in flight, mountain pine beetles prefer broad objects over narrow dark vertical objects, a trait that is likely to lead them to old trees (Shepherd 1966). Likewise, they are repelled by heavy resin flow during gallery initiation, and beetles orient away from trees with very high thresholds of resistance (Raffa and Berryman 1983b). Moreover, the most toxic allelochemicals to scolytids are generally the most repellent as well (Smith 1975, Bordasch and Berryman 1977, Raffa et al. 1985).

Although cooperative behavior incurs intraspecific competition, beetles can reduce this effect by preventing further entries once selected trees have been overcome and/or reducing oviposition to a fraction of the potential egg complement before reemerging to attack new hosts. These behaviors are governed by a complex integration of chemical, acoustic, and visual cues (Renwick and Vite 1970, Rudinsky 1969, 1973, Coulson et al. 1976).

7 FACTORS AFFECTING THE EXTENT OF INDUCED RESPONSES

7.1 Host Stress

Conifers show considerable intraspecific variation in their extent and rate of response to simulated and natural bark beetle–fungal attack. Induced responses are decreased by disease, mechanical damage, intraspecific competition, water deficit, natural and artificial defoliation, and poor site condition (Wright et al. 1979, 1984, Raffa and Berryman 1982b, Waring and Pitman 1983, Paine et al. 1985, Miller et al. 1986). These physiological relationships are ecologically significant because bark beetle outbreaks are correlated with each of the above factors (Keen 1938, Rudinsky 1962). They also support the view that almost any biotic or abiotic stress will lower tree resistance (Waring and Pitman 1983). For example, lodgepole pines with high growth rates have stronger resistant responses to controlled inoculations (Shrimpton 1973a,b) and are more likely to survive natural mountain pine beetle outbreaks than trees with slow growth rates (Mahoney 1978, Raffa and Berryman 1983).

7.2   Effects of Prior Exposure, Inoculum Load, and Site of Attack

Most of the available evidence suggests that conifer responses to subcortical invasion are short term and localized (Miller 1984, Paine et al. 1985). In this regard they seem to differ substantially from many induced responses to defoliators. However, Cook and Hain (1988) observed elevated and altered monoterpene composition in previously attacked and inoculated trees. Within a tree's capacity to respond, inoculum load does not affect the magnitude of host responses (Paine and Stephen 1985).

Experiments examining within-tree variation along different heights of trees have yielded mixed results. Reid and Shrimpton (1971) observed more resinous responses along the basal 20 ft than in the upper stem of lodgepole pines, but Paine et al. (1985) found no height-related variation in loblolly pines. No significant variation due to aspect was observed in *P. resinosa* or *P. banksiana* inoculated with *Ceratocystis ips*, *C. nigrocarpa*, or *L. terebrantis* (Raffa and Smalley 1988a,b).

7.3   Seasonal and Age Related Variation

Pronounced seasonal differences occur in the induced subcortical responses of conifers. However, the pattern of change for each host or fungal species is unique (Paine 1984, Raffa and Smalley 1988c).

Age-specific patterns (Section 6.5) in induced defensive capacity may allow for substantial host reproduction prior to attack and have important implications to conifer–bark beetle coexistence. In many systems trees produce cones for several decades before their resistance to bark beetles declines. In other systems mortality to bark beetles of aging trees may even be adaptive. If the host is a pioneer species likely to be replaced by understory successors, reduced resistance to bark beetles, bark beetle outbreak, fuel accumulation in the form of thousands of hectares of dead trees, intense large–scale fire, and rapid seed germination from long-lived serotinous cones provide the most likely route of insuring successful conifer reproduction (Amman 1977, Peterman 1978, Raffa and Berryman 1987).

7.4   Genetic and Environmental Components of Constitutive and Induced Defenses

Unfortunately, the degree to which variation in conifer defenses is under genetic vs environmental control is unknown. Available data suggest that the proportion of genetic and environmental regulation of various components of the constitutive and induced defense systems varies considerably (Table 6). For example, an extensive literature on chemotaxonomy, tree

TABLE 6 Proposed Relative Genetic and Environmental Contributions to Various Constitutive and Induced Parameters of Conifer Resistance to Bark Beetles

Parameter	Source of Variation		References
	Genetic	Environmental	
CONSTITUTIVE			
Monoterpene composition	High	Low	1–4
Resin acids	High	Low	5, 6
Total oleoresin content	Moderate	Moderate	7–10
Viscosity	High	Low	9, 11
Rate of crystallization	High	Low	12
Oleoresin exudation flow and pressure	Low	High	8, 13, 14
INDUCED			
Elicitation (occurrence)	High?	Low	15–19
Extent and rate of response	Moderate?	High	17–23

References. 1, Hanover 1966; 2, Hanover 1975; 3, Zavarin et al. 1969; 4, Tobolski and Hanover 1971; 5, Zinkel 1977; 6, Hanover 1966; 7, Mergen 1953; 8, Rudinsky 1966; 9, Mergen et al. 1955; 10, Squillace and Gansel 1974; 11, McReynolds and Lane 1971; 12, Buijtenen and von Santamour 1972; 13, Lorio and Hodges 1968; 14, Vite 1961; 15, Raffa and Smalley 1988a; 16, Raffa and Smalley 1986b; 17, Wright et al. 1979; 18, Wright et al. 1984; 19, Raffa and Berryman 1982a; 20, Raffa and Berryman 1982b; 21, Paine et al. 1985, Miller et al. 1986; 22, Paine and Stephen 1987b; 23, Shrimpton and Watson 1971.

breeding, and naval stores production indicates strong genetic control of the chemical and physical properties of constitutive conifer resin. Nevertheless, constitutive resin composition can be altered by stressful site conditions or severe trauma such as a lightning strike (Blanche et al. 1985). Other aspects of the constitutive defense system are largely determined by environmental conditions. Resin flow rate, for example, is strongly influenced by edaphic, climatic, biotic, and anthropogenic factors (Waters et al. 1985).

There is little intraspecific variation in the qualitative elicitation of induced responses, as all (or almost all) members of the host population respond to their normal beetle-vectored pathogens (Raffa and Berryman 1982a,b, 1983a,b, Paine 1984, Paine and Stephen 1987a,b, Lieutier and Berryman 1988, Raffa and Smalley 1988a–c). Quantitative aspects of the response vary enormously, however, and these are strongly influenced by environmental factors (see section 7.1). There is some circumstantial evidence for a genetic component to these parameters as well. Raffa and Berryman (1982a) assayed surviving lodgepole pines in a stand where mortality exceeded 99%. Although there were no microsite, canopy structure, age, or size differences between the few trees that survived and those that were killed, the former group had higher monoterpene accumulation rates during induced responses to controlled inoculations than other trees in the population.

Whitham (1983) has proposed that a combination of both genetic and environmental control of host defensive traits may partially explain how long-lived trees can remain resistant against insects with short generation times and high potential for evolutionary change. This appears to be true in conifer–scolytid systems. Selection may have favored beetle orientation to a particular physiological condition, rather than to a particular genotype. As a consequence, stable coexistence is common in such a potentially unstable system.

8 ROLE OF HOST RESISTANCE IN THE POPULATION DYNAMICS AND MANAGEMENT OF BARK BEETLES

8.1 Population Dynamics

Bark beetle populations are governed by host availability, which in turn is determined by the interface of tree resistance levels with beetle densities on an areawide basis (Rudinsky 1962, Berryman 1974, 1976, 1979, Coulson 1979, Coulson et al. 1983). During endemic periods, beetle populations rise and fall with the availability and size of stressed trees that are unable to resist attack. Each successful attack depletes the limited pool of stressed trees, and there are rarely sufficient beetles to kill healthy trees. Losses during dispersal offset reproductive gains, and populations can remain near a stable equilibrium density. If an areawide stress such as drought or age weakens many of the trees, however, the population can rise above a critical density, beyond which enough beetles are present to overcome vigorous tree defenses through mass attack. Each reproductive increment expands the range of susceptible trees, generating a positive feedback cycle. Even if the predisposing stress is alleviated, the population can continue to expand exponentially. Nearly all suitably sized trees within the host species are eventually killed. With their resource exhausted, beetle populations collapse. Such outbreaks have major effects on forest composition and succession (Borden 1971, Amman and Baker 1972, Amman 1977, Geiszler et al. 1980, Raffa and Berryman 1987).

At low population densities of beetles, mortality due to resinosis and other tree defenses might be expected to be great. However, because aversion from well-defended trees is adaptive and occurs frequently, beetle life tables typically reveal low numbers of beetles killed by resinosis, and very high losses during their search for acceptable hosts (Berryman 1973, 1979, Cole 1981, Wright et al. 1984).

The effects of induced plant responses on the population dynamics of bark beetles can be quite different from what typifies folivores. Whereas high folivore populations often result in decreased host suitability and subsequent population decline, high scolytid densities are only marginally influenced by induced defenses. Conversely, low folivore densities are unlikely to induce long-term responses and so their progeny are unaffected; at low scolytid densities, host defenses play a major role in population behavior. The rapid, lethal responses exhibited by both the herbivore and the plant have apparently led to more eruptive cycles in conifer–bark beetle systems as compared to less dynamic folivore–host interactions.

8.2 Management Considerations

The use of induced host resistance in the management of insect pests has much potential, but also poses several difficult problems. Of foremost importance is the trade-off between defensive capacity and desirable agronomic traits. Plants must allocate limited resources to growth, reproduction, maintenance and defense (Bazzaz et al. 1987); resistance can be correlated with reduced growth rate. There may also be disparities between crop defensive capabilities and palatability to humans, cropping convenience, or resistance to other biotic and abiotic agents (Raffa 1987).

Fortunately, these problems are largely alleviated in bark beetle management. First, the crop is not for human consumption, so noxious defensive chemicals are of less concern. Second, the correlation between defensive capacity (both constitutive and induced) and general plant vigor (see section 7.1) provides much flexibility in terms of cultural manipulation. There is a high degree of compatibility between enhancing host resistance to bark beetles and overall forest management objectives. For example, thinning a stand increases both growth rate and resistance by reducing crowding (Mitchell et al. 1983). Likewise, trees begin to lose resistance about the same time their growth rate slows (Shrimpton 1978, Paine and Stephen 1987b), an optimal time for harvest. Maintenance of high host vigor through stand management practices is the most effective means of preventing bark beetle outbreaks (Rudinsky 1962, Waters et al. 1985).

There are, however, several complicating factors that can reduce the compatibility of resistance enhancement with general forest management. Fluctuations in the value of wood products can make adherence to prescribed rotation and thinning regimens difficult. When these practices are not cost-effective in terms of the immediate return on the removed wood, their pest management value must offset operational costs. Also, cultural practices that disfavor one insect or pathogen may favor another. Thinning young stands, for example, can favor certain root-dwelling insects and fungi. Once established, these organisms can increase susceptibility to bark beetles. A greater understanding of tree physiology, stand dynamics, and forest economics is needed to resolve these issues.

The natural pattern of intertree variation in defensive capability also lends a cautionary note to some proposed methods of enhancing resistance. Specifically, the insertion of foreign genes that encode for an exotic form of resistance may strongly favor insect biotype evolution in this system. The availability of a small pool of weakened trees due to such inevitable events as lightning, wind, and drought is a major factor in competition among individual beetles and consequently the genetic makeup of the beetle population (Raffa 1987, 1988). Forest managers should determine and provide conditions under which those beetles that primarily orient to weakened trees outcompete less discriminating conspecifics. Conferring a resistant factor to all members of the host population, including stressed trees, could place intense selective pressures on bark beetles to overcome the resistance factor. A large pool of accumulated stressed trees would suddenly become available, and eruptive population cycles could result. There is no certainty that under these conditions the current compatibility between host vigor and host resistance would be preserved, or that currently available methods of utilizing host resistance would remain effective. This caution does not necessarily apply to all plant–herbivore systems to equivalent degrees, but stresses the unique features of bark beetle–conifer associations.

Addendum: The literature review was completed in November 1988.

ACKNOWLEDGMENT

This work was supported by the College of Agricultural and Life Sciences, University of Wisconsin, Madison.

REFERENCES

Albersheim, P., and S. S. Valent. 1974. Host-pathogen interactions. VII. Plant pathogens secrete proteins which inhibit enzymes of the host capable of attacking the pathogen. *Plant Physiol.* 53:684–687.

Amman, G. D. 1969. Mountain pine beetle emergence in relation to the depth of lodgepole pine bark. *USDA Forest Service Research Note INT-96.*

Amman, G. D. 1972. Mountain pine beetle brood production in relation to thickness of lodgepole pine phloem. *J. Econ. Entomol.* 65:138–140.

Amman, G. D. 1977. In W. J. Mattson (ed.), *Arthropods in Forest Ecosystems.* New York: Springer, pp. 3–18.

Amman, G. D., and B. H. Baker. 1972. Mountain pine beetle influence on lodgepole pine stand structure. *J. For.* 70:204–209.

Amman, G. D., and W. F. Cole. 1983. Mountain pine beetle dynamics in lodgepole pine forests. II. Population dynamics. *USDA Forest Service General Technical Report INT-145.*

Bannan, M. W. 1936. Vertical resin ducts in the secondary wood of the Abietineae. *New Phytol.* 35:11–46.

Barras, S. J., and T. Perry. 1971. *Leptographium terebrantis* sp. nov. associated with *Dendroctonus terebrans* in loblolly pine. *Mycopathol. Mycol. Appl.* 43:1–10.

Bazzaz, F. A., N. R. Chiariello, P. D. Coley, and L. F. Pitelka. 1987. Allocating resources to reproduction and defense. *BioScience* 37:58–67.

Bernard-Dagan, C. 1988. Seasonal variations in energy sources and biosynthesis of terpenes in Maritime Pine. In W. J. Mattson, J. Levieux, and C. Bernard-Dagan (eds.), *Mechanisms of Woody Plant Defenses against Insects*. New York: Springer, pp. 93–116.

Berryman, A. A. 1972. Resistance of conifers to invasion by bark beetle–fungal associations. *BioScience* 22:598–602.

Berryman, A. A. 1973. Population dynamics of the fir engraver *Scolytus ventralis* (Coleopera: Scolytidae). I. Analysis of population behavior and survival from 1964 to 1971. *Can. Entomol.* 105:1465–1488.

Berryman, A. A. 1974. Dyanmics of bark beetle populations: toward a general productivity model. *Environ. Entomol.* 3:579–585.

Berryman, A. A. 1976. Theoretical explanation of mountain pine beetle dynamics in lodgepole pine forests. *Environ. Entomol.* 5:1225–1233.

Berryman, A. A. 1979. Dynamics of bark beetle populations: analysis of dispersal and redistribution. *Mitt. Schweiz. Entomol. Ges.* 52:227–234.

Berryman, A. A. 1988. Towards a unified theory of plant defense. In W. J. Mattson, J. Levieux, and C. Bernard-Dagan (eds.), *Mechanisms of Woody Plant Defenses against Insects*. New York: Springer, pp. 39–56.

Blanche, C. A., J. D. Hodges, and T. E. Nebeker. 1985. Changes in bark beetle susceptibility indicators in a lightning-struck loblolly pine. *Can. J. Forest Res.* 15:397–399.

Bordash, R. P., and A. A. Berryman. 1977. Host resistance to the fir engraver beetle, *Scolytus ventralis* (Coleoptera: Scolytidae). 2. Repellency of *Abies grandis* resins and some monoterpenes. *Can. Entomol.* 109:95–100.

Borden, J. H. 1971. Changing philosophy in forest-insect management. *Bull. Entomol. Soc. Am.* 17:268–273.

Borden, J. H. 1984. Semiochemical-mediated aggregation and dispersion in the Coleoptera. In T. Lewis (ed.), *Insect Communication*. London: Academic, pp. 123–149.

Borden, J. H., D. W. A. Hunt, D. R. Miller, and K. N. Slesser. 1986. An uncertain outcome of response by individual beetles to variable stimuli. In T. L. Payne, M. C. Birch, and C. E. J. Kennedy (eds.), *Mechanisms in Insect Olfaction*. Oxford: Clarendon, pp. 98–109.

Brand, J. M., J. W. Bracke, A. J. Markovetz, D. L. Wood, and L. E. Browne. 1975. Production of verbenol pheromone by a bacterium isolated from bark beetles. *Nature* 254:137.

Brand, J. M., J. W. Bracke, L. N. Britton, A. J. Markovetz, and J. S. Barras. 1976. Bark beetle pheromones: production of verbenone by a mycangial fungus of *Dendroctonus frontalis*. *J. Chem. Ecol.* 2:195–199.

Brattsten, L. B. 1983. Cytochrome P-450 involvement in the interactions between plant terpenes and insect herbivores. In P. A. Hedin (ed.). *Plant Resistance to Insects*. Washington, D.C.: American Chemical Society, pp. 173–198.

Buijtenen, J. P., and F. Van Santamour. 1972. Resin crystallization related to weevil resistance in white pine (*Pinus strobus*). *Can. Entomol.* 104:215–218.

Carroll, C. R., and C. A. Hoffman. 1980. Chemical feeding deterrent mobilized in response to insect herbivory and counter adaptation by *Epilachna tredecimnotata*. *Science* 209:414–416.

Cheniclet, C., C. Bernard-Dagan, and G. Pauly. 1988. Terpene biosynthesis under pathological conditions. In W. J. Mattson, J. Levieux, and C. Bernard-Dagan (eds.), *Mechanisms of Woody Plant Defenses against Insects*. New York: Springer, pp. 117–130.

Christiansen, E., and A. Ericsson. 1986. Starch reserves in *Picea abies* in relation to defense reaction against a bark beetle transmitted blue-stain fungus, *Ceratocystis polinica*. *Can. J. Forest Res.* 16:78–83.

Christiansen, E., and R. Horntvedt. 1983. Combined *Ips Ceratocystis* attack on Norway spruce and defensive mechanisms of the trees. *Z. Angew. Entomol.* 96:110–118.

Cobb, F. W., K. Krstic, E. Zavarin, and H. W. Barber, Jr. 1968. Inhibitory effects of volatile oleoresin components on *Fomes annosus* and four *Ceratocystis* species. *Phytopathology* 58:1327–1335.

Cole, W. E. 1962. The effects of intraspecific competition within mountain pine beetle broods under lab conditions. *USDA Forest Service Research Note* 97.

Cole, W. E. 1981. Some risks and causes of mortality in mountain pine beetle populations: a long-term analysis. *Res. Popul. Ecol.* 23:116–144.

Cook, S. P., and F. P. Hain. 1985. Qualitative examination of the hypersensitive response of loblolly pine, *Pinus taeda* L., inoculated with two fungal associates of the southern pine beetle, *Dendroctonus frontalis* Zimmermann (Coleoptera:Scolytidae). *Environ. Entomol.* 14:396–400.

Cook, S. P., and F. P. Hain. 1986. Defensive mechanisms of loblolly and shortleaf pine against attack by southern pine beetle, *Dendroctonus frontalis* Zimmermann, and its fungal associate, *Ceratocystis minor* (Hedgecock) Hunt. *J. Chem. Ecol.* 12:1397–1406.

Cook, S. P., and F. P. Hain. 1988. Wound response of loblolly and shortleaf pine attacked or reattacked by *Dendroctonus frontalis* Zimmermann (Coleoptera:Scolytidae) or its fungal associate, *Ceratocystis minor* (Hedgecock) Hunt. *Can. J. Forest Res.* 18:33–37.

Coulson, R. N. 1979. Population dynamics of bark beetles. *Annu. Rev. Entomol.* 24:417–447.

Coulson, R. N., M. Mayyasi, J. L. Foltz, F. P. Hain, and W. C. Martin. 1976. Resource utilization by the southern pine beetle, *Dendroctonus frontalis* (Coleoptera:Scolytidae). *Can. Entomol.* 108:353–362.

Coulson, R. N., P. B. Hennier, R. O. Flamm, E. J. Rykiel, L. C. Hu, and T. L. Payne. 1983. The role of lightning in the epidemiology of the southern pine beetle. *Z. Angew. Entomol.* 96:182–193.

Coyne, J. F., and L. H. Lott. 1976. Toxicity of substances in pine oleoresin to southern pine beetle. *J. Ga. Entomol. Soc.* 11:301–305.

Croteau, R., A. J. Burbott, and W. D. Loomis. 1972. Apparent energy deficiency in mono- and sesqui-terpene biosynthesis in peppermint. *Phytochemistry* 11:2937–2948.

Dussourd, E. E., and T. Eisner. 1987. Vein-cutting behavior: insect counterplay to the latex defense of plants. *Science* 237:898–901.

Elkinton, J. S., D. L. Wood, and L. E. Brown. 1981. Feeding and boring behavior of the bark beetle, *Ips paraconfusus*, in extracts of ponderosa pine phloem. *J. Chem. Ecol.* 7:209–220.

Feeny, P. 1970. Seasonal changes in oak leaf tannins and nutrients as a cause of spring feeding by Winter Moth caterpillars. *Ecology* 51:565–581.

Ferrell, G. T. 1973. Weather logging, and tree growth associated with fir engraver attack scars in white fir. *USDA Forest Service Research Paper PSW-92.* 11 pp.

Ferrell, G. T. 1983. Host resistance to the fir engraver, *Scolytus ventralis* (Coleoptera: Scolytidae): frequency of attacks contacting cortical resin blisters and canals of *Abies concolor. Can. Entomol.* 115:1421–1428.

Ferrell, G. T. 1988. Wound-induced oleoresins of *Abies concolor*: is it part of host resistance to the fir-engraver, *Scolytus ventralis*? In. W. J. Mattson, J. Levieux, and C. Bernard-Dagan (eds.), *Mechanisms of Woody Plant Defenses against Insects.* New York: Springer, pp. 305–312.

Gambliel, H. A., R. D. Cates, M. Caffey-Moquin, and T. D. Paine. 1985. Variation in the chemistry of loblolly pine in relation to infection by the blue-stain fungus. In S. J. Branham and R. C. Thatcher (eds.). *Integrated Pest Management Research Symposium: The Proceedings*, pp. 177–185. USDA FS Gen. Tech. Rept. SO-56. Asheville, NC.

Gara, R. I., and J. E. Coster. 1968. Studies on the attack behavior of the southern pine beetle. III. Sequence of tree infestation within stands. *Contrib. Boyce Thompson Inst.* 24:69–79.

Geiszler, D. R., R. I. Gara, C. H. Driver, V. F. Gallucci, and R. E. Martin. 1980. Fire, fungi, and beetle influences on a lodgepole pine ecosystem of south-central Oregon. *Oecologia* 46:239–243.

Green, T. R., and C. A. Ryan. 1972. Wound-induced proteinase inhibitor in plant leaves: a possible defense against insects. *Science* 175:776–777.

Gregoire, J. C., J. C. Braekman, and A. Tondeur. 1982. Chemical communication between larvae of *Dendroctonus micans* Kug. (Coleoptera:Scolytidae). *Les Colloques de L'INRA.* 7. Les Mediateurs chemiques:16–20.

Hain, F. P., W. D. Mawby, S. P. Cook, and F. H. Arthur. 1983. Host conifer reaction to stem invasion. *Z. Angew. Entomol.* 96:247–256.

Hanover, J. A. 1975. Physiology of tree resistance to insects. *Annu. Rev. Entomol.* 20:75–95.

Hanover, J. W. 1966. Genetics of terpenes. 1. Gene control of monoterpenes levels in *Pinus monticola* Dougl. *Heredity* 21:73–84.

Hemingway, R. W., G. W. McGraw, and S. J. Barras. 1977. Polyphenols in *Ceratocystis minor*–infected *Pinus taeda*: fungal metabolites, phloem and xylem phenols. *J. Agric. Food Chem.* 25:717–720.

Hodges, J. D., W. W. Elam, W. F. Watson, and T. E. Nebeker. 1979. Oleoresin characteristics and susceptibility for four southern pines to southern pine beetle attacks. *Can. Entomol.* 11:889–896.

Hodges, J. D., T. E. Nebeker, J. D. DeAngelis, and C. A. Blanche. 1985. Host/beetle interactions: influence of associated microorganisms, tree disturbance, and host vigor. In S. J. Branham and R. C. Thatcher (eds.), *Integrated Pest Management Research Symposium: The Proceedings*, pp. 161–168. USDA FS Gen. Tech. Rept. SO-56. Asheville, NC.

Horntvedt, R., E. Christiansen, H. Solheim, and S. Wang. 1983. Artificial inoculation with *Ips typographus*-associated blue stain fungi can kill healthy Norway spruce trees. *Meddelelser Norsk Inst. Skogforskning* 38:1–20.

Hughes, P. R. 1973. *Dendroctonus*: production of pheromones and related compounds in response to host monoterpenes. *Z. Angew. Entomol.* 73:294–312.

Hynum, B. J., and A. A. Berryman. 1980. *Dendroctonus ponderosae* (Coleoptera:Scolytidae): pre-aggregation landing and gallery-initiation on lodgepole pine. *Can. Entomol.* 112:185–191.

Jones, D. A. 1972. Cyanogenic glycosides and their function. In J. B. Harborne (ed.), *Phytochemical Ecology*. New York: Academic, pp. 103–124.

Keen, F. P. 1938. Insect enemies of western forests. *USDA Misc. Pub. No. 273.*

Lieutier, F., and A. A. Berryman. 1988. Elicitation and defensive reactions in conifers. In W. J. Mattson, J. Levieux, and C. Bernard-Dagan (eds.), *Mechanisms of Woody Plant Defenses against Insects*. New York; Springer, pp. 313–320.

Lorio, P. L., Jr. 1986. Growth-differentiation balance: a basis for understanding southern pine beetle-tree interactions. *For. Ecol. Manag.* 14:259–273.

Lorio, P. L., Jr. and J. D. Hodges. 1968. Microsite effects on oleoresin exudation pressure of large loblolly pines. *Ecol.* 49:1207–1210.

Mahoney, R. L. 1978. Lodgepole pine/mountain pine beetle risk classification methods and their applications. In A. A. Berryman, G. D. Amman, R. W. Stark, and D. L. Kibbee (eds.). *Theory and Practice of Mountain Pine Beetle Management in Lodgepole Pine Forests*. Moscow: University of Idaho, pp. 106–113.

Mason, R. R. 1966. Dynamics of *Ips* populations after summer thinning in a loblolly pine plantation: with special reference to host tree resistance. Ph.D. dissertation, University of Michigan.

Mathre, D. E. 1964. Pathenogenicity of *Ceratocystis ips* and *Ceratocystis minor* to *Pinus ponderosa*. *Contrib. Boyce Thompson Inst.* 22:363–388.

McReynolds, R. D., and J. M. Lane. 1971. Adapting the bubble-time method for measuring viscosity of slash pine oleoresin. *USDA Forest Service Research Note 147.*

Mergen, F. 1953. Gum yields in longleaf pine are inherited. *USDA Forest Service SE Forest Experiment Station Research Notes 29.*

Mergen, F., P. E. Hoekstra, and R. M. Echols. 1955. Genetic control of oleoresin yield and viscosity in slash pine. *For. Sci.* 1:19–30.

Miller, R. A. 1984. Physiological responses of lodgepole pine (*Pinus contorta* Douglas var. *latifolia* Engelmann) to bark beetle attack (*Dendroctonus ponderosae* Hopkins) and artificial stress. Ph.D. Dissertation, Washington State University, Pullman.

Miller, R. A., A. A. Berryman, and C. A. Ryan. 1986. Biotic elicitors of defense reactions in lodgepole pine. *Phytochemistry* 25:611–612.

Miller, R. H., and A. A. Berryman. 1985. In. L. Safranyik (ed.). *The Role of the Host in the Population Dynamics of Forest Insects*. Victoria, BC: Candian Forest Service and USDA Forest Service, pp. 13–23.

Miller, R. H., H. S. Whitney, and A. A. Berryman. 1986. Effects of induced translocation stress and bark beetle attack (*Dentroctonus ponderosae*) on heat pulse velocity and the dynamic wound response of lodgepole pine (*Pinus contorta* var. *latifolia*). *Can. J. Bot.* 64:2669–2674.

Mitchell, R. G., R. H. Waring, and G. B. Pitman. 1983. Thinning lodgepole pine increases tree vigor and resistance to mountain pine beetle. *For. Sci.* 209:204–211.

Mitton, J. B., and K. B. Sturgeon. 1982. Biotic interactions and evolutionary changes. In J. B. Mitton and K. B. Sturgeon (eds.), *Bark Beetles in North American Forests: A System for the Study of Evolutionary Biology.* Austin: University of Texas Press, pp. 3–20.

Moeck, H. A., D. L. Wood, and K. Q. Lindahl, Jr. 1981. Host selection behavior of bark beetles (Coleoptera:Scolytidae) attacking *Pinus ponderosa* with special emphasis on the western pine beetle, *Dendroctonus brevicomis. J. Chem. Ecol.* 7:49–83.

Moser, J. C. 1985. Use of sporothecae by phoretic *Tarsonemus* mites to transport ascospores of coniferous bluestain fungi. *Trans. Br. Mycol. Soc.* 84:750–753.

Paine, T. D. 1984. Seasonal response of ponderosa pine to inoculation of the mycangial fungi from the western pine beetle. *Can. J. Bot.* 62:551–555.

Paine, T. D., and F. M. Stephen. 1985. Fungi associated with the southern pine beetle: avoidance of the induced response in pine hosts. *Oecologia* 74:377–379.

Paine, T. D., and F. M. Stephen. 1987a. Fungi associated with the southern pine beetle: avoidance of induced defense reponse in loblolly pine. *Oecologia* 74:337–379.

Paine, T. D., and F. M. Stephen. 1987b. Influence of tree stress and site quality on the induced defensive system of loblolly pine. *Can. J. Forest Res.* 17:569–571.

Paine, T. D., F. M. Stephen, and R. G. Cates. 1985. Induced defenses against *Dendroctonus frontalis* and associated fungi: variation in loblolly pine resistance. In S. J. Branham and R. C. Thatcher (eds.), *Integrated Pest Management Research Symposium: The Proceedings*, pp. 169–176. USDA FS Gen. Tech. Rept. SO-56. Asheville, NC.

Peterman, R. M. 1978. The ecological role of the mountain pine beetle in lodgepole pine forests. In A. A. Berryman, G. D. Amman, R. W. Stark, and D. L. Kibbee (eds.), *Theory and Practice of Mountain Pine Beetle Management in Lodgepole Pine Forests.* Moscow: College of Forest Resources, University of Idaho, pp. 16–26.

Popp, M. P., J. J. Johnson, and R. C. Wilkinson. 1988. Ethylene's role in the response of slash pine to beetle vectored fungi. *Entomological Society of America National Meeting, Dec. 4–8, Louisville, KY.*

Raffa, K. F. 1987. Devising pest management tactics-based on plant defense mechanisms: theoretical and practical considerations. In S. Ahmad and L. B. Brattsten (eds.), *Molecular Mechanisms in Insect–Plant Interactions.* New York: Plenum, pp. 303–329.

Raffa, K. F. 1988. Host orientation behavior of *Dendroctonus ponderosae*: integration of token stimuli and host defenses. In W. J. Mattson (ed.), *Mechanisms of Woody Plant Resistance to Insects and Pathogens.* New York: Springer, pp. 369–390.

Raffa, K. F., and A. A. Berryman. 1980. Flight responses and host selection by bark beetles. In A. A. Berryman and L. Safranyik (eds.), *Proceedings of the Second IUFRO Conference on Dispersal of Forest Insects: Evaluation, Theory, and Management Implications*. Pullman: Conference Office, Cooperative Extension Service, Washington State University, pp. 213–233.

Raffa, K. F., and A. A. Berryman. 1982a. Physiological differences between lodgepole pines resistant and susceptible to the mountain pine beetle and associated microorganisms. *Environ. Entomol.* 11:486–492.

Raffa, K. F., and A. A. Berryman. 1982b. Accumulation of monoterpenes and associated volatiles following fungal inoculation of grand fir with a fungus transmitted by the fir engraver *Scolytus ventralis* (Coleoptera:Scolytidae). *Can. Entomol.* 114:797–810.

Raffa, K. F., and A. A. Berryman. 1982c. Gustatory cues in the orientation of *Dendroctonus ponderosae* (Coleoptera:Scolytidae) to host trees. *Can. Entomol.* 114:97–104.

Raffa, K. F., and A. A. Berryman. 1983a. The role of host plant resistance in the colonization behavior and ecology of bark beetles (Coleoptera:Scolytidae). *Ecol. Monogr.* 53:27–49.

Raffa, K. F., and A. A. Berryman. 1983b. Physiological aspects of lodgepole pine wound responses to a fungal symbiont of the mountain pine beetle. *Can. Entomol.* 115:723–734.

Raffa, K. F., and A. A. Berryman. 1987. Interacting selective pressures in conifer-bark beetle systems: a basis for reciprocal adaptations? *Am. Natur.* 129:234–262.

Raffa, K. F., and E. B. Smalley. 1988a. Response of Red and Jack pines to inoculation with microbial associates of the pine engraver, *Ips pini. Can. J. Forest Res.* 18:581–586.

Raffa, K. F., and E. B. Smalley. 1988b. Host resistance to invasion by lower stem root infesting insects of pine: response to controlled inoculations with the fungal associate *Leptographium terebrantis. Can. J. Forest Res.* 18:675–681.

Raffa, K. F., and E. B. Smalley. 1988c. Seasonal and longterm responses of host trees to microbial associates of the pine engraver, *Ips pini. Can. J. Forest Res.* 18:1624–1634.

Raffa, K. R., A. A. Berryman, J. Simasko, W. Teal, and B. L. Wong. 1985. Effects of grand fir monoterpenes on the fir engraver beetle (Coleoptera; Scolytidae) and its symbiotic fungi. *Environ. Entomol.* 14:552–556.

Reid, R. W., and D. M. Shrimpton. 1971. Resistance response of lodgepole pine to inoculation with *Europhium clavigerum* in different months and at different heights on stem. *Can. J. Bot.* 49:349–351.

Reid, R. W., H. S. Whitney, and J. A. Watson. 1967. Reactions of lodgepole pine to attack by *Dendroctonus ponderosae* Hopkins and blue stain fungi. *Can. J. Bot.* 45:1115–1126.

Renwick, J. A. A., and J. P. Vite. 1970. Systems of chemical communication in *Dendroctonus. Contrib. Boyce Thompson Inst.* 24:283–292.

Rhoades, D. F. 1985. Offensive–defensive interactions between herbivores and plants: their relevance in herbivore population dynamics and ecological theory. *Am. Natur.* 125:205–238.

Rudinsky, J. A. 1962. Ecology of Scolytidae. *Annu. Rev. Entomol.* 7:327–348.

Rudinsky, J. A. 1966. Host selection and invasion by the Douglas fir beetle, *Dendroctonus pseudotsugae* Hopkins, in coastal Douglas-fir forests. *Can. Entomol.* 98:98–111.

Rudinsky, J. A. 1969. Masking of the aggregation pheromone in *Dendroctonus pseudotsugae* Hopk. *Science* 166:884–885.

Rudinsky, J. A. 1973. Multiple functions of the Douglas-fir beetle pheromone 3-methyl-2-cyclohexen-1-one. *Environ. Entomol.* 2:579–585.

Russell, C. E., and A. A. Berryman. 1976. Host resistance to the fir engraver beetle. I. Monoterpene composition of *Abies grandis* pitch blisters and fungus-infected wounds. *Can. J. Bot.* 54:14–18.

Safranyik, L., D. M. Shrimpton, and H. S. Whitney. 1975. An interpretation of the interaction between lodgepole pine, the mountain pine beetle and its associated blue stain fungi in western Canada. In D. M. Baumgartner (ed.), *Management of Lodgepole Pine Ecosystems*. Pullman: Washington State University Cooperative Extension Service, pp. 406–428.

Sequeira, L. 1980. Defenses triggered by the invader: recognition and compatibility phenomena. In J. G. Horsfall and E. B. Cowling (eds.), *Plant Disease: An Advanced Treatise*. New York: Academic, pp. 179–200.

Shepherd, R. F. 1966. Factors influencing the orientation and rats of activity of *Dendroctonus ponderosae* (Coleoptera:Scolytidae). *Can. Entomol.* 98:507–518.

Shrimpton, D. M. 1973a. Extractives associated with the wound response of lodgepole pine attacked by the mountain pine beetle and associated microorganisms. *Can. J. Bot.* 51:527–534.

Shrimpton, D. M. 1973b. Age- and size-related response of lodgepole pine to inoculation with *Europhium clavigerum*. *Can. J. Bot.* 51:1155–1160.

Shrimpton, D. M. 1978. Resistance of lodgepole pine to mountain pine beetle infestation. In A. A. Berryman, G. D. Amman, R. W. Stark, and D. L. Kibbee (eds.), *Theory and Practice of Mountain Pine Beetle Management in Lodgepole Pine Forests*. Moscow: College of Forest Resources, University of Idaho, pp. 64–76.

Shrimpton, D. M., and J. A. Watson. 1971. Response of lodgepole pine seedlings to inoculation with *Europhium clavigerum*, a blue stain fungus. *Can. J. Bot.* 49:373–375.

Smith, R. H. 1963. Toxicity of pine resin vapors to three species of *Dendroctonus* bark beetles. *J. Econ. Entomol.* 56:823–831.

Smith, R. H. 1965. Effects of monoterpene vapors on the western pine beetle. *J. Econ. Entomol.* 58:509–510.

Smith, R. H. 1975. Formula for describing effect of insect and host tree factors on resistance to western pine beetle attack. *J. Econ. Entomol.* 68:841–844.

Squillace, A. E., and C. R. Gansel. 1974. Juvenile: mature correlations in slash pine. *Forest Sci.* 20:225–229.

Stark, R. W. 1965. Recent trends in forest entomology. *Annu. Rev. Entomol.* 10:303–324.

Stephen, F. M., and T. D. Paine. 1985. Seasonal patterns of host tree resistance to fungal associates of the southern pine beetle. *Z. Angew. Entomol.* 99:113–122.

Struble, G. R. 1957. The fir engraver, a serious enemy of western true firs. *USDA Prod. Res. Rep. No. 11.*

Thalenhorst, W. 1958. Grundzuge der populations-dynamik des grossen Fichtenborkenkafers *Ips typographus* L. *Shriftrenreihe der Forstlichen Fakultar der Universitat Gottingen. No. 21.*

Tobolski, J. J., and J. W. Hanover. 1971. Genetic variation in the monoterpenes of Scotch pine. *Forest Sci.* 17:293–299.

Vite, J. P. 1961. The influence of water supply on oleoresin exudation pressure and resistance to bark beetle attack in *Pinus ponderosa. Contrib. Boyce Thompson Inst. Plant Prot.* 21:37–66.

Vite, J. P., and D. L. Wood. 1961. A study on the applicability of the measurement of oleoresin exudation pressure in determining susceptibility of second-growth ponderosa pine to bark beetle infestation. *Contrib. Boyce Thompson Inst.* 21:67–78.

Waring, R. H., and G. B. Pitman. 1983. Physiological stress in lodgepole pine as a precursor to mountain pine beetle attack. *Z. Angew. Entomol.* 96:265–270.

Waters, W. E., R. W. Stark, and D. L. Wood. 1985. *Integrated Pest Management in Pine Bark Beetle Ecosystems.* New York: Wiley.

Whitham, T. G. 1983. Host manipulation of parasites: within-plant variation as a defense against rapidly evolving pests. In R. F. Denno and M. S. McClure (eds.), *Variable Plants and Herbivores in Natural and Managed Systems.* New York: Academic, pp. 15–39.

Whitney, J. H. 1982. Relationships between bark beetles and symbiotic organisms. In J. B. Mitton and K. B. Sturgeon (eds.), *Bark Beetles in North American Forests: A System for the Study of Evoluationary Biology.* Austin: University of Texas Press, pp. 183–211.

Wong, B. L., and A. A. Berryman. 1977. Host resistance to the fir engraver beetle. 3. Lesion development and containment of infection by resistant *Abies grandis* inoculated with *Trichosporium symbioticum. Can. J. Bot.* 55:2358–2365.

Wood, D. L. 1972. Selection and colonization of ponderosa pine by bark beetles. In H. F. van Emden (ed.), *R.E.S. Symposium No. 6, Insect/Plant Relationships.* Oxford: Blackwell, pp. 101–107.

Wood, D. L. 1982. The bark and ambrosia beetles of North and Central America (Coleoptera:Scolytidae), a taxonomic monograph. *The Great Basin Nat. Mem.* 6, Provo, UT: Brigham Young University.

Wright, L. E., A. A. Berryman, and S. Gurusiddaiah. 1979. Host resistance to the fir engraver beetle, *Scolytus ventralis* (Coleoptera: Scolytidae). 4. Effect of defoliation on wound monoterpenes and inner bark carbohydrate concentrations. *Can. Entomol.* 111:1255–1261.

Wright, L. E., A. A. Berryman, and B. E. Wickman. 1984. Abundance of the fir engraver, *Scolytus ventralis*, and the Douglas-fir beetle, *Dendroctonus pseudotsugae*, following tree defoliation by the Douglas-fir tussock moth, *Orgyia pseudotsugata. Can. Entomol.* 116:293–305.

Zavarin, E., W. B. Critchfield, and K. Snazberk. 1969. Turpentine composition of *Pinus banksiana* hybrids and hybrid derivatives. *Can. J. Bot.* 47:1444–1453.

Zinkel, D. R. 1977. Pine resin acids as chemiotaxonomic indicators. In *TAPPI Conference Papers.* Forest Biology Wood Chemistry Conference, Madison, WI, pp. 53–56.

CHAPTER 12

THE EFFECTS OF INDUCIBLE RESISTANCE IN HOST FOLIAGE ON BIRCH-FEEDING HERBIVORES

S. NEUVONEN and E. HAUKIOJA
Department of Biology, University of Turku, SF-20500 Turku 50, Finland

1 Introduction
2 Inducible resistance in the birch–*Epirrita* system
   2.1 Effects of rapid inducible resistance on *Epirrita* performance
   2.2 Effects of delayed inducible resistance of *E. autumnata*
   2.3 Counteradaptations by *Epirrita* to inducible resistance
   2.4 The evolutionary origin of delayed inducible resistance
3 The variable role of inducible resistance in different birch–herbivore systems
4 Conclusion
   References

1 INTRODUCTION

Birches (*Betula* spp.) are widespread and abundant deciduous trees which in Britain alone harbor over 300 herbivorous insect species (Kennedy and Southwood 1984). These herbivores differ greatly in their average densities and in population dynamics. Large variation indicates differences in the factors regulating herbivore populations living on the same host plant. For example, host quality and inducible changes in it may have different effects on the population dynamics of different herbivore species.

In this chapter we describe the role of inducible resistance in the interactions between the mountain birch (*Betula pubescens* ssp. *tortuosa*) and the autumnal moth (*Epirrita* (= *Oporinia*) *autumnata*. We discuss also the variable effects of inducible resistance in other systems involving birch and its herbivores.

Inducible changes in foliage quality can be studied by using chemical

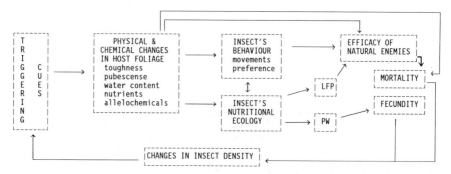

Figure 1 Potential causal links among induced changes in foliage quality (with some examples of relevant variables), the behavior and performance of insect herbivores (interacting with their natural enemies), and the cues triggering inducible responses. LFP, length of feeding (larval) period; PW, pupal weight.

analyses (Niemelä et al. 1979, Tuomi et al. 1984, Wratten et al. 1984, Bergelson et al. 1986, Hartley and Lawton, this volume, Chapter 5) or by conducting bioassays with insects. The basic difficulty with chemical analyses is that it is often impossible to know which, if any, of the plant traits changing after foliage damage are important for the herbivore. The more straightforward bioassays directly monitor effects on the herbivore but may be of limited value in revealing the role of inducible resistance in the population dynamics of the insect. The least time-consuming tests to make, like preference tests and measurement of consumption rates (Lewis and van Emden 1986, Kogan 1986), may be of little relevance. Clearly, the most appropriate variables to measure are survival, fecundity, and pupal weight, which are strongly correlated (Tingey 1986, Haukioja and Neuvonen 1987).

Figure 1 describes how inducible changes in foliage quality affect the behavior and performance of insect herbivores, and the cues triggering the inducible resistance. Note that the rate at which the different effects (shown by arrows) are realized is variable. The movements and foliage preferences of insects may be affected in a few hours (Wratten et al. 1984, Bergelson et al. 1986) while the effects on fecundity and subsequent insect density may not be fully realized until the next generation. Furthermore, various attributes of foliage quality respond to triggering cues with different tempos, and relaxation of inducible changes may take from months to years (Tuomi et al. 1984, Wratten et al. 1984).

The situation is actually more complex than in Figure 1 because of variation in the spatial scale. Damage to only one part of the foliage in a tree apparently induces the strongest responses at the site of damage, but some leaf traits may change even in distant parts of the tree (Edwards and Wratten 1983, Haukioja and Neuvonen 1985a, Neuvonen et al. 1987). Insect species with different mobility differ in their capacity to respond behaviorally to induced changes in food quality. Finally, before the next larval generation, many flying insects may redistribute their offspring among trees within a stand.

2 INDUCIBLE RESISTANCE IN THE BIRCH–*EPIRRITA* SYSTEM

The mountain birch forests in northwestern Europe are periodically defoliated by geometrid larvae. Several species (*E. autumnata*, *Operophtera* spp., and *Erannis defoliaria*) fluctuate in density drastically and fairly synchronously (Tenow 1972, Koponen 1983). These oscillations are cyclic, with peaks occurring at 9- to 10-year intervals (Tenow 1972, Haukioja et al. 1988b).

Insect damage to birch foliage triggers responses that may affect herbivores rapidly or the effects may be carried over to the subsequent herbivore generation(s). Distinguishing between rapid and delayed inducible resistance is essential since these two types of responses affect the population dynamics of a herbivore in fundamentally different ways. Rapid inducible resistance (RIR) tends to stabilize herbivore population dynamics. On the other hand, delayed inducible resistance (DIR) introduces a time lag to the negative feedbacks regulating the population dynamics of insects and may generate cyclic density fluctuations (Haukioja 1982, Berryman et al. 1987, Haukioja et al. 1988b).

2.1 Effects of Rapid Inducible Resistance on *Epirrita* Performance

Mechanically damaged birch foliage supports the growth of *Epirrita* larvae less well than undamaged control foliage (Haukioja and Niemelä 1977, Haukioja and Hanhimäki 1985), but quantifying the effects of RIR on the population dynamics of *E. autumnata* presents serious difficulties (Fowler and Lawton 1985, Neuvonen and Haukioja 1985, Haukioja and Neuvonen 1987).

Bioassays attempting to estimate the efficacy of RIR on fecundity and survivorship of an insect are sensitive to the methods used in triggering the response, ways of testing the leaves, and temporal changes in the strength of the response (Fig. 2). Methods used should mimic natural damage as well as possible, and bioassays should be conducted in the field. The basic problem with field tests is that feeding larvae trigger inducible changes both in experimental and in control trees (Fig. 2D). Thus, it is not easy to get an unequivocally pure control treatment. The induction of control foliage may be largely prevented by conducting the bioassays on detached leaves in the laboratory (Fig. 2A–C). In these cases the results depend on the frequency and cumulative effects of leaf damage.

All available estimates of the effects of rapid inducible resistance on the performance of *E. autumnata* are based on bioassays made with detached leaves. Damaging the leaves of mountain birch 2 days before the leaves were fed to larvae (one treatment/tree, different trees each day; see Fig. 2B) resulted in 10–30% lower larval weights on induced (=damaged and nearby

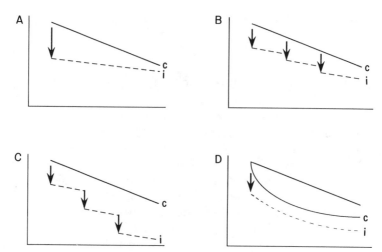

Figure 2 Different methods of damaging (inducing) leaves may produce disparate estimates on the efficacy of inducible responses. Foliage quality (measured by insect performance) is plotted on the y axis, and time is plotted on the x axis. c, control foliage (quality decreases seasonally); i, induced foliage. (*A*) A single induction (↓) reduces foliage quality, but the effect weakens with time. (*B*) The induction is repeated to prevent the relaxation of the response. In this case the effects of repeated treatments are not additive, or different host individuals should be used for each induction. (*C*) Repeated induction may have cumulative (additive) effects on foliage quality. (*D*) A single artificial induction is followed by responses induced by the bioassay procedure (insect feeding triggers responses in both control and induction treatments).

undamaged) leaves than on control leaves (Haukioja 1980, Haukioja and Hanhimäki 1985). Larvae grew for a longer time on induced than on control foliage and, consequently, pupal weights were only 0–14% lower on induced than on control foliage (Haukioja and Niemelä 1979, Haukioja et al. 1983b, Haukioja and Hanhimäki 1985). The responses triggered by foliage damage spread to undamaged leaves, too: *Epirrita* larvae grew better on control foliage than on intact leaves adjacent to damaged ones. However, the effects on larvae were slightly stronger when fed with damaged leaves than with intact nearby leaves (Haukioja 1982, Haukioja and Hanhimäki 1985).

Also, the survival of *E. autumnata* larvae may be slightly lower on induced than on control foliage (Haukioja and Niemelä 1977, 1979). When the total effects of RIR on the capacity of *Epirrita* to increase were estimated by combining the effects on survival and fecundity, the reduction ranged from 0 to 22% (summarized in Haukioja and Neuvonen 1987).

Annual variation in the intensity of RIR has received little attention, but it may be important because a temporal weakening of the induced response for example, due to weather conditions, may allow a more rapid increase in the population density of herbivores. Annual variation in RIR has been

TABLE 1   The Mean Pupal Weights (PW) of Female *E. autumnata* Reared on Control (Undamaged) and Induced (Damaged and Adjacent Intact) Birch[a] Leaves

Experiment	PW (mg)	SD	*n* (trees)
Kevo			
Control	83.3	3.9	22
Induced	80.4	6.7	22
		$F^b = 3.03, p < .10$	
Ruissalo			
Control	86.3	4.7	12
Induced	80.5	4.2	12
		$F^b = 8.45, p < .01$	

[a] In the Kevo experiment the host was *B. pubescens* ssp. *tortuosa* and in the Ruissalo experiment it was *B. pendula*.
[b] Statistical analyses based on tree specific means.

suggested in the birch–*Epirrita* system (Haukioja and Hanhimäki 1985; cf. also the results with *B. pendula* in Wratten et al. 1984 and in Edwards et al. 1986), but many confounding factors involved in bioassays (e.g., different sets of experimental trees and animals) have hampered a thorough analysis of variation in the efficacy of inducible resistance through time.

The previous experiments with only a single induction treatment may underestimate the efficacy of RIR. In two experiments where manual induction was applied repeatedly to the same trees during the entire larval period of *Epirrita* (see Fig. 2*C*), female pupal weights were 4–7% lower on induced than on control foliage (Table 1).

2.2   Effects of Delayed Inducible Resistance on *E. autumnata*

Delayed inducible resistance can be triggered by mechanical damage to leaves and by applying feces of *Epirrita* larvae (or small amounts of fertilizer) beneath the birches (Haukioja et al. 1985). The effects of these cues are additive. Insect damage causes a stronger reduction in the pupal weights of *Epirrita* in the following year than artificial damage alone (Haukioja and Neuvonen 1985a, Neuvonen et al. 1987). Foliage responses triggered by previous insect damage are different from the effects of physical foliage damage per se, and they are manifested also in the previously undamaged branches of the same trees (Haukioja and Neuvonen 1985a, Neuvonen et al. 1987). Whether insect saliva, implicated as a cue for the RIR by Hartley and Lawton (this volume, Chapter 5), is the causal factor for RIR is not known.

The foliage of defoliated mountain birches has lower nitrogen and higher phenolic content than that of control trees in the years following damage (Tuomi et al. 1984). The effects of these changes on the nutritional ecology of *Epirrita* were studied by short-term laboratory trials with 5th instar larvae

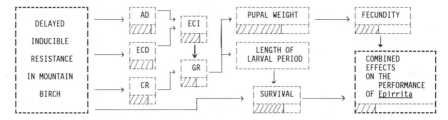

Figure 3   The effects of delayed inducible resistance on different variables measuring the performance of *E. autumnata*. The shaded bar shows the performance of *Epirrita* on induced foliage as a proportion of that on control foliage. AD, approximate digestibility; ECD, conversion efficiency of digested food; ECI, conversion efficiency of ingested food; CR, consumption rate; GR, growth rate. *Note:* AD*ECD = ECI; CR*ECI = GR. (From Neuvonen and Haukioja 1984, Haukioja et al. 1985.)

(Fig. 3) (for methodology, see Neuvonen and Haukioja 1984, Kogan 1986). The digestibility (AD) of leaves from defoliated trees was about 12% lower than in control trees (Neuvonen and Haukioja 1984). *Epirrita* larvae were not able to compensate for the reduced quality of foliage by increasing food intake: The consumption rate was significantly lower on leaves from defoliated than from control trees (Neuvonen and Haukioja 1984). Because of lower consumption rate and AD on mountain birches defoliated during the previous year, growth rates of *Epirrita* larvae were reduced (Fig. 3). Decreased growth rates of *Epirrita* on defoliated birches both prolonged larval periods and lowered pupal weights (Neuvonen and Haukioja 1984, Haukioja et al. 1985). The lengthening of the larval period caused by DIR in birch foliage may be detrimental for *Epirrita* in two ways: Larvae are forced to feed on foliage whose nutritive quality declines with the season (see Haukioja et al. 1978, Ayres and MacLean 1987) and they are exposed to natural enemies for a longer period.

The effects of foliage quality on the performance of *E. autumnata* have been measured in field trials by rearing larvae throughout their development within mesh bags (cf. Tingey 1986). The enclosure protects larvae from most natural enemies but some parasites and predatory spiders do attack *Epirrita* larvae through the net (personal observation). Thus, differences in mortality are due to both direct effects of food quality and interactions between the efficacy of natural enemies and food quality. These factors reduced the survival of *Epirrita* larvae from 40% on control trees to 20–30% on defoliated trees (Haukioja et al. 1985). Mean pupal weight of *Epirrita* reared on trees showing DIR has been about 30% lower than on untreated trees (Haukioja et al. 1985). The potential fecundity of female *Epirrita* strongly correlates with pupal mass (Haukioja and Neuvonen 1985b).

Summarizing (see Fig. 3), simulated partial defoliation (50% of each leaf torn off, insect frass to soil beneath the trees) in the previous year reduced the performance (combined effects on survival and fecundity) of *E. autum-*

nata 70–78% below that on control birches (Haukioja et al. 1985). Even 15% leaf removal (without frass application) reduced the performance of *Epirrita* by 40% in the following year (Haukioja and Neuvonen 1987).

The DIR provides a causal mechanism that produces time lags in the population dynamics of *Epirrita*. Its relaxation rate is crucial in determining the effect of the feedbacks and may contribute to intervals between successive outbreaks of *Epirrita*. Mountain birches that were artificially defoliated 2–4 years earlier still reduced *Epirrita* fecundity by 16–38% (Haukioja 1982, Haukioja and Neuvonen 1987).

2.3 Counteradaptations by *Epirrita* to Inducible Resistance

Because the inducible resistance of host trees has potentially drastic effects on the performance of *Epirrita*, it is natural to study whether *Epirrita* can in some way cope with these responses (cf. Rhoades 1985). The density (crowding) of larvae triggers phenotypic changes in *Epirrita* that can alleviate the consequences of induced deterioration in foliage quality. In laboratory trials crowded *Epirrita* has shorter larval periods than and similar pupal weights as their solitary siblings when reared on low-quality diets (previously defoliated birches or inferior food plants) (Haukioja 1980, Haukioja et al. 1988a). The reason for the better growth of crowded individuals on poor diets was their higher consumption rates (Haukioja et al. 1988a). Experiments with different types of poor-quality diets suggested that the group response in *E. autumnata* larvae may generally relieve the consequences of low-quality food, and not be a specific response against induced reactions in foliage (Haukioja et al. 1988a).

2.4 The Evolutionary Origin of Delayed Inducible Resistance

Two different explanations have been proposed for the evolutionary origin of the DIR in mountain birch. The first is that defoliation affects nutrient content and production of allelochemicals in foliage by causing a relative shortage of mineral nutrients (Tuomi et al. 1984). In its pure form this nutrient stress hypothesis does not require any adaptive interpretation (Haukioja and Neuvonen 1985a, Tuomi et al. this volume, Chapter 4). The second explanation is an adaptive one: The reduced foliage quality after defoliation is a specific evolutionary (defensive) response to pressures exerted by herbivores (Haukioja and Neuvonen 1985a).

Several observations are consistent with the defensive hypothesis but difficult to obtain from the nutrient stress hypothesis (Table 2). However, a DIR can be triggered by manual defoliation only in trees growing on nutrient-poor sites, which is consistent with the nutrient stress hypothesis (Tuomi et al. 1984, Haukioja and Neuvonen 1985a). We cannot, therefore, exclude the nutrient stress hypothesis as a partial explanation of a DIR. The incidental responses suggested by the nutrient stress hypothesis may have

TABLE 2   A Comparison of the Match[a] between Experimental Results and Two Alternative Hypotheses[b] Explaining the Occurrence of Delayed Inducible Resistance (DIR) in Mountain Birch

	Hypothesis		
Experimental Result	DH	NSH	References
DIR reduces the consumption rate of *Epirrita* larvae	+	0/−	1
Insect damage causes stronger DIR than manual damage	+	0	2, 3
Frass application triggers DIR	+	0/−	4
Fertilization does not mitigate DIR triggered by manual damage	0/+	−	2

[a] +, Consistent with the hypothesis; −, contradictory with the hypothesis.
[b] DH, defensive hypothesis; NSH, nutrient stress hypothesis.
References. 1, Neuvonen and Haukioja 1984; 2, Haukioja and Neuvonen 1985a; 3, Neuvonen et al., 1987; 4, Haukioja et al. 1985.

been molded in a defensive direction if such alterations have increased the fitness of individual birches (Haukioja and Neuvonen 1985a, Tuomi et al., this volume, Chapter 4). This is probable in areas where heavy defoliations occur (see also Mattson et al. 1988). On the other hand, birches outside the outbreak range of defoliating insects may not express the DIR (Haukioja et al. 1981).

3   THE VARIABLE ROLE OF INDUCIBLE RESISTANCE IN DIFFERENT BIRCH–HERBIVORE SYSTEMS

Most insect species apparently cause no significant damage to the host tree. *Epirrita autumnata* is an extreme case among birch-feeding insects because it can destroy trees over vast areas (Tenow 1972). Consequently, it is not possible to generalize the results obtained with mountain birch and *Epirrita* to all birch-feeding herbivores. Even *E. autumnata* does not cause regular defoliations in all parts of its shared range with birch (Tenow 1972, Haukioja et al. 1988b), suggesting geographical variation in the factors that regulate this insect. We now address the following questions:

1. Do the inducible responses in birch foliage have measurable effects on herbivore species other than *E. autumnata*?
2. Are there birch-feeding insects that are not affected by induced resistance?
3. Is it possible to find consistent patterns (e.g., with respect to insect taxonomy, feeding type, seasonal occurrence, or geographical area) in the occurrence and efficacy of inducible resistance?

4. Which generalizations can be drawn about the importance of inducible resistance in various birch–herbivore systems?

Table 3 shows that rapid inducible responses in birch foliage have negative effects on the performance of herbivores other than *E. autumnata*. In addition, lowered palatability to generalist herbivores (slugs and two species of polyphagous Lepidoptera) after foliage damage have been reported (Edwards and Wratten 1982, Wratten et al. 1984), although effects on palatability were not consistent in birches or insects from different areas (see Hartley and Lawton, this volume, Chapter 5). In short-term bioassays lepidopteran and sawfly larvae gained 11–24% less weight on induced than on control foliage in early and mid season. Most species seem to be able to compensate for this reduction in foliage quality by feeding for a longer time. As a result, the RIR in birch foliage usually had only slight effects on pupal weights. Although larval mortality was not often significantly affected in tests, the direction of the changes was fairly consistent: Survival has been over 10% lower on induced than on control foliage, excluding experiments with two late season sawflies (Table 3).

Rapid inducible resistance has negative effects on the performance of herbivores belonging to several taxonomic groups. In addition to the lepidopterans and sawflies listed in Table 2, Fowler and MacGarvin (1986) found reductions in the density of phloem feeders and adult *Phyllobius* weevils on damaged branches of white birch. The responses seem also to affect herbivores with different feeding types such as leaf chewers (most species in Table 2) and miners (*Coleophora serratella*; Bergelson et al. 1986). In sucking insects, the evidence is contradictory: Although Fowler and MacGarvin (1986) found that leaf damage reduced the numbers of phloem feeders, other bioassays showed no effects of RIR on the performance of *Euceraphis* aphids (Wratten et al. 1984, Neuvonen and Lindgren, personal observation).

The effects of RIR on a birch-feeding herbivore can be best predicted on the basis of its feeding season. Bioassays both in Finland and in Great Britain have shown that RIR can be triggered by damaging early but not late season birch foliage (Wratten et al. 1984, Fowler and Lawton 1985). This is apparently not due to methodological reasons since the procedures have been essentially similar to those with early season foliage (see Table 3).

Both rapid and delayed inducible resistance in birch foliage are easily triggered by artificially damaging foliage and affect a variety of leaf-chewing herbivores, though insect damage seems to cause stronger responses (Haukioja and Neuvonen 1985a, Neuvonen et al. 1987, Hartley and Lawton, this volume, Chapter 5). The occurrence of RIR only in early season foliage implies that the interactions it mediates are more likely to occur in early than in late season guilds. Foliage changes induced in early season persist at least two months (Wratten et al. 1984). Thus, early season inducible changes may have negative effects on late season herbivores (Fowler and MacGarvin 1986, Neuvonen et al. 1988). On the other hand, not all types

TABLE 3 Effects of Rapid Inducible Responses in the Foliage of Different *Betula* Species[a] on the Performance[b] of Various Herbivores[c]

Host	Herbivore	D	B	M	MD	LFP	Weight	Survival	References
mb	*Brephos parthenias*	e	m	B	35	+10% *	−4% ns	−12% ns	1
mb	*Pristiphora* sp.	e	m	C	36	+5% *	−0% ns	−21% ns	1
mb	*Pteronidae* sp.	m	m	C	24	Fixed	−11% *	?	1
		m	m	C	44	+5% *	−2% ns	−19% *	1
mb	*Dineura viriididorsata*	l	l	B, C	36	−1% ns	−0% ns	0% ns	1
mb	*Trichiosoma lucorum*	l	l	B, C	36	0% ns	−2% ns	−7% ns	1
mb	*Eriogaster lanestris*	e	m	C	50	Fixed	−11% *	?	1
		l	l	C	81	Fixed	−3% ns	?	1
wb	*Apocheima pilosaria*	e	e	D	28	Fixed	−24% *	−32% *	2
sb	*Coleophora serratella*	e	e	D	?	ca. +3% *	?	−14% ns	3
gb	*Lymantria dispar*	e	m	C/D?	?	?	−9% *	−15% *	4
mb	*Dineura viriididorsata*	e	l	D	17	Fixed	−14% *	?	5

[a] Host species: mb, *B. pubescens* ssp. *tortuosa*; wb, *B. pubescens*; sb, *B. pendula*; gb, *B. populifolia*.

[b] Percentage change in a variable due to inducible responses. LFP, length of feeding period (fixed: the experiment was finished after a certain time). Weight = pupal or final larval weight if LFP not fixed; if LFP is fixed then weight = larval weight at the end of the bioassay. * Significant ($p < .05$) difference; ns, difference not significant; based on analyses in the original articles.

[c] Column D: time when experimental damage was started (e, early; m, mid; l, late season). Column B: timing of the bioassay (see column D). Column M: method of inducing foliage (see Fig. 2). Column MD: mean duration (days) of the bioassay.

References. 1, Haukioja and Niemelä 1979; 2, Fowler and MacGarvin 1986; 3, Bergelson et al. 1986; 4, Wallner and Walton 1979; 5, Neuvonen et al. 1988.

of damage to birch trees cause subsequent deterioration in foliage quality. Increased densities of insects have been observed on birches browsed either artificially or by moose during the previous winter(s) (Danell and Huss-Danell 1985), although the performance of *Epirrita* larvae was not significantly affected (Neuvonen and Danell 1987). A total defoliation early in the season may also have positive effects on some birch-feeding herbivores. Although birches refoliating after an early season artificial defoliation supported the growth of a late season sawfly species less well than control trees, they harbored significantly more weevils (*Deporaus betulae*) (Neuvonen et al. 1988, cf. Faeth 1987).

In addition to seasonal variation, there may also be geographical or local variation in the importance of inducible resistance for birch-feeding herbivores. In Fennoscandia, extensive defoliations of birch are restricted to marginal growth sites in northern and mountainous areas, although both the host and the defoliating geometrid species occur over much larger areas. The effects of RIR on the performance of *E. autumnata* seem to be equally negative on birches from outbreak (northern Finland) and nonoutbreak (southern Finland) areas (Haukioja and Hanhimäki 1985). Furthermore, moth strains originating within and outside the outbreak area were equally susceptible to RIR (Haukioja and Hanhimäki 1985). Also, the verification of RIR in Finland, Great Britain, and the United States with different birch species (*B. pendula, B. populifolia, B. pubescens*; Wallner and Walton 1979, Edwards and Wratten 1982, Wratten et al. 1984, Haukioja and Hanhimäki 1985, Bergelson et al. 1986, Fowler and MacGarvin 1986) suggests that there are no fundamental geographical differences in these responses.

In addition to the mountain birch and the autumnal moth, the effects of DIR have also been shown in other birch–herbivore systems that are characterized by high levels of defoliation. The performance (combined effects on fecundity and larval survival) of gypsy moth (*Lymantria dispar*) was 18–41% lower on defoliated than on control gray birches (*B. populifolia*) (Wallner and Walton 1979, Valentine et al. 1983). Defoliation of *B. resinifera* had striking effects on the fecundity and survival of *Rheumaptera hastata* in the following year. Total defoliation reduced the performance of this geometrid by 37% and cumulative defoliations had additive effects: After two years of defoliation moth performance on treated trees was about 70% lower than on controls and after three total defoliations it was over 95% lower (Werner 1979, 1981).

4 CONCLUSION

This review shows that induced responses (both rapid and delayed) in birch foliage may have considerable effects on the success of herbivorous insects. The realism of efficacy estimates depends on several methodological details:

1. Realistic estimates of the efficacy of inducible resistance are possible only if the natural level of foliage damage is low and the bioassay procedure does not induce control foliage.
2. The induction treatment should simulate natural damage well. Specific cues may be more efficient than artificial damage of foliage in triggering induced resistance. The specificity of the effects on different herbivore species is poorly documented.
3. From the viewpoint of insect population dynamics, survival and fecundity of herbivores are much more informative parameters than preference trials or chemical analyses. Short-term bioassays may not reveal the full efficacy of inducible resistance.
4. The ability of insects to compensate for the effects of reduced foliage quality should be determined to ascertain the relevance of the measured variables for their population dynamics. Possible interactions among the effects of inducible resistance and other regulating factors may also be relevant.

Unfortunately, the procedures applied in studying the effects of inducible resistance on herbivores have been so variable that much of the observed variation in the effects of induced responses may be due to methodological differences. The experimental evidence shows, however, that RIR has effects on the performance and behavior of several birch-feeding insects, even at endemic levels of herbivory (Edwards and Wratten 1985, Bergelson et al. 1986, Fowler and MacGarvin 1986). The effects of RIR have not been as strong as those from DIR, although the difference may partly be caused by differing methodology.

When compared with other potential regulating factors, the importance of DIR is well established in birch–herbivore systems showing cyclic fluctuations in insect densities (Werner 1981, Haukioja et al. 1988b). The case for RIR is not as clear (Fowler and Lawton 1985, Hartley and Lawton, this volume, Chapter 5) and may not be solved before ingenious experiments overcome the practical problems discussed previously.

The importance of inducible resistance in herbivore communities is determined not only by the efficacy of the responses but also by how often and to what extent they are triggered. Inducible resistance (especially DIR) obviously has a greater role in birch–herbivore systems characterized by drastic density fluctuations of herbivores than in systems showing generally low levels of herbivory. DIR may be a key factor in the cyclic density fluctuations of birch defoliators in some areas, but this does not deny the importance of other factors (e.g., parasitoids, diseases) (Haukioja et al. 1983a, Haukioja et al. 1988b).

REFERENCES

Ayres, M. P., and S. F. MacLean, Jr. 1987. Development of birch leaves and the growth energetics of *Epirrita autumnata* (Geometridae). *Ecology* 68:558–568.

Bergelson, J., S. Fowler, and S. Hartley. 1986. The effects of foliage damage on casebearing moth larvae, *Coleophora serratella*, feeding on birch. *Ecol. Entomol.* 11:241–250.

Berryman, A. A., N. C. Stenseth, and A. S. Isaev. 1987. Natural regulation of herbivorous forest insect populations. *Oecologia (Berlin)* 71:174-184.

Danell, K., and K. Huss-Danell. 1985. Feeding by insects and hares on birches earlier affected by moose browsing. *Oikos* 44:75–81.

Edwards, P. J., and S. D. Wratten. 1982. Wound-induced changes in palatability in birch (*Betula pubescens* Ehrh. ssp. *pubescens*). *Am. Natur.* 120:816–818.

Edwards, P. J., and S. D. Wratten. 1983. Wound induced defenses in plants and their consequences for patterns of insect grazing. *Oecologia (Berlin)* 59:88–93.

Edwards, P. J., and S. D. Wratten. 1985. Induced plant defenses against insect grazing: fact or artefact? *Oikos* 44:70–74.

Edwards, P. J., S. D. Wratten, and S. Greenwood. 1986. Palatability of British trees to insects: constitutive and induced defenses. *Oecologia (Berlin)* 69:316–319.

Faeth, S. H. 1987. Community structure and folivorous insect outbreaks: the roles of vertical and horizontal interactions. In P. Barbosa and J. C. Schultz (eds.), *Insect Outbreaks*. San Diego: Academic, pp. 135–171.

Fowler, S. V., and J. H. Lawton. 1985. Rapidly induced defenses and talking trees: the devil's advocate position. *Am. Natur.* 126:181–195.

Fowler, S. V., and M. MacGarvin. 1986. The effects of leaf damage on the performance of insect herbivores on birch, *Betula pubescens*. *J. Anim. Ecol.* 55:565–573.

Hartley, S. E., and J. H. Lawton. 1991. The biochemical basis, and significance of rapidly induced changes in birch foliage. In D. W. Tallamy and M. J. Raupp (eds.), *Phytochemical Induction by Herbivores*. New York: Wiley, pp. 105–132.

Haukioja, E. 1980. On the role of plant defenses in the fluctuation of herbivore populations. *Oikos* 35:202–213.

Haukioja, E. 1982. Inducible defenses of white birch to a geometrid defoliator, *Epirrita autumnata*. In *Proc. 5th Int. Symp. Insect–Plant Relationships*. Wageningen: Pudoc, pp. 199–203.

Haukioja, E., and S. Hanhimäki. 1985. Rapid wound-induced resistance in white birch (*Betula pubescens*) foliage to the geometrid *Epirrita autumnata*: a comparison of trees and moths within and outside the outbreak range of the moth. *Oecologia (Berlin)* 65:223–228.

Haukioja, E., and S. Neuvonen. 1985a. Induced long-term resistance of birch foliage against defoliators: defensive or incidental? *Ecology* 66:1303–1308.

Haukioja, E., and S. Neuvonen. 1985b. The relationship between size and reproductive potential in male and female *Epirrita autumnata* (Lepidoptera:Geometridae). *Ecol. Entomol.* 10:267–270.

Haukioja, E., and S. Neuvonen. 1987. Insect population dynamics and induction of plant resistance: the testing of hypotheses. In P. Barbosa and J. C. Schultz (eds.), *Insect Outbreaks*. San Diego: Academic, pp. 411–432.

Haukioja, E., and P. Niemelä. 1977. Retarded growth of a geometrid larva after mechanical damage to leaves of its host tree. *Ann. Zool. Fennici* 14:48–52.

Haukioja, E., and P. Niemelä. 1979. Birch leaves as a resource for herbivores: seasonal occurrence of increased resistance in foliage after mechanical damage of adjacent leaves. *Oecologia (Berlin)* 39:151–159.

Haukioja, E., P. Niemelä, L. Iso-Iivari, H. Ojala, and E. M. Aro. 1978. Birch leaves as a resource for herbivores. I. Variation in the suitability of leaves. *Rep. Kevo Subarctic Res. Sta.* 14:5–12.

Haukioja, E., P. Niemelä, L. Iso-Iivari, S. Sirén, K. Kapiainen, K. J. Laine, S. Hanhimäki, and M. Jokinen. 1981. Koivun merkitys tuntu rimittarin kannanvaihtelussa. *Luonnon Tutkija.* 85:127–140.

Haukioja, E., K. Kapiainen, P. Niemelä, and J. Tuomi. 1983a. Plant availability hypothesis and other explanations of herbivore cycles: complementary or exclusive alternatives? *Oikos* 40:419–432.

Haukioja, E., P. Niemelä, and K. Kapiainen. 1983b. Herbivory and tree line birches. *Proc. Northern Quebec Tree-Line Conference 1981.* Nouveau Quebec: Poste-de-la-Balaine, pp. 151–159.

Haukioja, E., J. Suomela, and S. Neuvonen. 1985. Long-term inducible resistance in birch foliage: triggering cues and efficacy on a defoliator. *Oecologia (Berlin)* 65:363–369.

Haukioja, E., E. Pakarinen, P. Niemelä, and L. Iso-Iivari. 1988a. Crowding-triggered phenotypic responses alleviate consequences of crowding in *Epirrita autumnata* (Lepidoptera:Geometridae). *Oecologia (Berlin)* 75:549–558.

Haukioja, E., S. Neuvonen, S. Hanhimäki, and P. Niemelä. 1988b. The autumnal moth in Fennoscandia. In A. A. Berryman (ed.), *Dynamics of Forest Insect Populations: Patterns, Causes, and Management Strategies.* New York: Plenum, pp. 163–178.

Kennedy, C. E. J., and T. R. E. Southwood. 1984. The number of species of insects associated with British trees: a re-analysis. *J. Anim. Ecol.* 53:455–478.

Kogan, M. 1986. Bioassays for measuring quality of insect food. In J. R. Miller and T. A. Miller (eds.), *Insect–Plant Interactions.* New York, Springer, pp. 155–189.

Koponen, S. 1983. Phytophagous insects of birch foliage in northernmost woodlands of Europe and eastern North America. *Proc. Northern Quebec Tree-Line Conference 1981.* Nouveau Quebec: Poste-de-la-Balaine, pp. 165–176.

Lewis, A. C., and H. F. van Emden. 1986. Assays for insect feeding. In J. R. Miller and T. A. Miller (eds.), *Insect–Plant Interactions.* New York: Springer, pp. 95–119.

Mattson, W. J., R. K. Lawrence, R. A. Haack, D. A. Herms, and P. J. Charles. 1988. Defensive strategies of woody plants against different insect feeding guilds in relation to plant ecological strategies and intimacy of association with insects. In W. J. Mattson, J. Levieus, and C. Bernard-Dagan (eds.). *Mechanisms of Woody Plant Defenses against Insects: Search for Pattern.* New York: Springer, pp. 3–38.

Neuvonen, S., and K. Danell. 1987. Does browsing modify the quality of birch foliage for *Epirrita autumnata* larvae? *Oikos* 49:156–160.

Neuvonen, S., and E. Haukioja. 1984. Low nutritive quality as defense against herbivores: induced responses in birch. *Oecologia (Berlin)* 63:71–74.

Neuvonen, S., and E. Haukioja. 1985. How to study induced plant resistance? *Oecologia (Berlin)* 66:456–457.

Neuvonen, S., E. Haukioja, and A. Molarius. 1987. Delayed inducible resistance in four deciduous tree species. *Oecologia (Berlin)* 74:363–369.

Neuvonen, S., S. Hanhimäki, J. Suomela, and E. Haukioja. 1988. Early season damage to birch foliage affects the performance of a late season herbivore. *J. Appl. Entomol.* 105:182–189.

Niemelä, E., M. Aro, and E. Haukioja. 1979. Birch leaves as a resource for herbivores. Damage-induced increase in leaf phenolics with trypsin inhibiting effects. *Rep. Kevo Subarctic Res. Stat.* 15:37–40.

Rhoades, D. F. 1985. Offensive-defensive interactions between herbivores and plants: their relevance in herbivore population dynamics and ecological theory. *Am. Natur.* 125:205–238.

Tenow, O. 1972. The outbreaks of *Oporinia autumnata* Bkh. and *Operophthera* spp. (Lepidoptera:Geometridae) in the Scandinavian mountain chain and northern Finland 1862–1968. *Zool. Bidrag Uppsala, Suppl.* 2:1–107.

Tingey, W. M. 1986. Techniques for evaluating plant resistance to insects. In J. R. Miller and T. A. Miller (eds.), *Insect–Plant Interactions*. New York: Springer, pp. 251–284.

Tuomi, J., P. Niemelä, E. Haukioja, S. Sirén, and S. Neuvonen. 1984. Nutrient stress: an explanation for plant anti-herbivore responses to defoliation. *Oecologia (Berlin)* 61:208–210.

Tuomi, J., P. Niemelä, and T. Fagerström. 1991. Carbon allocation, phenotypic plasticity, and induced defenses. In D. W. Tallamy and M. J. Raupp (eds.), *Phytochemical Induction by Herbivores*. New York: Wiley, pp. 85–104.

Valentine, H. T., W. E. Wallner, and P. M. Wargo. 1983. Nutritional changes in host foliage during and after defoliation and their relation to the weight of gypsy moth pupae. *Oecologia (Berlin)* 57:298–302.

Wallner, W. E., and G. S. Walton. 1979. Host defoliation : a possible determinant of gypsy moth population quality. *Ann. Entomol. Soc. Am.* 72:62–67.

Werner, R. A. 1979. Influence of host foliage on development, survival, fecundity and oviposition of the spear-marked black moth, *Rheumaptera hastata* (Lepidoptera:Geometridae). *Can. Entomol.* 111:317–322.

Werner, R. A. 1981. Advantages and disadvantages of insect defoliation in the taiga ecosystem. *Proc. 32nd Alaska Sci. Conf.*, Fairbanks, AK.

Wratten, S. D., P. J. Edwards, and I. Dunn. 1984. Wound-induced changes in the palatability of *Betula pubescens* and *B. pendula*. *Oecologia (Berlin)* 61:372–375.

CHAPTER 13

VARIABLE INDUCED RESPONSES: DIRECT AND INDIRECT EFFECTS ON OAK FOLIVORES

STANLEY H. FAETH

Department of Zoology, Arizona State University, Tempe, AZ 85287-1501

1   Introduction
    1.1   Variable consequences of induced responses
2   Mitigating factors in induced responses
    2.1   Variation in chemical changes
    2.2   Variation in nutritional changes
    2.3   Variation in phenological and morphological changes
    2.4   Are phenological changes defensive?
3   Other complicating factors
    3.1   Life histories of participating insect species
    3.2   Behavior of phytophagous insects
    3.3   Induced responses and intraspecific competition
    3.4   Induced responses and the third trophic level
4   Future directions
    4.1   Know thy effects
    4.2   Know thy response
5   Conclusion
    References

1   INTRODUCTION

Plants often undergo chemical changes following herbivory (e.g., Green and Ryan 1972, Carroll and Hoffman 1980, Edwards and Wratten 1982, Schultz and Baldwin 1982, Fowler and Lawton 1985). Preliminary evidence suggested that these chemical changes inhibit herbivore growth and development, decrease herbivore population growth, and thus reduce herbivore pressure on the plant. Consequently, such changes were viewed as a previously unrecognized line of evolved plant defenses above and beyond constitutive defenses. The general acceptance of induced chemical changes as plant defenses is evidenced by attempts to incorporate induced plant responses into programs of control for insect pests of agricultural and silvacultural systems (Kogan and Paxton 1983, Karban, this volume, Chapter 17). Despite its intuitive appeal, Fowler and Lawton (1985), in a review of purported cases of inducible defenses, found problems with evidence that postinjury changes have detrimental effects on individual growth and development, much less on population dynamics of herbivorous insects. In fact, induced changes may have no effect on individuals or populations and may even benefit (in terms of growth, development, or fecundity) herbivores (Fowler and Lawton 1985, Myers 1988, Coleman and Jones, this volume, Chapter 1).

In many ways, the development of ideas concerning induced defenses stems from coevolutionary arguments. Ehrlich and Raven (1965) proposed that the proliferation of allelochemicals in plants and the diversity of phytophagous insects were mutually explanatory: Each had reciprocally evolved in a coevolutionary arms race. Later workers questioned the importance of reciprocal coevolution in structuring phytophagous insect communities (Janzen 1980, Fox 1981, Thompson 1982, Futuyma 1983, Strong et al. 1984, Jermy 1984, Bernays and Graham 1988), and alternative explanations were proposed (Strong et al. 1984, Bernays and Graham 1988). Many of the criticisms arose from the realization that interactions between plants and insects are extraordinarily complex, confounded by temporal and spatial variation in secondary chemistry, abiotic and biotic factors, and other plant features, such as genotype, age, phenology, and architecture, that may interact with or even override variation in secondary chemistry (Denno and McClure 1983, Barbosa 1988, Thompson 1988). Likewise, the popular notion that induced chemical changes are universally defensive has been challenged (Fowler and Lawton 1985, Myers and Williams 1987, Faeth 1988, Myers 1988). Researchers have recently recognized that variation in these factors may also influence how induced changes affect individuals and populations of herbivorous insects (Faeth 1987, 1988, Karban, this volume, Chapter 17, Raffa this volume, Chapter 11). Furthermore, in addition to phytochemistry, herbivory causes change in plant nutrition, phenology, and architecture, all of which may mask any effects by phytochemical induction. This chapter focuses on the complexity of induced responses and how many factors contribute to variable ecological effects on individuals and populations of phytophagous insects.

1.1 Variable Consequences of Induced Responses

The repercussions of changes induced by chewing and sucking insects on subsequent herbivores are highly variable (Table 1). Some studies show negative effects (Myers and Williams 1984, 1987, Niemelä et al. 1984, Karban and Carey 1984, Raupp and Denno 1984, Karban 1985, Tallamy 1985, Fowler and McGarvin 1986, Karban et al. 1987), some show positive effects (Rock-

TABLE 1 Studies Involving Induced Plant Responses and Their Effects on Invertebrate Herbivores

Reference	Plant/Herbivore	Time After Damage Measured[a]	Chemical Response[b]	Phenological Response[c]	Indirect Effects Measured[d]	Amount of Damage[e]	Type of Damage[f]	Effect on Herbivores[g]
Bergelson et al. (1986)	Birch/casebearer	1-8 days	Yes	No	No	Partial	Artificial (pin-pricking)	Movement and larv. devel. increased; no increase in mortality
Bergelson & Lawton (1988)	birch/casebearer/caterpillars	hrs.-days	Yes	No	Yes	Partial	Artificial (hole-punched)	Casebearer - slight increase in movement; caterpillar - no increased movement; neither had increased predation
Carroll & Hoffman (1980)	Cucurbita/beetles	40 min.	No	No	No	Partial	Artificial (1 cm slits)	Feeding rate increased - 1 spp. Feeding rate decreased - 1 spp.
Dixon & Barlow (1979, Barlow & Dixon (1980)	Lime trees/lime aphids	Recent infest. days?	No	No	No	Partial	Sucking insects	Smaller, longer development, increased mortality
Edwards & Wratten (1982)	Trees/snails	4-17 days	No	No	No	Partial	Artificial (1/4 leaf cut)	Feeding rate decreased - 1 spp. No effect -1 spp.
Faeth (1986)	Oaks/leafminers	days-months	Yes	Yes	Yes	Partial	Leafchewer/Artificial (hole-punched)	Adult oviposition decreased; larv. mortality increased; parasitism increased
Faeth (1987, 1988)	Oaks/leafminers	days-months	Yes	No	Yes	Complete	Artificial (manually-removed)	Adult oviposition increased; larv. and pupal mortality decreased; abscission death decreased; local extinct. - 1 spp.
Fowler & Lawton (1985)	Birch/folivores	4 weeks	No	No	No	Partial (5 or 25%)	Artificial (1/2 leaf torn)	No effects on subsequent folivore feeding
Fowler & MacGarvin (1986)	Birch/sucking insects/caterpillars	2 days-4 weeks	No	No	No	Partial	Artificial (hole-punched)	Larv. mass reduced; larv. mortality increased; no change in parasitism or amount of subsequent damage
Harrison & Karban (1986)	Cotton/mites	14-19 days	No	No	No	Partial	Mites	Avoidance by subsequent mites
Hartley (1988)	Birch/caterpillars	1-8 days	Yes	No	No	Partial	Artificial (10%	Caterpillars show no preference for

TABLE 1 (*Continued*)

Reference	Plant/ Herbivore	Time After Damage Measured[a]	Chemical Response[b]	Phenological Response[c]	Indirect Effects Measured[d]	Amount of Damage[e]	Type of Damage[f]	Effect on Herbivores[g]
Haukioja & Hanhimaki (1985)	Birch/caterpillars	days	No	No	No	Partial	leaf cut. insect (10% leaf area)	damaged/undamaged leaves or leaves with reduced phenolic synthesis
Haukioja & Niemelä (1977); Niemelä et al. (1979)	Birch/caterpillars	2 days	Yes	No	No	Partial	Artificial(Leaf torn/amt. not spec.)	Reduced growth and pupal mass; larval devel. time increased
Haukioja & Niemelä (1979)	Birch/moth & sawfly caterpillars	2 days	No	No	No	Partial	Artificial(1/2 leaf torn)	Larval development time increased
Hawkins (1988)	Birch/caterpillars/ leafminers	days	No	Yes	No	Partial	Artificial; Artificial (10-20% leaf torn) Leafchewers	Larv. devel. time increased - 4 spp. Larv. devel. time decreased - 1 spp. Larval mass decreased - 1 spp. Larval weight increased - 1 spp.; Parasitism of leafminers increased for specialists; no change for generalist parasites
Hunter (1987)	Oak/caterpillars	days-months	No	No	No	Partial/ Complete	Leafchewers	Reduced larval survival and pupal mass but abundances greater on damaged trees
Karban & Carey (1984)	Cotton/mites	12-26 days	No	No	No	Partial	Chewing Insects	Reduced population size
Karban (1985)	Cotton/mites	14-28 days	No	No	No	Partial	Mites/ mechanical abrasion	Reduced population growth
Karban (1987)	Cotton/mites	14-28 days	No	No	No	Partial	Mites	Reduced population growth, but effect strongest when least damage
Karban et al. (1987)	Cotton/mites/ verticillium wilt	3-14 days	No	No	No	Partial	Chewing insects/fungi	Reduced infestation of wilt; Reduced population of mites
Lawton (1986)	Birch/caterpillars	days	Yes	No	No	Partial	Leafminers	Caterpillars avoid damaged leaves

Reference	System	Time			Complete?	Treatment	Results
Leather et al. (1987)	Pine/moth caterpillars	previous yr.	Yes	No	Complete?	Chewing insects (50% foliage)	Adult oviposition decreased; Larv. mass and survival decreased
Myers & Williams (1984)	Alder/tent caterpillars	previous yrs.	No	No	Complete?	Leafchewers	Reduced larv. growth only after repeated, severe defoliation
Myers & Williams (1987)	Alder/tent caterpillars	1 day	Yes (nutrients only)	No	Partial	Artificial (1/2 leaf torn)	No effects
		4 previous			Partial	Chewing insects (adjacent leaves)	No effects
					Complete?	Chewing insects (4 prev. yrs.)	Larv. devel. time reduced; pupal and adult mass reduced
Niemelä et al. (1984)	Pine/sawflies	days? prev. yr.	No	No	Partial	Sawflies	Faster growth, lower mortality - 2 spp. No effect on growth - 2 spp. but higher mortality - 1 spp.
Neuvonen & Haukioja (1984)	Birch/geometrid	Prev. yr.	No	No	Complete	Artificial (100% of leaves removed)	Consumption rate reduced
Neuvonen et al. (1987)	four tree spp.	Prev. yr.	No	No	Partial	Artificial (1/2 leaf torn) caterpillars	Larval growth and survival reduced on 3 of 4 tree spp.
Pullin (1987)	Nettles/nymphalid caterpillars	days-months	Yes (nutrients only)	Yes	Complete	Artifical (mowing)	Growth rates and pupal mass greater on regrowth foliage
Raupp & Denno (1984)	Willow/beetles	4-7 days	No	No	Partial	Artificial (1/2 leaf torn) Nonspecific herbivore (> 5% of leaflet)	Adult reduced fecundity Larv. devel. time decreased
Renaud (1986)	Walnut/leafchewers	months	Yes	No	Complete	Artificial (50% of leaves removed)	No decline in subsequent herbivory
Rhoades (1983a)	Alder/tent caterpillars	weeks	Yes	No	Partial?	Leafchewers	Mortality increased; larv. devel. decreased; adult fecundity decreased
	Willow/webworm/tent caterpillars	weeks	Yes	No	Partial?	Leafchewers	Webworm larv. growth reduced; tent caterpillars - no effects 1979; larv. growth reduced 1981; larv. mortality increased

TABLE 1 (*Continued*)

Reference	Plant/Herbivore	Time After Damage Measured[a]	Chemical Response[b]	Pheno-logical Response[c]	Indirect Effects Measured[d]	Amount of Damage[e]	Type of Damage[f]	Effect on Herbivore[g]
Rockwood (1974)	Calabash tree	weeks/months	No	Yes	No	Complete	Artificial (100% of leaves)	Damage by, and apparently populations of, flea beetles increased
Roland & Myers (1987)	Oak/apple/winter moth caterpillars	previous yr./current yr.	No	No	No	Partial? Complete?	Caterpillars (0-100% leaf area)	Pupal mass reduced with increased defoliation in current year; pupal mass increased with defoliation in previous year
Schultz & Baldwin (1982)	Oaks/gypsy moth caterpillars	previous yr./current yr.	Yes	Yes	No	Complete (past yrs.) Partial? (curr. seas.)	Gypsy moth caterpillars	Effects on gypsy moth were ot determined
Silkstone (1987)	Birch/leafchewing insects	months	No	No	No	Partial	Artificial (hole punch 5 or 15% leaf area)	Less subsequent grazing on damaged leaves
Stiling & Simberloff	Oaks/leafminers	days-months	No	Yes	Yes	Partial	Leafminers	Slight increase in mortality due to premature leaf abscission
Valentine et al. (1983)	Oak/Gypsy moth caterpillars	prev. yr. or yrs. & days?	Yes (nutrients only)	No	No	Complete (past yrs.)	Gypsy moth caterpillars	Pupal mass reduced
	Birch/gypsy moth caterpillars	prev. yr. or yrs. & days?	Yes (nutrients only)	No	No	Partial (curr. seas.)	Artificial (1/3 leaf cut)	No effects
Wallner & Walton (1979)	Oak/gypsy moth caterpillars	prev. yr. or yrs. & days?	No	No	No	Complete (past yrs.) Partial (curr. seas.)	Gypsy moth caterpillars Artificial (1/3 leaf cut)	Pupal mass reduced; larv. devel. time increased; larv. mortality increased
West (1985)	Oaks/caterpillars/leafminers	days-months	Yes (nutrients only)	Yes	Yes	Partial	Artificial (1/4 leaf cut) Leafchewers	Leafminer mortality increased due to leaf abscission and unknown causes

Williams & Myers (1984)	Alder/Fall webworm caterpillars	prev. yr. or yrs. (6)	No	Yes	No	Partial?	Tent caterpillars/webworms (0-100% of leaves)	Greater pupal mass, faster larv. devel. on trees defol. for 2 yrs. Decreased pupal mass, slower larv. devel. on trees defol. for 3 yrs.
Williams & Whitham (1986)	Cottonwood/aphids	days	No	Yes	No	Partial	Aphid gallers	Aphid mortality and movement increased due to leaf abscission
Wratten et al. (1984)	Birch/caterpillars/aphids	6 hr.-28 days	Yes	No	No	Partial	Artificial 1/4 leaf cut on 50% of leaves	Feeding rate reduced; no effect on population size

[a] Closest approximation of the time interval between damage and when the effect on folivores was measured. Some time periods are estimated where time periods are only vaguely noted in the studies. Question marks (?) appear when time periods can only be inferred from the study.

[b] "Yes" indicates that phytochemical changes following damage were actually measured, rather than inferred form changes in the folivores themselves. "No" indicates no phytochemical changes were measured in this study, although in some cases, phytochemistry was documented in a related study.

[c] Phenological changes refer to phenological or morphological changes in abscission rates, leaf size, number, shape, thickness, time to refoliation or any other alteration in growth, development or morphology of the plant.

[d] Indirect effects refer to whether plant-mediated changes in attack by natural enemies, or intra- or interspecific competition between folivores was documented.

[e] "Partial" damage indicates that only parts of leaves were removed or damaged. "Complete" damage indicates whole leaves were removed, but not necessarily all leaves on a plant. In some cases, it was unclear if damage (particularly in past years) was the result of partial or complete damage. For example, in some studies, it was mentioned only that trees in past years were "severely defoliated." These are marked with question marks (?).

[f] "Type of damage" is generally categorized as "artificial" (i.e., damage was manually inflicted) or by organisms themselves. In each case, the method, organism, and amount of damage inflicted by each mode is listed when specified in the study.

[g] Effects on herbivores are summarized as best as can be discerned by information given in the study. No attempt was made to ascertain if the effects were validly determined or results interpreted in a statistically correct fashion (see Fowler and Lawton (1985)) for a critique of experimental design and purported results and conclusions for some these studies).

299

wood 1974, Myers and Williams 1984, 1987, Niemelä et al. 1984, Faeth and Bultman 1986), and some demonstrate no effect at all on herbivores (Wratten et al. 1984, Fowler and Lawton 1985, Bergelson et al. 1986). Such variation in plant-mediated effects, however, is expected if one closely examines confounding factors among studies. In Fowler and Lawton's review of purported cases of rapidly induced defenses, the induction time frames ranged from 40 min to 28 days, a difference of more than three orders of magnitude (Table 1). Inclusion of more recent studies inflates the time differential to more than four orders of magnitude (40 min to several years), although these studies do not claim to examine "rapid" induced defenses. Less than half of the cited studies actually determined chemical changes, their magnitude, or their duration. Furthermore, the inducing agent varied widely, from actual herbivory to artificial tearing or pin-pricking. In some studies whole leaves were removed; in others only parts of leaves were consumed or manually torn. Complicating factors such as phenological changes (i.e., increased leaf abscission, or changes in leaf structure), indirect influences of natural enemies, seasonal effects, and differences in the life histories of participating arthropod species (e.g., timing of induced responses with windows of oviposition or feeding), are generally not accounted for (Table 1).

Coleman and Jones (this volume, Chapter 1) state that despite these differences, the majority of studies show negative effects of induced responses. Indeed, of the insect species listed in Table 1, about 31 were negatively affected, 9 positively affected, and 10 showed no effect of induced responses. One must be cautious, however, of these tallies. These studies may not reflect actual frequencies because results contrary to the prevailing notion that induced responses are defensive may go disproportionately unpublished. Second, variation in factors influencing the impact of induced responses make comparisons and generalizations from past studies difficult. Unless complicating factors are carefully controlled in future studies, efforts to derive any general theory of induced plant responses are unlikely to prove fruitful. To illustrate these points, induced responses and their variable effects are examined in oaks and their folivores.

2   MITIGATING FACTORS IN INDUCED RESPONSES

2.1   Variation in Chemical Changes

Generally, concentrations of phenolic compounds increase following folivory of oaks and other trees (Niemelä et al. 1979, Bryant 1981, Bergelson et al. 1986, Puttick 1986, Tuomi et al. 1988, Hartley and Lawton, this volume, Chapter 5). Schultz and Baldwin (1982) reported that leaves on red oak trees subjected to previous and current gypsy moth defoliations were higher in total phenolics and tannins than those with lesser amounts of defoliation. Also, newly flushed leaves, produced after defoliation, had greater hydro-

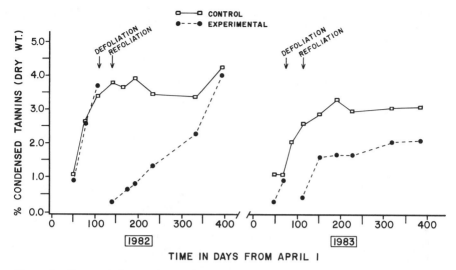

Figure 1   Seasonal changes in condensed tannin content of control and experimental (defoliated) trees in 1982–83 and 1983–84. Each point is the mean of six trees. See Faeth (1986) for phytochemical methods.

lyzable tannin content than leaves remaining on trees with low or severe defoliation. Change in foliar phenolics of Emory oak, *Quercus emoryi*, in Arizona after artificial folivory was quite different from that found in Schultz and Baldwin's (1982) study. After trees were completely defoliated, replacement leaves were higher in hydrolyzable tannin content, but these levels quickly returned to concentrations found in mature leaves on undefoliated trees (Faeth 1987, 1988).

Increases in phenolics in partially damaged leaves have also been reported in various oaks (Wratten et al. 1984, Bergelson et al. 1986, Faeth 1986, Hartley and Lawton, this volume, Chapter 5). Partially damaged leaves and replacement leaves of *Q. emoryi* are very different chemically even though the same mechanism, folivory, produced the effect. Partially damaged leaves are higher in condensed tannins than undamaged leaves, while the opposite is true for replacement leaves (Faeth 1987, 1988, Fig. 1). Thus, oak leaves respond differently depending on the severity of folivory. Based on induced chemical changes alone, one is left with the unconventional conclusion that less folivory (partial vs. complete leaf consumption) produces a greater plant response. Interaction between early and late season herbivores could therefore be inversely density dependent, at least when damage is severe enough that replacement leaves are produced.

A hypothetical depiction of the inversely density-dependent effects of oak responses on herbivore populations is shown in Figure 2. This relationship depends on two assumptions. First, increases in phenolics adversely affect herbivore populations (Feeny 1970, 1976, Levin 1971, Rhoades and Cates

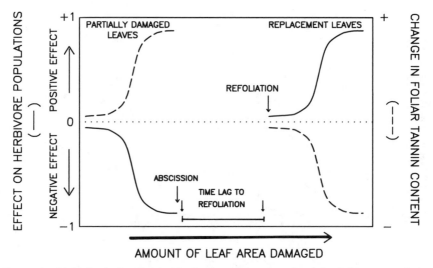

Figure 2 Hypothetical relationship between amount of oak leaf damage and the effects via changes in phenolics on a population of herbivorous insects. As partial damage to leaves on a tree increases, induction of chemical responses increases and herbivore population growth is increasingly hindered. However, once damage reaches the point where leaves begin to abscise and increasing numbers of replacement leaves are produced, population growth is enhanced by replacement leaves of superior quality (lower in phenolics and less likely to abscise). See text for assumptions. The relationship does not include indirect effects via natural enemies and is based upon results from the *Quercus emoryi*–herbivore–leaf miner system.

1976, Rhoades 1983b, 1985). More recent evidence, however, indicates that some insects are unaffected by phenolics and may even prefer foliage high in phenolics (Bernays 1978, 1981, Berenbaum 1980, Fox and MacCauley 1977, Zucker 1983, Martin and Martin 1984, Faeth and Bultman 1986, Lawton 1986). Second, increasing partial damage results in monotonic increases in negative effects on herbivore populations. Few controlled studies have examined this assumption. Karban (1987), for example, found that intraspecific competition between mite populations via induced responses was most intense when the initial mite population was smallest. Thus, the effects of partial damage via induced responses on herbivores may also be inversely density dependent, similar to the contrast between partial and complete leaf consumption.

If folivory increases phenolics, the impact on insect populations also depends on variability in both the duration of the chemical response and when folivory occurs. Some chemical changes after folivory are short-lived, while others persist for the duration of the growing season. For example, hydrolyzable tannin is higher in replacement leaves shortly after refoliation, but returns to that in mature, primary leaves within 30 days (Faeth 1986, 1987,

1988). It is likely that these changes simply reflect developmental changes in new leaves as they expand, since chemical changes mimic those of newly flushed spring leaves. Nevertheless, such alterations in the resource could affect insects feeding at that time. Replacement leaves do differ, however, from new spring leaves in condensed tannin content. Condensed tannin content is initially low in both spring leaves and replacement leaves, but spring leaves gradually increase in condensed tannin content, while that in replacement leaves remains low throughout the growing season (Fig. 1). Partially damaged leaves, alternatively, have higher condensed tannin content than intact leaves, and this difference persists throughout the growing season (Faeth 1986).

The seasonal timing of folivory may also alter induced responses and their effects on herbivore populations. One expects long-lived plants to be particularly variable in their response to folivory through the growing season as light, temperature, nutrient and moisture availability, and developmental state of the plant, as well as biological agents such as fungi and bacteria, change seasonally. There have been few systematic studies of variability in chemical responses and the impact of this variation on folivorous insects as a function of timing during the growing season. Tuomi et al. (1988) found that early and late defoliations of birch branches both result in increases in phenolics in the next growing season compared to undamaged branches; those defoliated earlier showed stronger induced responses. Replacement leaves produced after early and late season defoliations of *Quercus emoryi* differ in condensed tannin content from intact leaves as well as from each other (Fig.1). One might therefore predict that the effect of herbivory via induced chemical responses on insect populations in either the current or in future growing seasons depends on timing of defoliation.

At any given time, then, an individual tree may be a mosaic of waxing and waning chemical responses based on within-tree variation in intensity and timing of folivory. This mosaic becomes even more complex when such factors as the capacity to respond chemically based on tree genotype, age, and nutrient status are superimposed (Coleman and Jones, this volume, Chapter 1). How these complex responses and their distribution in time and space will affect current and future sites of folivorous insect species will be difficult to decipher.

2.2 Variation in Nutritional Changes

The effects of induced changes in plant secondary chemicals on herbivores are further complicated by changes in other physical and nutritional attributes of plants known to affect herbivore performance. For example, in oaks, leaves flushed after removal of primary leaves are generally lower in water content, greater in toughness, and lower in protein content (Schultz and Baldwin 1982, Faeth 1987, 1988). However, this trend depends on when replacement leaves are sampled. Faeth (1987, 1988) found that water content was higher in replacement leaves immediately after flushing, but declined to lower levels relative to mature, primary leaves on undefoliated trees. Defoliation in previous years and partial damage in the current season results in the lowering of sugar and nitrogen content in black oak (Valentine et al. 1983) and in alder (Myers and Williams 1987). Similarly, partial damage to Emory oak leaves results in lower water and protein content in remaining leaf parts (Faeth 1986). Clearly, if herbivory affects subsequent insects, it becomes difficult to disentangle the effects of induced secondary chemistry from changes in nutritional content, or from the interaction between the two (Neuvonen and Haukioja 1984, Duffey et al. 1986).

2.3 Variation in Phenological and Morphological Changes

Phenological and morphological features on plants are recognized as traits that can strongly influence folivory (e.g., leaf abscission, Faeth et al. 1981, Risley 1986, Escudero and del Arco 1987; stem length, Price et al. 1987a, b; foliar pubescence, Obrycki 1986, Ezcurra et al. 1987; leaflet position, Gall 1987). Phenological and morphological changes typically follow folivory, yet few studies have included these changes in examining induced responses (but see Myers and Bazely, this volume, Chapter 14) (Table 1). The most obvious phenological change is the temporary or permanent absence of some or all leaves, assuming that folivores are consuming entire leaves or that severely damaged leaves are abscised. If and when trees refoliate following folivory within a growing season depends on the type of normal plant growth patterns (determinate versus indeterminate growth), the extent and history of defoliation (Heichel and Turner 1976, Ericsson et al. 1980), the amount of root reserves and nutrients available (Crawley 1983), weather conditions, and competition with other plants (Lee and Bazzaz 1980).

Interacting with these factors is the frequency and timing of defoliation. If trees are defoliated late in the growing season, flushes of replacement leaves may not occur at all or may be delayed compared to early season defoliations. For example, in *Q. emoryi*, replacement leaves appeared 18 ± 3 (mean $\pm$ SE) days after defoliations on 8 June, but this interval lengthened to 32 ± 5 days when defoliations occurred on 14 July (Faeth 1987, 1988). Repeated defoliations in a single season or between seasons could deplete root reserves such that the lag to refoliation is delayed or does not occur at all. Furthermore, the effects of defoliation in one season may be carried over to the next. When larch (Benz 1977) and birch (Haukioja and Niemelä 1977, Haukioja 1980, Tuomi et al. 1989) are defoliated in the previous year, flushing is delayed or more variable the following spring.

Not only may defoliation affect availability of leaves to insect folivores, but reflushed leaves may differ morphologically and phenologically from primary leaves. Reflushed leaves are usually smaller than primary leaves (Tuomi et al. 1989, Figure 3) and differ in shape (Silva-Bohorques 1986, Faeth 1987, 1988). Replacement leaves may be less prone to leaf abscission,

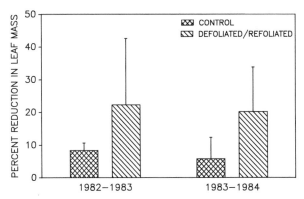

Figure 3 Mean difference in the percent leaf wet mass ($\pm$ SD) between primary, mature leaves (controls, $N = 6$ trees) on *Quercus emoryi* and replacement leaves produced after manual defoliation in 1982–83 ($N = 6$ trees). Mass of control leaves declines seasonally as water content decreases. The same experimental trees were defoliated again in a second season (1983–84). Mean size reduction of replacement leaves is significantly greater than primary leaves in 1983–84 ($t = 2.32$, df $= 10$, $p < .05$) but not in 1982–83 ($t = 1.66$, df $= 10$, $p > .10$).

a critical factor in the survival of endophagous insects such as leaf miners (Askew and Shaw 1979, Faeth 1985a, 1986, 1987, 1988, Stiling and Simberloff 1989).

Under most circumstances, herbivores consume only small parts of leaves (Edwards and Wratten 1983, Faeth 1985b). Partially damaged leaves differ phenologically from both primary and replacement leaves and are more likely to abscise than primary leaves (Faeth 1985a, 1986, Simberloff and Stiling 1987, Williams and Whitham 1986). This contrasts with phenological changes accompanying replacement leaves.

2.4   Are Phenological Changes Defensive?

Much like induction of chemical and nutritional changes, alterations in phenology, particularly leaf abscission, have been proposed as induced plant defenses (Owen 1978, Rhoades 1983b, Williams and Whitham 1986). Although induced abscission may increase mortality for some folivorous insects (Faeth et al. 1981, Williams and Whitham 1986, Simberloff and Stiling 1987, Stiling and Simberloff 1989, but see Kahn and Cornell 1983, Pritchard and James 1984), at present there is little evidence to suggest that premature leaf abscission is an evolved defense against herbivores (Bultman and Faeth 1986a, Stiling and Simberloff 1989). The most parsimonious explanation for premature leaf abscission is that it is a simple damage response that occasionally can affect some insect species on some host plants (Faeth et al. 1981, Stiling and Simberloff 1989). Research that documents whether plants can discriminate physical from insect damage and respond accordingly would be critical in deciphering whether phenological or chemical inductions are indeed defensive or simply incidental wound responses that may or may not have an impact on insect herbivores. Some inroads have been made toward that end (see Hartley and Lawton, this volume, Chapter 5), but at this point it is premature to conclude that such responses are active defenses, at least in the evolutionary sense.

Likewise, defoliator-induced delay in budbreak in the next growing season has been postulated as an induced defense (Haukioja et al. 1985). However, Tuomi et al. (1988, 1989) concluded that while such delays may affect insect populations, the most parsimonious explanation for such delays is resource limitation rather than an evolved defense against herbivores. Until future research indicates otherwise, phenological changes that accompany herbivory should be considered consequences of altered plant metabolism, rather than specific defenses against insect herbivores.

3 OTHER COMPLICATING FACTORS

Coleman and Jones (this volume, Chapter 1) review numerous plant factors (genotype, age, and resource status) that influence induction and its consequences. Many factors involving the life histories, behavior, intra- and interspecific interactions, and natural enemies of insect herbivores may also alter the effects of induction.

3.1 Life Histories of Participating Insect Species

The insect species that are available for plant colonization as well as their life histories are critical factors molding the outcome of induced responses. Auerbach and Simberloff (1985) provided an example with a study of fertilized and drought-stressed water oak trees (*Q. nigra*). Although these treatments did not include folivory, response of the oak trees were similar to what might occur following defoliation, perhaps because both herbivory and environmental stress affect resource limitation. Trees dropped leaves prematurely and then atypically reflushed leaves. Densities of two species of leaf miners that are restricted to oviposit and feed on supple, new foliage responded positively by increasing on these replacement leaves. However, when leaf flush occurred too late in the growing season, densities of these leaf miners were unaffected simply because ovipositing adults were not present. Similarly, I found that leaf miners of *Q. emoryi* oviposit preferentially on new leaves produced by defoliation, although these leaf miners generally do not encounter this resource; the leaf miners appear late in the growing season when leaves are largely mature. Coupled with this preference is an increase in survival on replacement leaves relative to that on mature leaves of undefoliated trees. However, if defoliation occurs late in the growing season and the lag to refoliation is increased, one dominant species (*Stilbosis juvanti*) goes extinct locally because its window of oviposition coincides with the time when leaves are absent from defoliated trees (Faeth 1987).

Generally, most folivorous insects are highly susceptible to even slight variations in host phenology because emergence, mating, oviposition, feeding, and pupation are typically dependent on various aspects of host quality associated with temporal change (Strong et al. 1984). For example, a localized induced change in phenology or in chemistry that slows or speeds development of a minority of the larval population may mean that those individuals emerge asynchronously with the majority of the population; successful mating is therefore unlikely. Although many studies have examined changes in development relative to induced changes (Table 1), I know of none that have examined the consequences in terms of mating success. This could have important effects on population dynamics and should prove an interesting area of future research.

Predicting when and how insect species might respond to induced changes based upon their life histories is difficult. However, one might speculate under what conditions the fitness of herbivorous insects is increased or decreased. Insect species that frequently encounter induced changes in evolutionary time might adapt to such changes and exploit them. This scenario would most likely occur when folivory and associated plant responses are predictable in time and space, and the insect species in question shows some fidelity or dependency to the particular plant species. Confirmation of this prediction is not yet available but some evidence supports this hypothesis. Opler (1974) reports that a leaf miner, *Neurobathra bohartiella*, oviposits and feeds only upon young, supple foliage of *Q. agrifolia*. Such leaves are plentiful at spring budbreak, but are only produced later in the season following physical damage or severe defoliation. Yet adults can be collected virtually all year round. Defoliation is caused primarily by the dioptid oak moth, *Phryganidia californica*, a monophagous specialist on *Quercus* in California that regularly defoliates oaks, particularly *Q. agrifolia* (Puttick 1986). Thus, it is tempting to speculate that success of second or third generations of the leaf miner within a season is partially dependent upon induced changes produced by the oak moth caterpillars. Damage by the oak moth caterpillar may be more predictable in space and time than secondary leaf flushes caused by more stochastic events such as physical damage from wind, hail, or fire. Variability in voltinism of this moth could be maintained via periodic defoliations by caterpillars. Auerbach and Simberloff (1985) also suggest that multivoltinism of the leaf miner *Neurobathra* on *Q. nigra* is maintained by secondary leaf flushes caused by early season defoliations by notodontid caterpillars.

It is possible that polyphagous or monophagous insects exapted to specific chemical profiles of foliage might actually benefit from induced changes. For example, some polyphagous insect species that normally encounter tannins in their diets tolerate and may even benefit from the presence of foliar tannins (Bernays 1981). If an induced response includes increased tannins, then fitness of these species should increase. Of course, this prediction ignores the effects of induced changes on other trophic levels which may either enhance or cancel beneficial effects (Price et al. 1980, Duffey et al. 1986, Price 1986). It is interesting to note that many late season insect species feeding on trees are generalists (Niemelä 1983), perhaps, in part, because they must contend with a wide and unpredictable array of induced changes precipitated by early feeding species.

3.2 Behavior of Phytophagous Insects

Feeding behavior of insect folivores is widely variable, with some insects taking only small bites of individual leaves and others consuming entire leaves or at least enough to cause abscission. Leaf area removed may be equivalent among or within trees but the distribution of tissue damage may be quite different. The distribution of herbivore damage as a function of insect species' feeding behaviors should have a strong effect on the outcome of plant-induced responses on other insect folivores. For example, Silkstone (1987) showed that artificial damage aggregated in the canopy of a tree produced a stronger deterrent to grazing than the same amount of damage dispersed throughout the tree.

Not only can insect feeding behaviors modify the plant response but the converse is also true. Edwards and Wratten (1983) and Edwards et al. (1988, this volume, Chapter 9) proposed that most partial leaf damage is overdispersed because insects behaviorally avoid local phytochemical induction. Others have shown that insect feeding behaviors may have evolved as a response to induced changes (e.g., Tallamy 1985, Dussourd and Eisner 1987, Edwards and Wanjura 1989). Fowler and Lawton (1985), however, criticized the general conclusion of Edwards and Wratten (1983) suggesting other mechanisms such as avoidance of natural enemies that cue to leaf damage, overdispersed oviposition by female insects, or sequential feeding by insects on leaves of different ages. Although image analyses and additional experiments suggest that feeding insects avoid a halo around previous feeding scars, adequate statistics are not yet available to test if insect damage is generally overdispersed in nature (Edwards et al., this volume, Chapter 9).

3.3 Induced Responses and Intraspecific Competition

Seasonal interactions among herbivore species as mediated by induced responses have most commonly been considered asymmetric interspecific competition (Lawton and Hassell 1981, Faeth 1986). Asymmetrical competition can occur due to direct effects of induced responses on growth, development, survival, and fecundity or by indirect alterations of intraspecific interactions. In the simplest sense, if defoliation results in a significant reduction in the amount of foliage on a tree, or in fewer or smaller replacement leaves, foliage could become limiting. However, given that most consider folivores not to be food limited in the strictest sense (Lawton and Strong 1981, Connell 1983, Orians and Paine 1983, Price 1983, Strong et al. 1984), this scenario may occur only under extreme conditions of folivory. In the more complex case where replacement leaves are of inferior quality because of induced changes in chemistry, phenology, or morphology, intraspecific competition could be more intense than when the damaging herbivore is absent. It should be emphasized, however, that relative quality of replacement leaves need not always decline, and instead may improve (Faeth 1987, 1988, Fig. 1), in which case the expectations for intraspecific competition would reverse.

Partial damage to leaves could also alter the probability of intraspecific interactions, and thus cause indirect effects on populations of folivorous insects. Several studies have shown that insects avoid partially damaged leaves or, alternatively, prefer undamaged leaves (Raupp and Denno 1984, Faeth 1985a, 1986, Hartley and Lawton, this volume, Chapter 5, Raupp and Sadof, this volume, Chapter 8, but see Hunter 1987), reducing the resource base and increasing the potential for intraspecific interactions. I have shown that several species of ovipositing leafminers avoid leaves that are damaged manually or by insects (Faeth 1985a, 1986). However, densities of these leaf miners are typically very low (<10 mines per 1000 leaves) so interspecific interactions were not significantly increased (although mortality from other sources, such as premature leaf abscission, is increased).

In contrast, another leaf-mining species, *Cameraria* sp. nov., *agrifoliella* group (Lepidoptera:Gracillaridae) that typically occurs at much higher densities (>100 per 1000 leaves), is highly clumped between trees, within trees, within terminal twigs, and within leaves (Faeth 1990a). Clumping has important consequences for survival and fecundity. Cannibalism increases and pupal mass, an indicator of fecundity (Marks 1976, Hough and Pimentel 1978), decreases with greater intra-leaf densities (Faeth, 1990a). Generally, clumping of this leaf miner at certain spatial scales can be explained by inherent variation in the propensity for certain leaves to abscise early, and in leaf size (Bultman and Faeth 1986a, b) but not in constitutive tannin or protein levels (Faeth 1990a). Induced responses, however, appear to be responsible for clumping at one spatial scale—clumping within twigs on leaves 4, 5, and 6. A common and abundant chewing species, *Epinotia* sp. (Lepidoptera:Tortricidae, near *emarginata*), ties apical leaves (1, 2, 3) together with silk and damages these leaves just before *Cameraria* adults oviposit. The distributions of the leaf tier and leaf miner are discordant (Fig. 4). Apparently, the leaf miner avoids these leaves for oviposition, possibly because of increased leaf abscission, smaller leaf areas for mining, structural and chemical changes that increase attack from parasitoids (section 4, Faeth 1985, 1986, Faeth and Bultman 1986), or predation from spiders that inhabit tied leaves (Corrigan and Bennett 1987). This avoidance may, in turn, cause

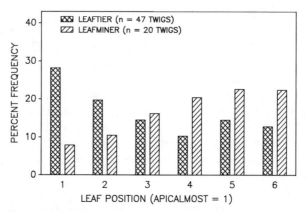

Figure 4 Distributions of the leaf miner *Cameraria* sp. nov. and the leaf tier *Epinotia* (near *emarginata*) on leaves within twigs of *Q. emoryi*. The distributions of the leaf tier and leaf miner are significantly different ($\chi^2 = 14.54$, df = 5, $p < .05$).

aggregated distributions on leaves 4, 5, and 6 where the risk of intraspecific interactions increases.

Because of the sedentary feeding habits of leaf miners, one might expect that interactions between leaf chewers like *Epinotia* and leaf miners are asymmetrical. However, Hartley and Lawton (1987, this volume, Chapter 5) have suggested that eriocraniid leaf miners induce higher phenolic levels to prevent leaf chewers from feeding on mined leaves and killing the larval leaf miner. There is no evidence that *Cameraria* on *Q. emoryi* induces chemical changes to ward off leaf-chewing insects, despite its long larval period within the leaf (about 11 months). Within trees, condensed tannin and protein content are not significantly different between mined and unmined leaves shortly after mining has begun, or later in the season after mines have expanded (Fig. 5). In fact, densities of this leaf miner are positively and negatively correlated with hydrolyzable tannin and protein content respectively early in the growing season, at least among trees (Faeth 1990a). Elevated tannin or lowered protein levels in mined leaves may result from habitat selection by ovipositing females rather than from induction.

3.4 Induced Responses and the Third Trophic level

Induced responses in plants may indirectly affect herbivores by altering attack by natural enemies (Price et al. 1980, 1986, Price 1986, Niemelä and Tuomi 1987). Plant-mediated changes in attacks by natural enemies can be a form of inter- or intraspecific competition via natural enemies (Holt 1977, Jeffries and Lawton 1984, Price et al. 1986). Induced defenses via natural enemies are thought to operate in several ways. First, natural enemies may use induced chemical and physical cues associated with damage to locate

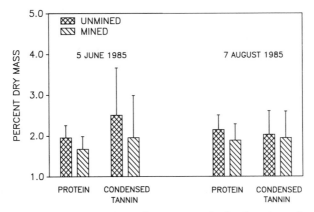

Figure 5 Protein and condensed tannin content of mined and unmined leaves of *Quercus emoryi* leaves when mining first begins (5 June, 1st instars) and later in the growing season (7 Aug, 4th instars). At time of first leaf collection, larvae had mined < 0.1% of upper surface leaf area, making any phytochemical induction at that time unlikely. Unmined and mined leaves do not differ significantly (univariate F tests on log transformed data) in protein or tannin content at either date (5 June 85, protein, $F = 1.12$, df = 7,19, $p = .24$; condensed tannin, $F = 1.46$, df = 7,19, $p = .39$; 7 Aug 85, protein, $F = 0.81$). See Faeth (1986) for phytochemical methods.

their hosts (Vinson 1976, Heinrich 1979, Weseloh 1981, Heinrich and Collins 1983, Duffey et al. 1986, Price 1986, Niemelä and Tuomi 1987, Hawkins 1988). Second, avoidance of damaged leaves may cause increased movement of insects searching for suitable feeding sites such that exposure and apparency to natural enemies is increased (Whitham 1983, Schultz 1983a, b). Third, induced chemical responses such as tannins, may prolong development time so that probability of discovery by natural enemies is enhanced (Price et al. 1980, Price 1986, but see Clancy and Price 1987).

There have been few rigorous experimental tests of the purported effects of induced responses on the third trophic level. Much like direct effects of induced responses on herbivores, results from studies to date are highly variable. Caterpillars feeding on damaged leaves increased movement and had longer developmental times, but attack by natural enemies did not increase (Bergelson et al. 1986, Fowler and MacGarvin 1986, Bergelson and Lawton 1988). West (1985) found that caterpillar and artificial damage to mined leaves increased mortality from premature leaf abscission and other unknown causes but not apparently from increased parasitism. Hawkins (1988) showed that damage to mined leaves did not affect parasitism of a casebearer on birch. However, Faeth (1985a, 1986) reported that either artificial or insect damage increased attack from parasitoids. Further studies showed that experimentally elevated tannins increased parasitism but also protected leaf miners from microbial attack such that overall survival was not altered (Faeth and Bultman 1986). Tannins have bactericidal and fungicidal properties (Swain 1979, Luthy et al. 1985). Thus, any negative effects of induced responses on endophagous insects via macroscopic parasites may be nullified by reduction in attack by microparasites (Taper and Case 1987, Taper et al. 1986). However, recent experiments where sealed, damaged leaves were attached to mined leaves show that physical alteration without chemical changes increases attack by natural enemies and decreases survival of *Cameraria* (Faeth 1990b). The net effect of damage by other herbivores on *Cameraria* may be negative and may be one factor in the behavioral avoidance of damaged leaves by ovipositing females (Faeth 1990b).

Price (1986) proposed that plants manipulate the third trophic level via variation in allelochemistry to which predators and parasites respond. Niemelä and Tuomi (1987) hypothesized that some plants have evolved leaves that mimic herbivore damage to attract enemies of herbivores. It is unlikely, however, that plants are able to manipulate the third trophic level via induced responses. It is clear that induced responses and their effects on the third trophic level are highly variable in space and time and depend on many extrinsic factors. A plant manipulating the third trophic level via induced responses could be likened to a flexible chain pushing an object across the floor; slight deviations in the direction of force would cause the chain to weave and buckle at the proximal links so that little or no force is transferred to distal ones, and movement of the object is frustrated. Systems with many links (i.e., within and between trophic level interactions) further reduce the probability of tight evolution or coevolution between plant responses and enemies of folivores. This is not to say that induced plant responses do not have an effect on either folivorous insects directly or attack by their enemies. It is unlikely that the selective chain is sufficiently rigid in time and space to affect evolutionary manipulations of higher tropic levels.

4 FUTURE DIRECTIONS

Guidelines on conducting research in ecology are rampant, no doubt because it is relatively easy to suggest what should be done and much more difficult in practice to comply with these suggestions. At the risk of tipping the balance even further, the following suggestions for studies of induced responses are offered.

4.1 Know Thy Effects

Paramount to the development of an ecological and evolutionary understanding of induced responses is determining if and under what circumstances induced changes affect populations of herbivores. The evidence to date is equivocal but not unexpected since comparing results of studies is much like comparing proverbial apples and oranges: Too many factors vary between studies for comparisons to be valid (Table 1). Extrapolation of laboratory studies of variation in development times, larval or pupal weights, survival, and fecundity to consequences at the population or community level is a large step. Clearly, in addition to controlling the many variables influencing induced effects on folivorous insects, it is necessary to demonstrate that effects at the level of the individual are translated into population changes. A measurable decrease in survival due to an induced response, for example, may be swamped by massive mortality from some other factor (compensatory mortality, Karban, this volume, Chapter 17). Yet there is no reason to preclude the possibility that other factors, such as enhanced efficiency of natural enemies, increase the impact of the induced response at the population level. Few studies include consideration of other trophic levels in examining the impact of induced changes (Price et al. 1980, Duffey et al. 1986, Price 1986). Some that do (Faeth 1985a, 1986, Taper et al. 1986, Taper and Case 1987) indicate that natural enemies can alter the influence of induced changes on insect populations, often in unexpected directions.

Community-level effects, such as changes in species diversity, composition, or organization, are more difficult to document than population-level events. Limited evidence, however, indicates that the impact of induced responses can be detected at the community level. For example, seasonality, local extinction, and change in dominance hierarchies of the leaf-mining guilds on oaks appear to be mediated by changes induced by the leaf-chewing guilds (West 1985, Faeth 1987, 1988), although these are probably not related to phytochemical induction. To what extent induced responses have been responsible for community patterns in other insect guilds is uncertain. It is doubtful that induced responses have served as a primary evolutionary force organizing many folivorous insect communities, given the complicating factors discussed in this and other chapters in this volume.

Even if induced responses are evident at the community level, it may not be appropriate to use conventional theories to predict or explain their effects on herbivore communities. For example, competitive interactions via induced responses may occur between distantly related species rather than between closely related species (Karban et al. 1987). Similarly, plant-mediated competitive or amensalistic interactions may be inversely density dependent (Faeth 1986, Karban 1987), reversing conventional expectations. Moreover, since these effects are often idiosyncracies of a particular system, generalizations and unifying theories are inappropriate. Nevertheless, recognition of the enormous variation in induced responses and their effects is a major advance and should provide fertile ground for future studies, just as the awareness of variability in constitutive chemistry has stimulated many lines of research (Denno and McClure 1983).

In any single study, it obviously is very difficult, if not impossible, to incorporate and control for all possible mitigating factors. It would seem, however, that multiple studies of the same three trophic level system would be a productive avenue of future research. Ecologists working at higher levels of organization can tell us if induced responses do indeed affect population dynamics and community organization and under what conditions. Because of the broad nature of such studies, the mechanisms for observed changes are often concealed in a black box. Toxicologists, pathologists, and biochemists can peek inside the box to reconcile mechanistic phenomena with effects observed at higher levels of organization.

4.2 Know Thy Response

Relatively few studies (Table 1) actually determine chemical, phenological, or structural changes in the host plant following folivory, and fewer still consider more than one of these responses simultaneously. Yet it is imperative to document when such changes occur in the plant and how long they last. Otherwise, there is the risk of drawing the right conclusion for the wrong reason. Most ecological studies use the insects themselves to assay effects by using choice experiments of damaged and undamaged foliage, or by determining growth rates, survival, or fecundity of insects on damaged and undamaged foliage. The presence or absence of an effect via assays does not alone prove or disprove the significance of the induced response in nature (Myers and Williams 1987). If an effect is documented through a laboratory assay, is it because of a chemical or phenological change? Is the change nutritional or allelochemical, or some interaction between the two (e.g., Duffey et al. 1986)? Would such an effect occur in the field given timing, extent of damage, type of damage, nature of response, and phenology of the insect species?

5   CONCLUSION

Folivory often results in chemical, phenological, and structural changes in plants that may affect current or future folivores. Less attention has been focused on the latter two categories, although induced phenological and structural changes in plants can affect folivorous insects, and sometimes override chemical induction, either directly (e.g., mortality through premature leaf abscission) or indirectly (e.g., altering attack by natural enemies). Presently, evidence of the actual effects of induced changes on folivorous insects is highly variable, ranging from positive, negative, and none at all. Variation in the nature, duration, and timing of the response; in the amount, type, and distribution of insect damage; in life histories of both insects; and in indirect effects via natural enemies or intraspecific competition, may alter the magnitude and direction of the effect of induced responses. Because of these confounding variables, it is unlikely that induced responses are defensive plant adaptations maintained by selective pressure from folivorous insects. Induced responses clearly impact some folivorous insects at the individual, population, and community level. Variability in these effects, however, prevent generalizations of how induced responses affect phytophagous insects.

ACKNOWLEDGMENTS

I thank B. Axelrod, M. Axelrod, P. Cheng, C. Febus, K. Hammon, S. Reeser, R. Rooney, and R. Sinnott for assistance in the lab or field. M. Auerbach, K. Hammon, R. Rooney, P. Stiling, and an anonymous reviewer made helpful comments on the manuscript. D. Tallamy and M. Raupp displayed extraordinary patience in editing this chapter. The research described herein was supported by NSF Grants BSR-8415616 and BSR-8715743.

REFERENCES

Askew, R. R., and M. R. Shaw. 1979. Mortality factors affecting the leaf-mining stages of *Phyllonorycter* (Lepidoptera:Gracillariidae) on oak and birch. *Zool. J. Linnean Soc.* 67:31–49.

Auerbach, M. J., and D. Simberloff. 1985. Responses of leaf miners to atypical leaf production patterns. *Ecol. Entomol.* 9:361–367.

Barbosa, P. 1988. Some thoughts on "the evolution of host range." *Ecology* 69:912–915.

Barlow, N. D., and A. F. G. Dixon. 1980. *Simulation of Lime Aphid Population Dynamics.* Wageningen, Netherlands: Centre for Agricultural Publishing and Documentation.

Benz, G. 1977. Insect-induced resistance as a means of self-defense in plants. *Eucarpia/IOBC Work. Group Breed. Resistance Insects Mites, SROP*, pp. 155–159.

Berenbaum, M. R. 1980. Adaptive significance of midgut pH in larval Lepidoptera. *Am. Natur.* 115:138–146.

Bergelson, J. M., and J. H. Lawton. 1988. Does foliar damage influence predation on the insect herbivores of birch? *Ecology* 69:434–445.

Bergelson, J. M., S. Fowler, and S. Hartley. 1986. The effects of foliage damage on the casebearing moth larvae, *Coleophora serratella*, feeding on birch. *Ecol. Entomol.* 11:241–250.

Bernays, E. A. 1978. Tannins: an alternative viewpoint. *Entomol. Exp. Applic.* 24:244–253.

Bernays, E. A. 1981. Plant tannins and insect herbivores: an appraisal. *Ecol. Entomol.* 6:353–360.

Bernays, E., and M. Graham. 1988. On the evolution of host specificity in phytophagous arthropods. *Ecology* 69:886–892.

Bryant, J. P. 1981. Phytochemical deterrence of snowshoe hare browsing by adventitious shoots of four Alaskan trees. *Science* 213:889–890.

Bultman, T. L., and S. H. Faeth. 1986a. Selective oviposition by a leaf miner in response to temporal variation in abscission. *Oecologia (Berlin)* 69:117–120.

Bultman, T. L., and S. H. Faeth. 1986b. Leaf size selection by leaf-mining insects on *Quercus emoryi* (Fagaceae). *Oikos* 46:311–316.

Carroll, C. R., and C. A. Hoffman. 1980. Chemical feeding deterrent mobilized in response to insect herbivory and counteradaptations by *Epilachna tredecimnotata*. *Science* 209:414–416.

Clancy, K. M., and P. W. Price. 1987. Rapid herbivore growth enhances enemy attack: sublethal plant defenses remain a paradox. *Ecology* 68:733–737.

Coleman, J. S., and C. Jones. 1991. A phytocentric prespective of phytochemical induction by herbivores. In D. W. Tallamy and M. J. Raupp (eds.), *Phytochemical Induction by Herbivores*. New York: Wiley, pp. 3–45.

Connell, J. H. 1983. On the prevalence and relative importance of interspecific competition: evidence from field experiments. *Am. Natur.* 122:661–696.

Corrigan, J. E., and R. G. Bennett. 1987. Predation by *Cheiracanthium mildei* (Araneae, Clubionidae) on larval *Phyllonorycter blancardella* (Lepidoptera: Gracillariidae) in a greenhouse. *J. Arachnol* 15:132–134.

Crawley, M. J. 1983. *Herbivory: The Dynamics of Animal–Plant Interactions*. Oxford, England: Blackwell.

Denno, R. F., and M. S. McClure (eds.). 1983. *Variable Plants and Herbivores in Natural and Managed Systems*. New York: Academic.

Dixon, A. F. G., and N. D. Barlow. 1979. Population regulation in the lime aphid. *Zool. J. Linnean Soc.* 67:225–237.

Duffey, S. S., K. A. Bloem, and B. C. Campbell. 1986. Consequences of sequestration of plant natural products in plant–insect–parasitoid interactions. In D. J. Boethel and R. D. Eikenbarry (eds.), *Interactions of Plant Resistance and Parasitoids and Predators of Insects*. Chicester, England: Ellis Horwood, pp. 31–60.

Dussourd, D. E., and T. Eisner. 1987. Vein-cutting behavior: insect counterploy to the latex defense of plants. *Science* 237:898–901.

Edwards, P. B., and W. J. Wanjura. 1989. Eucalypt-feeding insects bite off more than they can chew: sabotage of induced defenses? *Oikos* 54:246–248.

Edwards, P. J., and S. D. Wratten. 1982. Wound-induced changes in palatability in birch (*Betula pubescens* Ehrh. spp. *pubescens*). *Am. Natur.* 120:816–818.

Edwards, P. J., and S. D. Wratten. 1983. Wound-induced defenses in plants and their consequences for patterns of insect grazing. *Oecologia (Berlin)* 59:88–93.

Edwards, P. J., S. D. Wratten, and L. Winder. 1988. Insect herbivory in relation to dynamic changes in host plant quality. *Biol. J. Linnean Soc.* 35:339–350.

Edwards, P. J., S. D. Wratten, and R. M. Gibberd. 1991. The impact of inducible phytochemicals on food selection by insect herbivores and its consequences for the distribution of grazing damage. In D. W. Tallamy and M. J. Raupp (eds.), *Phytochemical Induction by Herbivores*. New York: Wiley, pp. 205–221.

Ehrlich, P. R., and P. H. Raven. 1965. Butterflies and plants: a study in coevolution. *Evolution* 19:586–608.

Ericsson, A., S. Larsson, and O. Tenow. 1980. Effects of early and late defoliation on growth and carbohydrate dynamics in Scots pine. *J. Appl. Ecol.* 17:747–769.

Escudero, A., and J. M. del Arco. 1987. Ecological significance of the phenology of leaf abscission. *Oikos* 49:11–14.

Ezcurra, E., J. C. Gomez, and J. Becerra. 1987. Diverging patterns of host use by phytophagous insects in relation to leaf pubescence in *Arbutus xalapensis* (Ericaceae). *Oecologia (Berlin)* 72:479–480.

Faeth, S. H. 1985a. Host leaf selection by leaf miners: interactions among three trophic levels. *Ecology* 66:870–875.

Faeth, S. H. 1985b. Quantitative defense theory and patterns of feeding by oak insects. *Oecologia (Berlin)* 68:34–40.

Faeth, S. H. 1986. Indirect interactions between temporally-separated herbivores mediated by the host plant. *Ecology* 67:479–494.

Faeth, S. H. 1987. Community structure and folivorous insect outbreaks: the roles of vertical and horizontal interactions. In P. Barbosa and J. C. Schultz (eds.), *Insect Outbreaks*. New York: Academic, pp. 135–171.

Faeth, S. H. 1988. Plant-mediated interactions between seasonal herbivores: enough for evolution or coevolution? In K. C. Spencer (ed.), *Chemical Mediation of Coevolution. AIBS Symposium Volume.* New York: Academic, pp. 391–414.

Faeth, S. H. 1990a. Aggregation of a leafminer, *Cameraria* sp. nov. (Davis): consequences and causes. *J. Anim. Ecol.* 59:569–586.

Faeth, S. H. 1990b. Structural damage to oak leaves alters natural enemy attack on a leafminer. *Entomol. Exp. Appl.* 57:57–63.

Faeth, S. H., and T. L. Bultman. 1986. Interacting effects of increased tannin levels on leaf-mining insects. *Entomol. Exp. Applic.* 40:297–300.

Faeth, S. H., E. F. Connor, and D. Simberloff. 1981. Early leaf abscission: a neglected source of mortality for folivores. *Am. Natur.* 177:409–415.

Feeny, P. 1970. Seasonal changes in oak leaf tannins and nutrients as a cause of spring feeding by winter moth caterpillars. *Ecology* 51:565–581.

Feeny, P. P. 1976. Plant apparency and chemical defense. *Rec. Adv. Phytochem.* 10:1–40.

Fowler, S. V., and J. H. Lawton. 1985. Rapidly induced defenses and talking trees: the devil's advocate position. *Am. Natur.* 126:181–195.

Fowler, S. V., and M. MacGarvin. 1986. The effects of leaf damage on the performance of insect herbivores on birch. *J. Anim. Ecol.* 55:565–573.

Fox, L. R. 1981. Defense and dynamics in plant-herbivore systems. *Am. Zool.* 21:853–854.

Fox, L. R., and B. J. Macauley. 1977. Insect grazing on *Eucalyptus* in response to variation in leaf tannins and nitrogen. *Oecologia (Berlin)* 29:145–162.

Futuyma, D. J. 1983. Evolutionary interactions among herbivorous insects and plants. In D. J. Futuyma and M. Slatkin (eds.), *Coevolution*. Sunderland, MA: Sinauer, pp. 207–231.

Gall, L. F. 1987. Leaflet position influences caterpillar feeding and development. *Oikos* 49:172–176.

Green, T. R., and C. A. Ryan. 1972. Wound-induced proteinase inhibitor in plant leaves: a possible defense mechanism against insects. *Science* 175:776–777.

Harrison, S., and R. Karban. 1986. Behavioural response of spider mites (*Tetranychus urticae*) to induced resistance of cotton plants. *Ecol. Entomol.* 11:181–188.

Hartley, S. E. 1988. The inhibition of phenolic biosynthesis in damaged and undamaged birch foliage and its effect on insect herbivores. *Oecologia (Berlin)* 76:65–70.

Hartley, S. E., and J. H. Lawton. 1987. Effects of different types of damage on the chemistry of birch foliage, and the responses of birch feeding insects. *Oecologia (Berlin)* 74:432–437.

Hartley, S. E., and J. H. Lawton. 1991. The biochemical basis and significance of rapidly induced changes in birch foliage. In D. W. Tallamy and M. J. Raupp (eds.), *Phytochemical Induction by Herbivores*. New York: Wiley, pp. 105–132.

Haukioja, E. 1980. On the role of plant defenses in the fluctuation of herbivore populations. *Oikos* 35:202–213.

Haukioja, E., and S. Hanhimaki. 1985. Rapid wound-induced resistance in white birch (*Betula pubescens*) foliage to the geometrid *Epirrita autumnata*: a comparison of trees and moths within and outside the outbreak range of the moth. *Oecologia (Berlin)* 65:223–228.

Haukioja, E., and P. Niemelä. 1977. Retarded growth of a geometrid larva after mechanical damage to leaves of its host tree. *Ann. Zool. Fennici* 14:48–52.

Haukioja, E., and P. Niemelä. 1979. Birch leaves as a resource for herbivores. Seasonal occurrence of increased resistance in foliage after mechanical damage of adjacent leaves. *Oecologia (Berlin)* 39:151–159.

Haukioja, E., P. Niemelä, and S. Siren. 1985. Foliage phenols and nitrogen in relation to growth, insect damage, and ability to recover after defoliation in the mountain birch, *Betula pubescens* spp. *tortusa*. *Oecologia (Berlin)* 65:214–222.

Hawkins, B. A. 1988. Foliar damage, parasitoids and indirect competition: a test using herbivores of birch. *Ecol. Entomol.* 13:301–308.

Heichel, G. H., and N. C. Turner. 1976. Phenology and leaf growth of defoliated hardwood trees. In J. F. Anderson and H. K. Kaya (eds.), *Perspectives in Forest Entomology*. New York: Academic, pp. 31–40.

Heinrich, B. 1979. Foraging strategies of caterpillars: leaf damage and possible predator avoidance strategies. *Oecologia (Berlin)* 42:325–337.

Heinrich, B., and S. L. Collins. 1983. Caterpillar leaf damage and the game of hide-and-seek with birds. *Ecology* 64:592–602.

Holt, R. D. 1977. Predation, apparent competition, and the structure of prey communities. *Theoret. Population Biol.* 12:197–229.

Hough, J. A., and D. Pimentel. 1978. Influence of host foliage on development, survival, and fecundity of the gypsy moth. *Environ. Entomol.* 7:97–102.

Hunter, M. D. 1987. Opposing effects of spring defoliation on late season oak caterpillars. *Ecol. Entomol.* 12:373–382.

Janzen, D. H. 1980. When is it coevolution? *Evolution* 34:611–612.

Jeffries, M. J., and J. H. Lawton. 1984. Enemy free space and the structure of ecological communities. *Biol. J. Linnean Soc.* 23:269–286.

Jermy, T. 1984. Evolution of insect/host plant relationships. *Am. Natur.* 124:609–630.

Kahn, D. M., and H. V. Cornell. 1983. Early leaf abscission and folivores: comments and considerations. *Am. Natur.* 122:428–432.

Karban, R. 1985. Resistance against spider mites in cotton induced by mechanical abrasion. *Entomol. Exp. Applic.* 37:137–141.

Karban, R. 1987. Environmental conditions affecting the strength of induced resistance against mites in cotton. *Oecologia (Berlin)* 73:414–419.

Karban, R. 1991. Inducible resistance in agricultural systems. In D. W. Tallamy and M. J. Raupp (eds.), *Phytochemical Induction by Herbivores.* New York: Wiley, pp. 403–419.

Karban, R., and J. R. Carey. 1984. Induced resistance of cotton seedlings to mites. *Science* 225:53–54.

Karban, R., R. Adamchak, and W. C. Schanhorst. 1987. Induced resistance and interspecific competition between spider mites and a vascular wilt fungus. *Science* 235:678–680.

Kogan, M., and J. Paxton. 1983. Natural inducers of plant resistance to insects. In P. A. Hedin (ed.), Washington, DC: American Chemical Society.

Lawton, J. H. 1986. Food-shortage in the midst of plenty? The case for birch-feeding insects. *Proc. 3rd European Congr. Entomol.*, pp. 219–228.

Lawton, J. H., and M. P. Hassell. 1981. Asymmetrical competition in insects. *Nature* 289:793–795.

Lawton, J. H., and D. R. Strong, Jr. 1981. Community patterns and competition in folivorous insects. *Am. Natur.* 118:317–338.

Leather, S. R., A. D. Watt, and G. I. Forrest. 1987. Insect-induced chemical changes in young lodgepole pine (*Pinus contorta*): the effect of previous defoliation on oviposition, growth and survival of the pine beauty moth, *Panolis flammea. Ecol. Entomol.* 12:275–281.

Lee, T. D., and F. A. Bazzaz. 1980. Effects of defoliation and competition on growth and reproduction in the annual plant. *Abutilon theophrasti. J. Ecol.* 68: 813–821.

Levin, D. A. 1971. Plant phenolics: an ecological perspective. *Am. Natur.* 105:157–181.

Luthy, P., C. Hofmann, and F. Jaquet. 1985. Inactivation of delta-endotoxin of *Bacillus thuringiensis* by tannin. *FEMS Microbiol. Lett.* 28:31–33.

Marks, M. L. 1976. Mating behavior and fecundity of the red bollworm *Diparapis cantanea* Hmps. (Lepidoptera:Noctuidae). *Bull. Entomol. Res.* 66:145–158.

Martin, M. M., and J. S. Martin. 1984. Does tent caterpillar attack reduce the food quality of red alder foliage? *Oecologia (Berlin)* 62:74–79.

Myers, J. H. 1988. The induced defense hypothesis: does it apply to the population dynamics of insects? In K. C. Spencer (ed.), *Chemical Mediation of Coevolution.* New York: Academic, pp. 345–366.

Myers, J. H., and D. Bazely. 1991. Thorns, spines, prickles and hairs: are they stimulated by herbivory and do they deter herbivores? In D. W. Tallamy and M. J. Raupp (eds.), *Phytochemical Induction by Herbivores.* New York: Wiley, pp. 325–344.

Myers, J. H., and K. S. Williams. 1984. Does tent caterpillar attack reduce the food quality of red alder foliage? *Oecologia (Berlin)* 62:74–79.

Myers, J. H., and K. S. Williams. 1987. Lack of short or long term inducible defenses in the red alder–western tent caterpillar system. *Oikos* 48:73–78.

Neuvonen, S., and J. Haukioja. 1984. Low nutritive quality as a defense against herbivores: induced responses in birch. *Oecologia (Berlin)* 63:71–74.

Neuvonen, S., E. Haukioja, and A. Molarius. 1987. Delayed inducible resistance against a leaf-chewing insect in four deciduous trees. *Oecologia (Berlin)* 74:363–369.

Niemelä, P. 1983. Seasonal patterns in the incidence of specialism: Macrolepidopteran larvae on Finnish deciduous trees. *Ann. Zool. Fennici* 20:199–202.

Niemelä, P., and J. Tuomi. 1987. Does the leaf morphology of some plants mimic caterpillar damage? *Oikos* 50:256–257.

Niemelä, P., E. M. Aro, and E. Haukioja. 1979. Birch leaves as a resource for herbivores. Damaged-induced increase in leaf phenolics with trypsin-inhibiting effects. *Rept. Kevo Subarctic Res. Sta.* 15:37–40.

Niemelä, P., J. Tuomi, R. Mannila, and P. Ojala. 1984. The effect of previous damage on the quality of Scots pine foliage as food for Diprionid sawflies. *Z. angew. Entomol.* 98:33–43.

Obrycki, J. J. 1986. The influence of foliar pubescence on entomophagous insects. In D. J. Boethel and R. D. Eikenbarry (eds.), *Interactions of Plant Resistance and Parasitoids and Predators of Insects.* Chichester, England: Ellis Horwood.

Opler, P. A. 1974. *Biology, Ecology, and Host Specificity of Microlepidoptera Associated with Quercus agrifolia (Fagaceae).* Berkeley: University of California Press.

Orians, G. H., and R. T. Paine. 1983. Convergent evolution at the community level. In D. J. Futumya and M. Slatkin (eds.), *Coevolution.* Sunderland, MA: Sinauer, pp. 431–458.

Owen, D. F. 1978. The effect of a consumer, *Phytomyza ilicis,* on seasonal leaf-fall in the holly, *Ilex aquifolium. Oikos* 31:268–271.

Price, P. W. 1983. Hypotheses on the organization and evolution of herbivore communities. In R. F. Denno and M. S. McClure (eds.), *Variable Plants and Herbivores in Natural and Managed Systems.* New York: Academic, pp. 559–596.

Price, P. W. 1986. Ecological aspects of host plant resistance and biological control: interactions among three trophic levels. In D. J. Boethel and R. D. Eikenbarry (eds.), *Interactions of Plant Resistance and Parasitoid and Predators of Insects.* Chicester, England: Ellis Horwood.

Price, P. W., C. E. Bouton, P. Gross, B. A. McPheron, J. N. Thompson, and A. E. Weis. 1980. Interactions among three trophic levels: influence of plants on interactions between insect herbivores and natural enemies. *Annu. Rev. Ecol. Syst.* 11:41–65.

Price, P. W., M. Westoby, B. Rice, P. R. Atsatt, R. S. Fritz, J. N. Thompson, and K. Mobley. 1986. Parasite mediation in ecological interactions. *Annu. Rev. Ecol. Syst.* 17:487–506.

Price, P. W., H. Roininen, and J. Tahvanainen. 1987a. Plant age and attack by the bud galler, *Euura mucronata. Oecologia (Berlin)* 73:334–337.

Price, P. W., H. Roininen, and J. Tahvanainen. 1987b. Why does the bud-galling sawfly, *Euura mucronata,* attack long shoots? *Oecologia (Berlin)* 74:1–6.

Pritchard, I. M., and R. James. 1984. Leaf fall as a source of miner mortality. *Oecologia (Berlin)* 64:140–141.

Pullin, A. S. 1987. Changes in leaf quality following clipping and regrowth of *Urtica dioica*, and consequences for a specialist insect herbivore, *Agalis urticae*. *Oikos* 49:39–45.

Puttick, G. M. 1986. Utilization of evergreen and deciduous oaks by the California oak moth *Phryganidia californica*. *Oecologia* (Berlin) 68:589–594.

Raffa, K. F. 1991. Induced defensive reactions in conifer–bark beetle systems. In D. W. Tallamy and M. J. Raupp (eds.), *Phytochemical Induction by Herbivores*. New York: Wiley, pp. 245–276.

Raupp, M. J., and R. F. Denno. 1984. The suitability of damaged willow leaves as food for the leaf beetle, *Plagiodera versicolora*. *Ecol. Entomol.* 9:443–448.

Raupp, M. J., and C. S. Sadof. 1991. Responses of leaf beetles to injury related changes in their salicaceous hosts. In D. W. Tallamy and M. J. Raupp (eds.), *Phytochemical Induction by Herbivores*. New York: Wiley, pp. 183–204.

Renaud, D. E. S. 1986. Impact of herbivory on defensive and reproductive allocation of Arizona walnut. MS Thesis, Arizona State University.

Rhoades, D. F. 1983a. Responses of alder and willow to attack by tent caterpillars and webworms: evidence for pheromonal sensitivity of willows. *Am. Chem. Soc. Symp. Ser.* 208:55–68.

Rhoades, D. F. 1983b. Herbivore population dynamics and plant chemistry. In R. F. Denno and M. S. McClure (eds.), *Variable Plants and Herbivores in Natural and Managed Systems*. New York: Academic, pp. 155–220.

Rhoades, D. F. 1985. Offensive–defensive interactions between herbivores and plants: their relevance in herbivore population dynamics and ecological theory. *Am. Natur.* 125:205–238.

Rhoades, D. F., and R. G. Cates. 1976. Towards a general theory of plant antiherbivore chemistry. *Rec. Adv. Phytochem.* 10:168–213.

Risley, L. S. 1986. The influence of herbivores on seasonal leaf-fall: premature leaf abscission and petiole clipping. *J. Agric. Entomol.* 3:152–162.

Rockwood, L. L. 1974. Seasonal changes in the susceptibility of *Crescentia alata* leaves to the flea beetle *Oedionychus* sp. *Ecology* 55:142–148.

Roland, J., and J. H. Myers. 1987. Improved insect performance from host–plant defoliation: winter moth on oak and apple. *Ecol. Entomol.* 12:409–414.

Schoener, T. W. 1983. Field experiments on interspecific competition. *Am. Natur.* 122:240–285.

Schultz, J. C. 1983a. Impact of variable plant defensive chemistry on susceptibility of insects to natural enemies. In P. Hedin (ed.), *Mechanisms of Plant Resistance to Insects. Symp. Am. Chem. Soc.*, pp. 37–54.

Schultz, J. C. 1983b. Habitat selection and foraging tactics of caterpillars in heterogenous trees. In R. F. Denno and M. S. McClure (eds.), *Variable Plants and Herbivores in Natural and Managed Systems*. New York: Academic, pp. 61–90.

Schultz, J. C., and I. T. Baldwin. 1982. Oak leaf quality declines in response to defoliation by gypsy moth larvae. *Science* 217:149–151.

Silkstone, B. E. 1987. The consequences of leaf damage for subsequent insect grazing on birch (*Betula* spp.). *Oecologia (Berlin)* 74:149–152.

Silva-Bohorques, I. 1986. Interspecific interactions between insects on oak trees, with special reference to defoliators and the oak aphid. Ph.D. Thesis, University of Oxford.

Simberloff, D., and P. Stiling. 1987. Larval dispersion and survivorship in a leaf-mining moth. *Ecology* 67:1647–1657.

Stiling, P. D., and D. Simberloff. 1989. Leaf abscission: induced defense against pests or response to damage? *Oikos* 55:43–49.

Strong, D. R., Jr., J. H. Lawton, and T. R. E. Southwood. 1984. *Insects on Plants. Community Patterns and Mechanisms.* London: Blackwell Scientific.

Swain, T. 1979. Tannins and lignins. In G. A. Rosenthal and D. H. Janzen (eds.), *Herbivores. Their Interactions with Secondary Plant Metabolites.* New York: Academic, pp. 657–682.

Tallamy, D. W. 1985. Squash beetle feeding behavior: an adaptation against induced cucurbit defenses. *Ecology* 66:1574–1579.

Taper, M. L., and T. J. Case. 1987. Interactions between oak tannins and parasite community structure: unexpected benefits of tannins to cynipid gall-wasps. *Oecologia (Berlin)* 71:254–261.

Taper, M. L., E. M. Zimmerman, and T. J. Case. 1986. Sources of mortality for a cynipid gall-wasp (*Dryocomus dubiosus* (Hymenoptera:Cynipidae)): the importance of the tannin/fungus interaction. *Oecologia (Berlin)* 68:437–445.

Thompson, J. N. 1982. *Interaction and Coevolution.* New York: Wiley.

Thompson, J. N. 1988. Coevolution and alternative hypotheses on insect/plant interations. *Ecology* 69:893–895.

Tuomi, J., P. Niemelä, M. Rousi, S. Siren, and T. Vuorisalo. 1988. Induced accumulation of foliage phenols in mountain birch: branch response to defoliation? *Am. Natur.* 132:602–608.

Tuomi, J., P. Niemelä, I. Jussila, T. Vuorisalo, and J. Jormalainen. 1989. Delayed budbreak: a defensive response of mountain birch to early season defoliation? *Oikos* 54:87–91.

Valentine, H. T., W. E. Wallner, and P. M. Wargo. 1983. Nutritional changes in host foliage during and after defoliation, and their relation to weight of gypsy moth pupae. *Oecologia (Berlin)* 57:298–302.

Vinson, S. B. 1976. Host selection by insect parasitoids. *Annu. Rev. Entomol.* 21:109–138.

Wallner, W. E., and G. S. Walton. 1979. Host defoliation: a possible determinant of gypsy moth population quality. *Annu. Entomol. Soc. Am.* 72:62–67.

Weseloh, R. M. 1981. Host location by parasitoids. In D. A. Nordlund and W. J. Lewis (eds.), *Semiochemicals: Their Role in Pest Control.* New York: Wiley, pp. 79–95.

West, C. 1985. Factors underlying the late seasonal appearance of the lepidopterous leaf-mining guild on oak. *Ecol. Entomol.* 10:111–120.

Whitham, T. G. 1983. Host manipulation of parasites: within-plant variation as a defense against rapidly evolving pests. In R. F. Denno and M. S. McClure (eds.), *Variable Plants and Herbivores in Natural and Managed Systems.* New York: Academic, pp. 15–41.

Williams, A. G., and T. G. Whitham. 1986. Premature leaf abscission: an induced plant defense against gall aphids. *Ecology* 67:1619–1627.

Williams, K. S., and J. H. Myers. 1984. Previous herbivore attack of red alder may improve food quality for fall webworm. *Oecologia (Berlin)* 63:166–170.

Wratten, S. D., P. J. Edwards, and I. Dunn. 1984. Wound-induced changes in the palatability of *Betula pubescens* and *B. pendula*. *Oecologia (Berlin)* 61:372–375.

Zucker, W. V. 1983. Tannins: does structure determine function? *Am. Natur.* 121:335–365.

CHAPTER 14

THORNS, SPINES, PRICKLES, AND HAIRS: ARE THEY STIMULATED BY HERBIVORY AND DO THEY DETER HERBIVORES?

JUDITH H. MYERS
The Ecology Group, Departments of Zoology and Plant Science, University of British Columbia, Vancouver, Canada V6T 1Z4

DAWN BAZELY
Department of Biology, York University, North York, Ontario, Canada M3J 1P3

1   Introduction
2   Types of physical defenses
    2.1   Hairs as plant defenses
    2.2   Thorns, spines, and prickles
3   Changes in plant physical characteristics following herbivory
4   Environmental stress and physical "defense" structures
5   Discussion
    References

1   INTRODUCTION

Interpreting defensive strategies of plants is exceedingly difficult. For a plant, defense against herbivores is likely to result from a fine balance between herbivore pressure, the cost of adaptations for deterring herbivores, and the value of leaves and other plant parts as related to the potential growth, survival, and reproduction of the individual. The equilibrium point of these interrelated processes and pressures can change over both ecological and evolutionary time (Grime 1977, Coley et al. 1985). Plants that maintain the flexibility to respond to herbivores directly by increasing their defense following damage would seem best able to track the moving equilibrium (Myers 1988).

But can one interpret changes in plants following herbivore damage to be adaptations against herbivores? The response of plants to herbivory might be a generalized reaction to stress resulting from the removal of photosynthetic material, or it might be adaptation against the invasion of pathological microorganisms into damaged tissue. Leaves of stress-tolerant plants tend to be less palatable to herbivores (Grime 1977, Coley et al. 1985, Bryant et al. 1988). Some plants that lack alkaloid defenses compensate for herbivore damage by regrowth while some with alkaloids do not compensate for lost tissue (van der Meijden et al. 1988). Plant growth and foliage quality for herbivores seem to be closely related and it is difficult to know how selection has acted on these patterns. Do well-defended plants not need to have compensatory mechanisms to deal with herbivore damage? Does an inability to respond to damage select for greater chemical defense? Are plants living in areas where resources are limited unattractive to herbivores and unable to regrow following damage because nutrients are unavailable? Disentangling these related processes is very difficult.

Physical structures of plants may serve as defenses against herbivores and their expression may be less closely related to metabolic processes than are some of the chemical changes observed in plants following damage. If mechanical defenses of plants are independent of plant growth and environmental stress, they could serve as independent measures of the defensive response of plants to herbivore damage. In this chapter we ask if plants respond to herbivore damage by increasing physical defenses such as hairiness, thorniness, or spininess, and if the expression of these morphological characteristics is related to plant growth and/or environmental stress.

2 TYPES OF PHYSICAL DEFENSES

A variety of morphological characteristics and structures could help to defend plants against herbivores. These include leaf shape and texture, hairs, spines, and thorns. Before we consider the physical responses of plants to herbivore damage, we must try to determine if these physical qualities influence the feeding of herbivores.

2.1 Hairs as Plant Defenses

Trichomes or hairs occur on the leaves of most species of angiosperms. Hairs are formed by the plant epidermis and Cronquist (cited in Johnson 1975) has suggested that the almost universal tendency for plants to produce hairs may be related to genetically controlled developmental processes leading to the production of root hairs. The trichome density on plants varies from sparse to dense and woolly. Glandular hairs produce chemicals and the role of these as insect defenses of plants has been reviewed (Levin 1973, Johnson 1975, Webster 1975). Glandular hairs reduce the survival and reproduction of many small insects and mites by impeding their movement as well as engulfing them in toxic chemicals.

Of 33 insect– or mite–plant relationships reviewed by Webster (1975), increased hairiness improved resistance to attack in 20 cases, decreased resistance in seven cases, and had no effect in six cases. In particular, leaf hairiness frequently increased attack by Lepidoptera and reduced attack and survival of beetles, aphids, mites, and white flies. We have not included the resistance of solanaceous plants to mites and aphids in this survey because the glandular hairs of this family act as chemical defenses. Hairy leaves are sometimes associated with increased attack by mites (Harvey and Martin 1980, Peters and Berry 1980) or have no influence on mite populations, as in the case of red mite on apples (Paiva and Janick 1980). In most cases of increased susceptibility, the insects prefer to oviposit on hairy plants (Norris and Kogan 1980).

Hairs on leaves may also reduce the effectiveness of predators and parasites in attacking insect herbivores, and in this way herbivore damage may increase with leaf hairiness (Norris and Kogan 1980). Therefore, an increase in plant hairiness does not always reduce insect damage, but it frequently does for small sucking insects and mites.

Hairs may also deter large herbivores, and the stinging hairs of nettles, *Urtica dioica*, are a good example. Here again it is the combination of a physical and chemical process that makes this defense so effective. Pollard and Briggs (1984) review studies that have considered the palatability of nettles to grazing animals and report that while some vertebrates such as the European bison and mountain gorillas are undeterred by plants with stinging hairs, other grazers such as cattle and rabbits avoid nettles. Some horses can eat nettles with no obvious adverse effects (Myers, personal observations in Oxfordshire, England). Experiments by Pollard and Briggs (1984) showed that rabbits and sheep grazed more on plants with lower hair density. Snails, slugs, and numerous insect species graze on nettles and seem to be uninhibited by the hairy leaves.

Pollard (1986) compared the density of hairs on the petioles of the nettle, *Cnidoscolus texanus*, in 12 sites in Oklahoma. Nettles at the sites heavily grazed by cattle tended to have more hairs on their petioles. Whether these differences were genetically or phenotypically determined has not been reported nor has a study been done to determine the biological significance of these differences for herbivore behavior.

2.2 Thorns, Spines, and Prickles

Thorns and spines are often interpreted as adaptations to protect plants from large herbivores, but the effectiveness of these structures has rarely been tested. Thorns are modified branches, and spines are modified leaves. Both contain vascular tissue. Prickles, like hairs, are projections from the epidermis and subepidermal tissues (Wilson and Loomis 1962). However, these terms are frequently interchanged. Roses and blackberries actually have prickles instead of thorns, but we will not attempt to modify the poetic tradition of talking about rose thorns in this paper.

Geographical concurrence between thorny plants and large herbivores could provide evidence for thorns or spines being responses of plants to herbivory. For example, many xeric areas in Africa and other tropical savannah areas are characterized by large browsing mammals and thorny plants (Richards 1952, Brown 1960, Huntley 1982). In South Africa the vegetation of the nutrient-poor fynbos community lacks stem spines, and large vertebrate herbivores are uncommon in these areas. In nonfynbos plant communities, spinescent plants comprise a larger portion of the total plant cover, are characterized by stem spines, have higher nutrient value, and are thought to be exposed to greater herbivore pressure (Campbell 1986). Acacias in Australia generally lack thorns and large browsing mammals are also absent here (Foster and Dagg 1972). However, one species, *Acacia victoriae*, has the common name prickly wattle and is characterized by paired prickles on the branches (King 1986). Savannahs in Queensland, Australia, with fertile soil that are now overgrazed by introduced mammals have become dominated by spinescent taxa, *Maytenus*, *Canthium*, and *Carissa* (Huntley 1982). In Hawaii, where browsing mammals are absent, *Urtica* and *Smilax* both lack thorns (Karban, personal communication). This same pattern also occurs in Micronesia (Bryant, personal communication).

Large mammalian herbivores are not always present in habitats characterized by spiny plants. Janzen (1986) interprets the diversity of Cactaceae in the Chihuahuan Desert of Mexico as reflecting the "ghosts of herbivory past." Islands tend to lack both spiny plants and large herbivores (Carlquist 1974) but five *Rubus* species native to New Zealand have prickly petioles and leaves, and one of these also has prickly stems (Poole and Adams 1980).

Large mammalian herbivores are lacking from the evolutionary history of plants in New Zealand. However, browsing by moas, large flightless birds indigenous to New Zealand, may have influenced the evolution of the morphology of trees on this island (Greenwood and Atkinson 1977). Approximately 10% of woody species of plants in New Zealand show divaricating growth, branching at a wide angle, which results from reduced apical growth and increased lateral branching. Outer branches have fewer and smaller leaves and the interlaced branches make penetration to the inner branches difficult. In some species the divaricating growth occurs only in juvenile tissue, not current growth of mature tissue. These morphological characteristics can be interpreted as adaptations against the browsing moas whose horny bills would be impervious to spines and thorns. The moas have been extinct for 300 years but the divaricating growth pattern, which is genetically determined, has persisted (Greenwood and Atkinson 1977). The association between divaricating growth form and juvenile tissue in some woody species means that any damage that stimulates reversion to juvenile growth would also increase the "herbivore defense" of the plants.

Holly leaves, *Ilex opaca*, are well known for their spinose teeth on leaf edges. Supnick (1983) found that trees in southern areas had significantly more leaf spines than those in northern areas (11.47 vs. 6.97 spines/leaf) and herbivory was higher in southern localities, Virginia and North Carolina (43%), than in northern areas, New Jersey (21%) and Cape Cod (2%). Herbivores fed on leaves with statistically fewer spines (8.6 spines on leaves with herbivore damage vs. 9.0 spines for leaves without damage), but whether this variation is sufficient to be biologically relevant is unknown. More spinose teeth on leaves in areas with greater herbivory could indicate that herbivores select for increased spininess or that increased spininess has little impact on herbivore damage. The number of spines on holly leaves may also be influenced by other environmental characteristics such as temperature, humidity, or day length, and the association with herbivory could be spurious. Experimental blunting of spinose teeth of *Ilex opaca* did not influence feeding by fall webworm caterpillars, *Hyphantria cunea*, but caterpillar feeding increased greatly following removal of the tough leaf margin (Potter and Kimmerer 1988). While little herbivory of holly was observed in the field, captive rabbits and deer did not discriminate between leaves with and without spinose teeth. Lower leaves of holly trees have more teeth than upper leaves (Potter and Kimmerer 1988), which may be an ontogenetic phenomenon.

The impact of plant spinescence on the feeding behavior of large mammalian herbivores was studied by Cooper and Owen-Smith (1986). They concluded that thorns and spines primarily reduce the size of bites taken by browsing herbivores, and thorns in conjunction with small leaves are particularly effective in reducing the feeding rate of ungulates. However, ungulates compensate for reduced feeding rates by feeding longer. Smaller impala and goats are less influenced by thorns than are larger kudu. Branches of various *Acacia* species from which thorns are removed receive noticeably more browsing than control branches. The thorny acacias in this study were higher in nutrient value but about equally acceptable to browsing ungulates as unprotected plants of lower nutritional value.

Some vertebrate browsers appear to have adapted to prickly and thorny plants. Rates of ingestion by giraffe, *Giraffa camelopardalis*, in the Serengeti were not influenced by the spinescence of *Acacia* (Foster and Dagg 1972, Pellew 1984), and camels, *Camelus dromedarius*, sometimes prefer thorny branches over green leaves even though chewing the branches takes longer (Gauthier-Pilters and Dagg 1981). The gerenuk, *Litocranius walleri*, is another African mammal that feeds mainly on thorny vegetation (Leuthold 1970). Pointed muzzles and mobile lips seem to be adaptations for feeding on thorny vegetation.

Experimental plots in Wytham Woods, Oxfordshire, England that had been exposed to limited grazing by sheep in the spring or autumn for 3 years (Gibson et al. 1987), had approximately a third as many evergreen blackberry plants as ungrazed plots, and these plants had only a third as many reproductive shoots as did those on the ungrazed plots (Myers and Bazely, unpublished results). Prickliness did not prevent sheep from having an impact on the numbers of blackberry plants that were colonizing these old fields, but whether the prickles reduced damage to some extent cannot be determined.

Herbivores eat the leaves of blackberry leaving the prickly petioles, and if they can penetrate the thickets, they can defoliate the plants almost totally (see below). The thorns on the branches of blackberries may have their greatest impact by acting like Velcro and holding the canes together, thus making penetration of the patch by large vertebrates impossible. Because blackberries reproduce so successfully by vegetative growth, periferal canes may defend central canes of the same plant. While prickles may not effectively defend plants at low density, at high density, blackberries are almost immune from damage from large mammals. Another possible role of blackberry prickles might be to protect fruits and seeds from small mammals (Steve Courtney, personal communication). This is only speculation but deserves testing.

Thorns and spines on palm trees can reduce the growth of vines and epiphytes and therefore act as defensive structures (Maier 1982). They may also prevent mammals from climbing trees to eat leaves and fruits.

The diversity of patterns of spines on cacti is striking. While the role of spines in reducing herbivory seems obvious, the importance of herbivores in the evolution of spinescence is less clear. Cactus spines protect plants from high damaging radiation but also can reduce photosynthetically active radiation. Nobel (cited in Gibson and Nobel 1986) showed that periodic removal of spines from *Opuntia bigelovii* increased the growth of the plants. Cacti living in forests are usually spineless (Gibson and Nobel 1986), while plants growing in the direct sun have stronger spines (Kupper and Roshardt 1960). This could be interpreted as the cost of spines being greater for plants growing in the shade where further reduction of photosynthetically active radiation would be particularly detrimental. In contrast, herbivore damage might be less in the woods, reducing the need for spines in these areas.

Spines influence the surface of cacti by affecting the convection coefficient, by shading, and by emitting long-wave radiation (Nobel 1978). In a two-species comparison, the elevation range of the species with the densest coverage of spines extended 400 m higher than that of the species with few spines. Nobel (1980) successfully predicted the distribution of three species of cactus based on their spininess and temperature relationships. If spines are adaptations for thermoregulation, spininess may change with the habitat of the plants.

The development of spines in cacti is a trade-off between costs, benefits, and environmental conditions. It is interesting that spininess is still a characteristic of cacti in the Chihuahuan Desert even 10 000 years after the extinction of large mammalian herbivores (Janzen 1986). This indicates that the cost of producing spines must not be very great, and selection against their production small, or they are maintained by other positive selection pressures. There is a danger in considering all types of spines to have the same evolutionary cause, and for the cacti in particular, the physiological roles of spines must be taken into account when interpreting geographical distributions.

To summarize, while thorns do not stop feeding by herbivores, thorns and spines can influence the acceptability of plants as food to some large mammals. Caution is required, however, when assuming that increased thorniness will be translated directly into reduced herbivory. We now consider if herbivore damage stimulates changes in plant morphology that may reduce further herbivore damage.

3 CHANGES IN PLANT PHYSICAL CHARACTERISTICS FOLLOWING HERBIVORY

Changes in the structure of plants following damage by herbivores may influence the acceptability of the plants to herbivores in the future and could act as defense mechanisms. Observing these changes is easier than quantifying their impact on the herbivores. Although physical structures of plants change following damage from herbivores (Table 1), there are few examples in which the selective advantage of the change has been quantified. While the increase in spininess or thorniness may be statistically significant, a biological significance usually remains to be shown. The increase in thorniness and spininess following damage is often interpreted as a defense strategy, but evidence for this is rare.

TABLE 1   Examples of Herbivores Influencing the Physical Defenses of Plants

Plant	Herbivore	Effect	Reference
Acacia depranolobium	Goats	Increased thorn length	1
Grass	Ungulates	Increased silica	2
Glochidion obovatum	Deer	More thorns/cm branch	3
Opuntia	Artificial	More spines produced	4
Bull nettle	Cattle	Higher hair density	5
Acacia raddiana	Goats	More spines	6
Rubus trivalis	Cattle	Larger, heavier prickles	7
Blackbrush	Goats	Spines further apart	8
Rubus	Deer	More prickles/cm branch and Prickles/internode	9
Rubus	Artificial	More prickles/cm branch	9

References. 1, Young 1987; 2, McNaughton and Tarrants 1983; 3, Okuda 1987; 4, Myers 1987 and Monro and Myers, unpublished results; 5, Pollard 1986; 6, Seif El Din and Obeid 1971; 7, Abrahamson 1975; 8, Provenza and Malechek 1984; 9, Bazely, Myers, and DaSilva 1991.

A good example of damage inducing a physical response in plants that may be related to herbivore defense is the leaf-folding behavior of sensitive plants such as *Schrankia microphylla* and *Mimosa pudica* (Eisner 1981). In this case mechanical stimulation causes the leaves to fold and almost disappear, thus exposing the sharp recurved thorns along the stems for maximum pricking potential.

Another type of physical defense is the silica content of grass which influences the toughness of the plants. McNaughton and Tarrants (1983) found that when grown in a common environment, African grasses native to areas heavily grazed accumulated more silica in their leaves than plants from less heavily grazed areas. Silica content was higher in plants that had been defoliated. If herbivores can discriminate plants high in silica, this could serve as an induced defense. Grasses also respond to fungal pathogens by modifying cell wall structure to physically prevent development of fungal hyphae and reduce the movement of nutrients to the fungus (Hargreaves and Keon 1986).

That the production of spines by the cactus, *Opuntia stricta*, might be stimulated by herbivore damage was first suggested by the observation that cacti on an island used for cattle grazing had significantly more plants with spines than cacti on two nearby islands that lacked cows. Those plants with spines on the island with cows had an average of over three times as many spines as spined plants on the other islands (Fig. 1). Further observations of cactus in open woodland and adjacent pastures at five sites in Queensland and New South Wales supported this initial observation (Fig. 2). Experimental cutting of plants stimulated the growth of new cladodes with increased numbers of spines both in a field experiment (Fig. 3) and in a lab-

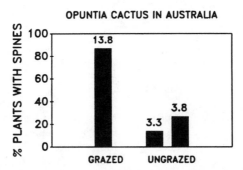

Figure 1   Percentage of *Opuntia stricta* plants with spines on three islands off the coast of Queensland, Australia. Cattle were present on one island and grazing damage was observed. Cattle were absent from the other two islands. The mean number of spines on plants with spines is given on top of the bars. Both the proportion of plants with spines (chi-squared test, $p < .05$) and mean number spines (t test, $p < .05$) were significantly different between grazed and ungrazed sites). (Monro and Myers, unpublished results.)

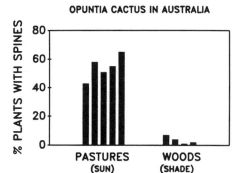

Figure 2 Percentage of *Opuntia stricta* plants with spines in four pastures and adjacent open woodlands and an additional pasture site in Queensland and New South Wales Australia. Differences between the means of percent plants with spines were significant with a *t* test ($p < .01$). (Myers and Monro, unpublished results.)

oratory experiment (Myers 1987). However, we know nothing about the influence of the spines on grazers or how effective they might be as defense structures (Gibson and Nobel 1986).

The development of spines in cacti is stimulated by gibberellin, the activity of which is influenced by concentrations of auxin and cytokinin (Mauseth and Halperin 1975). Damage to the plants could activate both new growth and spine production through its effect on the balance of plant hormones. Cacti are excellent plants for further studies of spininess and changes induced by herbivore damage.

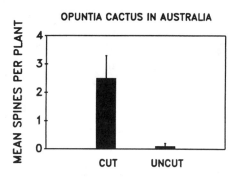

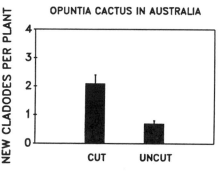

Figure 3 Mean number of spines and new cladodes on 22 pairs of *Opuntia stricta* plants of which one member of each pair had the top half of the newest cladode cut off to simulate cattle damage on 8 Oct 79. Plants were scored on 25 Feb 80 following the summer growth season. Bars are 1 SE and difference in mean number of spines was significant with *t* tests ($p < .01$). (Monro and Myers, unpublished results.)

In a study of the reproductive strategies of two species of *Rubus*, Abrahamson (1975) observed that prickles of grazed plants were larger and sharper than those of plants in an ungrazed area. No data were given and it was not clear if this was a response to grazing directly or reduced competition from other vegetation in the grazed site. Similarly, ramets of *Aralia spinosa* arising after injury had more dense and longer prickles than ramets that were undamaged (White 1988).

Juvenile growth of some trees is characterized by thorns and small, glabrous leaves. Fruit trees such as apples, plums, and citrus in particular have thorns on juvenile growth that are lacking on mature branches (Westwood 1978, Janick 1986). Pruning of top growth can stimulate juvenility in many plants as can damage that stimulates growth from adventitious buds (Poincelot 1980). Pruning roots stimulates the development of mature growth in upper branches in some situations (Poincelot 1980). New growth from adventitious buds stimulated by cutting or browsing damage will have juvenile characteristics. Thus, damage by herbivores can stimulate the development of new growth characterized by thorns, and this induced change could be interpreted as herbivore defense. This pattern is similar to the increased concentrations of phenols in juvenile tree tissue that can be stimulated by heavy browsing (Bryant et al., this volume, Chapter 6). The physical and chemical characteristics of juvenile tissues of trees could be interpreted as defenses against herbivores.

Because trees are modular in their structure, branches can act as integrated physiological units (Watson and Casper 1984). If herbivores remove leaves and the apical buds of lower branches, branches may react through reduced flower or fruit production or reduced growth, or actually die while other parts of the tree continue quite normally. Therefore, the modified structure of trees following herbivore damage may result from local nutrient limitation. If branch growth is reduced, the density of thorns (thorns/centimeter) may increase. What would appear to be an induced defense against herbivores, may be an induced growth response following damage. With this in mind we can look at the influence of browsing on the structure of trees.

Browsing can change the gross morphology of trees as well. Sika deer (*Cervus nippon*) on Miyajima Island, Japan, feed heavily on a deciduous, shrubby plant, *Glochidion obovatum*. Plants that are more than 200 cm tall show an hourglass shape: Leaves on the lower part of the tree are smaller, the number of branches is greater, and the percentage of branches with thorns is higher (Okuda 1987). Branch length and the number of thorns per centimeter branch length are highly correlated. Deer browsing stunts the growth of branches and leaves which, in turn, increases the density of thorns, but not necessarily the number of thorns. Upper branches of *G. obovatum* shed their thorns as they grow older, and new stems in the upper parts of the trees have significantly fewer thorns per 10 cm branch length than branches in the lower, browsed part of the trees. Branches are also longer in the upper parts of the tree which are out of reach of browsing deer. How the morphological changes in the structure of the trees influence deer browsing is not considered in the study.

A very different response by grazed plants is seen in blackbrush, *Coliogyne ramosissima*. Apical dominance causes ungrazed plants to be compact and spiny. Heavy grazing by goats removes the apical dominance and plants grow with a more open form so that the spiny tips of their leaves are farther apart. This allows goats to have access to more nutritious vegetation in the center of the plant which is also high in tannin (Provenza and Malechek 1984). This example shows how the control of development influences the response of plants to herbivore damage, but does not necessarily result in changes that can be easily interpreted as adaptive. However, goats may be better able to overcome the defenses of blackbrush than the herbivores that provided the evolutionary selection pressure.

In Senegal, goat browsing of *Acacia raddiana* resulted in stunted branches with stiff thorns. *Acacia senegal* growing in the same area did not produce thorns and was replaced by *A. raddiana* in areas of heavy browsing by goats (Seif el Din and Obeid 1971). Thorns on branches of *Acacia depronolobium* browsed by goats in Kenya were significantly longer than those on unbrowsed branches (Young 1987). Whether the longer thorns (2.34 vs. 2.27 cm) deterred goats or other herbivores to a greater extent was not determined.

Probably the best examples of induced physical defenses are the cases of plankton, colonial protozoans, rotifers, cladocerans, and snails producing keels, spines, helmets, crests, or thicker shells in response to waterborne cues released by predators (Gilbert 1966, Barry and Bayly 1985, Palmer 1985, Havel 1986). In these examples, the induced change in morphology is not dependent on the prey being damaged or losing tissue. Thus, the response cannot be caused by a shift in nutrient availability or environmental stress, but is cued by potential damage due to predation. The costs and benefits of inducible defenses in the marine bryozoan *Membranspora membranacea* have been measured by Harvell (1986). Colonies that grow spines enjoy protection from their predators, nudibranchs, but they grow more slowly, and this reduces their fecundity. Unfortunately, relationships between physical defensive structures and herbivory are not so clear.

A well-known example of a prickly plant is the evergreen blackberry, *Rubus vestitus*. Blackberry is heavily browsed in Wytham Woods, Oxfordshire, England, by a small deer that has been introduced to Britain, the muntjack (*Muntiacus reevesi*). In spring 1987 we observed that plants growing in shaded areas formed open thickets that the deer could penetrate. In May, 63% of the leaves of plants growing in five shaded patches had been removed, leaving only the petioles. In sunny areas, dense thickets prevented deer from reaching leaves beyond the edge of the patch and herbivore damage was restricted to the periphery. An average of 16% of the leaves had been eaten in five blackberry patches exposed to the sun (Bazely, Myers, and DaSilva 1991). In this situation sun exposure and herbivore damage were negatively correlated, but both the number of prickles per internode and number per centimeter of blackberry branch were significantly higher for plants growing in the shaded areas where herbivore damage was extensive. These results are consistent with the hypothesis that herbivory stimulates the production of prickles, but because the effects of grazing and sun exposure cannot be separated, we carried out experiments to test the impact of sun and simulated browsing on prickle production by blackberries.

First-year bramble shoots were dug and potted in early May 1987. After establishment, in late June, node length and the number of prickles were counted on the plants and the plants were randomly assigned to a treatment (3 blocks and 3 replicates): exposed to sun or shaded by plastic screening, leaves removed or unbrowsed, and fertilized or unfertilized. After 1 and 2 months, the length of internodes and the number of prickles on new growth were measured. Comparisons were made of the difference between internode length and number of prickles for the internodes measured before and after treatment. Simulated browsing had no influence on the number of prickles/internode, but because internodes were shorter on plants from which leaves were removed, the number of prickles per centimeter increased on browsed plants (Fig. 4). Nitrogen fertilization reduced the influence of the simulated browsing, and observations made 2 months after treatment were the same as those made 1 month after treatment. Thus, in Wytham Woods the impact of browsers appears to be greater than the influence of shading as indicated by the higher density of prickles on blackberry canes in the heavily browsed shaded patches. Spineless varieties of blackberry have been selected (Jennings 1986) and these would provide good material for experiments on the selective advantages of prickles in blackberries.

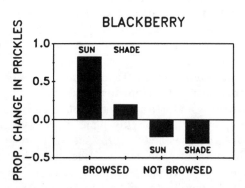

Figure 4 Proportional difference between the mean number of prickles per centimeter cane internode of transplanted blackberry plants prior to and 1 month after treatment. Treatments were leaves removed (browsed) or undisturbed (not browsed), and plants exposed to direct sun or shaded with black plastic netting. Differences between light ($p < .02$) and grazing ($p < .01$) are statistically significant by two-factor anova. Interaction between treatments approaches significance ($p = .08$). (Bazely, Myers, and DaSilva 1991)

4 ENVIRONMENTAL STRESS AND PHYSICAL "DEFENSE" STRUCTURES

Though the density of thorns, spines, or hairs can increase following herbivore damage, it is unknown whether such changes are adaptive responses by the plants to the herbivores or passive responses mediated by processes of growth and development. Since so little is known about the response of herbivores to these changes, we cannot evaluate selective pressures or adaptive responses. But we can ask if characteristics interpreted as being mechanical defenses are related to environmental stresses other than those imposed by herbivores.

The degree of hairiness of some plants varies geographically (Levin 1973). Early studies reported a positive association between harsh environments and pubescence (Johnson 1975). The type of hair may determine whether a relationship between environmental harshness and leaf pubescence occurs. Selection by herbivores is likely to explain the pattern of glandular leaf hairs (Levin 1973, Becerra and Ezcurra 1986) better than nonglandular or shag-type hairs, which are more likely to have been considered in early studies of leaf hairiness (Johnson 1975). In a comparison of plant communities covering a moisture gradient from a sandy beach to a red maple swamp, Johnson (1975) found similar proportions of pubescent species but more dense pubescence on mature foliage of plants in dryer environments. In another study of six different communities, pubescent plants in deserts were more densely covered than those of other habitats (Johnson 1975).

A relationship between leaf pubescence and xeric conditions could be explained if hairs reduce transpiration. However, not all experiments support this explanation. Johnson (1975) concluded that there are no definitive answers to the question of which selective factors determine leaf hairiness and suggested that complex interactions between physical and biotic factors will make simple patterns difficult to show. Interpretations of experiments on induced changes in leaf pubescence must consider adaptations to physiological conditions as well as herbivore avoidance.

Like blackberries, nettles growing in the shade have longer internodes and fewer hairs than those growing in the sun (Pollard and Briggs 1984). The classic studies of Clausen et al. (1940) on *Potentilla* showed that clonal material grown in sunny, dry conditions was more hairy than material planted in the shade. This difference was partly associated with smaller leaf size, but in some cases, the number of hairs increased.

Changes in environmental conditions can modify the hairiness of plants (Table 2). In general, the hairiness of leaves increases with environmental stress (drought, cold, damage, or zinc deficiency). Therefore, characteristics that appear to be responses to herbivory are similar to those induced by stressful environments. In many cases, differences in hairiness among species of plants living in different environments are genetically controlled. However, phenotypic variation can also be associated with the environment.

TABLE 2   Examples of Phenotypic Variation in Foliar Hairs and Thorns Associated with Variation in the Environment

Plant	Environment	Response	Reference
Phaseolus vulgaris	Dry	More hairs	1
Zea mays	Zinc deficient	Fewer hairs	2
Triticum aestivum	High humidity	More hairs	3
Rubus	Damage, freezing	Thorny adventitious shoots	4
Urtica	Sun	More hairs	5
Tillandsia	Xeric	More hairs	6
Cactus	Shade, moist	Few spines	7, 8
Rubus	Sun	More prickles/cm	9
Potentilla	Sun, dry	More hairs	10

References. De Fluiter and Ankesmit 1948; 2, Hagen and Anderson 1967; 3, Smith et al. 1971; 4, Hall et al. 1986; 5, Pollard and Briggs 1984; 6, Gilmartin 1983; 7, Kupper and Roshardt 1960; 8, Gibson and Nobel 1986; 9, Bazely, Myers, and DaSilva 1991; 10, Clausen et al. 1940.

5   DISCUSSION

While thorns, spines, prickles, and hairs are generally assumed to be adaptations for defense against herbivores, little evidence exists to strongly support this assumption. Few studies have determined whether an increase in mechanical defenses alters herbivore behavior, and to our knowledge, no study has assessed the value of physical defenses to plant fitness. Such experiments should be possible in field experiments in which the reproduction and survival of varieties of plants that lack thorns are compared to thorny varieties in the presence of natural herbivores. However, such comparisons might require years of observation to obtain results.

In this review four patterns have become apparent in regard to physical defenses of plants:

1. Physically defended plants are most common in dry savanna areas with good soil quality and either current or historically high densities of large mammalian herbivores. Defense is greatest where herbivory is high.

2. Of physically defended plant species, individuals growing in sunny, open areas frequently have higher numbers or densities of thorns and similar structures than plants in shaded areas.

3. Physically defended plants exposed to natural or simulated herbivory respond by regrowth which frequently has smaller leaves, shorter branches, and greater densities of the physical defense structures.

4. Physical defenses occur more frequently in the juvenile stage of the

plant and do not occur on mature growth, and/or are more common on lower branches than on upper branches of trees or shrubs. Damage that causes a reversion to the juvenile stage or that stimulates adventitious regrowth can result in stems that have greater physical defenses.

These observations can be used to test several hypotheses about plant defenses. First, Janzen (1974) proposed that plants growing in areas with poor soil quality should make a higher investment in defense because their metabolic tissues are more costly to replace and are therefore more valuable. Coley et al. (1985) expanded this nutrient-availability hypothesis to explain evolutionary patterns of chemical defenses of trees. Trees growing in nutrient-poor soils are characterized by low herbivore damage and high levels of secondary chemicals thought to act as herbivore defenses. However, the pattern of physical defenses of trees in savanna areas does not support the nutrient-availability hypothesis. Here areas of high soil nutrients have many large herbivores and many thorny plants (Huntley 1982, Owen-Smith 1982, Campbell 1986), resulting in a pattern just the opposite of that predicted by the nutrient-availability hypothesis.

Another hypothesis to explain the patterns of herbivory and chemical defenses is the carbon/nitrogen balance hypothesis (Bryant et al. 1983). In nutrient-limited environments, chemical defenses are largely carbon-based (tannins, phenols, or resins). Trees growing in high-nutrient or low-carbon environments will have nitrogen-based defenses, and juvenile plants may be better protected by these defensive chemicals than mature trees that have escaped herbivory through their height. Physical structures such as thorns can be thought of as carbon-based defenses since their production requires carbohydrates. Trees growing in sunny and arid savannas with good soil quality should be capable of regrowth following damage and may have an excess of carbohydrate. Therefore, thorns and so on may provide a mechanism for using excess carbohydrates while at the same time reducing the feeding efficiency of herbivores (Bryant, personal communication). The increased thorniness of plants in sunny areas is consistent with the C/N balance hypothesis. If thorn production is influenced by the C/N ratio, nitrogen fertilization should reduce thorniness. In two experiments in which plants were damaged and fertilized, the addition of nitrogen had no influence on the production of spines but reduced the production of prickles (Myers 1987, Bazely, Myers, and DaSilva 1991). Production of thorns and spines might be related to carbohydrate availability (exposure to sun) but in these experiments it was not always related to the ratio of carbon to nitrogen.

We observed a second contradiction to the C/N balance hypothesis in our experiments on blackberry. Removing leaves stimulated regrowth of blackberry canes, but canes in the sun were shorter than those in the shade. The number of prickles per internode remained the same in sun exposed and shaded plants, but the density was greater on the shorter internodes of sun-exposed plants. It is not clear why growth of blackberry is less in the sun than the shade, but water loss may be involved. Internode length did not vary significantly in natural patches of blackberry in exposed and shaded sites, however (Bazely, Myers, and DaSilva 1991).

An important difference between chemical defenses and physical defenses is the dependence of the latter on regrowth for expression while the chemical composition of leaves can change almost immediately following damage. The morphology of new growth will be controlled by balances in growth hormones that are influenced by sun exposure, maturity of the tissue, availability of nutrients, and apical dominance. The modular nature of trees allows for variation in growth, and the structure of different parts of the same individual may vary with the history of damage. While reversion to juvenile tissue or increased density of thorns on shorter branches may appear to be induced defensive responses, the selection pressures that have led to these patterns of growth following damage are probably more general. Age-specific selection for increased defense of juvenile trees is not the same as an induced defensive response (Bryant et al., this volume, Chapter 6). The value of changes in the morphology of plants to individual fitness must be measured before their role can be understood. It is unlikely that thorns, prickles, spines, and hairs are adaptations to only herbivore damage. It is far more likely that they are complex adaptive responses to an array of biotic and abiotic factors.

ACKNOWLEDGMENTS

We thank John Monro who collaborated with J.M. in the cactus studies reported here, and Karen Burke Da Silva for her help with the blackberry experiments. We thank Rick Karban, Hugh Daubenny, George Eaton, Liz Bernays, and Gerald Straley for useful comments and for suggesting references. Unpublished research presented here was supported by the Natural Sciences and Engineering Research Council of Canada.

REFERENCES

Abrahamson, W. G. 1975. Reproductive strategies in dewberries. *Ecology* 56:721–726.

Barry, M. J., and I. A. E. Bayly. 1985. Further studies on predator induction of crests in Australian *Daphnia* and the effects of crests on predation. *Aust. J. Mar. Freshwater Res.* 36:519–535.

Bazely, D., J. H. Myers, and K. DaSilva. 1991. The response of numbers of bramble prickles to herbivory and depressed resource availability. *Oikos* (In Press).

Becerra, J., and E. Ezcurra. 1986. Glandular hairs in the *Arbutus–Xalapensis* complex in relation to herbivory. *Am. J. Bot.* 73:1427–1430.

Brown, W. L. 1960. Ants, acacias and browsing mammals. *Ecology* 41:587–592.

Bryant, J. P., F. S. Chapin III, and D. R. Klein. 1983. Carbon/nutrient balance of boreal plants in relation to vertebrate herbivory. *Oikos* 40:357–368.

Bryant, J. P., D. J. Tuomi, and P. Niemela. 1988. Environmental constraint of constitutive and long-term inducible defenses in woody plants. In K. Spencer (ed.), *Chemical Mediation of Coevolution*. San Diego: Academic, pp. 367–391.

Bryant, J., K. Danell, F. Provenza, P. Reichardt, and T. Clausen. 1991. Effects of mammal browsing on the chemistry of deciduous woody plants. In D. W. Tallamy and M. J. Raupp (eds.), *Phytochemical Induction by Herbivores*. New York: Wiley, pp 135–154.

Campbell, B. M. 1986. Plant spinescence and herbivory in a nutrient poor ecosystem. *Oikos* 47:168–172.

Carlquist, S. 1974. *Island Biology*. New York: Columbia University Press.

Clausen, J. D., D. Keck, and W. M. Hiesey. 1940. Experimental studies on the nature of species. I. The effect of varied environments of Western North American plants. *Carnegie Inst. Washington Publ. 52*.

Coley, P. D., J. P. Bryant, and F. S. Chapin. 1985. Resource availability and plant anti-herbivore defense. *Science* 230:895–899.

Cooper, S. M., and N. Owen-Smith. 1986. Effects of plant spinescence on large mammalian herbivores. *Oecologia (Berlin)* 68:446–455.

De Fluiter, H. J., and G. W. Ankersmit. 1948. Data on the infestation of bean by *Aphis fabae*. *Tigdschr Planttenziekten* 54:1–13 (cited in Webester 1975).

Eisner, T. 1981. Leaf folding in a sensitive plant: a defensive thorn-exposure mechanism? *Proc. Nat. Acad. Sci. U.S.A.* 78:402–404.

Foster, J. B., and A. I. Dagg. 1972. Notes on the biology of giraffe. *E. Afr. Wildl. J.* 10:1–16.

Gauthier-Pilters, H., and A. I. Dagg. 1981. *The Camel*. Chicago: University of Chicago Press.

Gibson, A. C., and P. S. Nobel. 1986. *The Cactus Primer*. Cambridge, MA: Harvard University Press.

Gibson, C. W. D., H. C. Dawkins, V. K. Brown, and M. Jepsen. 1987. Spring grazing by sheep: effects on seasonal changes during early old field succession. *Vegetatio* 70:33–43.

Gilbert, J. 1966. Rotifer ecology and embryological induction. *Science* 209:1234–1237.

Gilmartin, A. J. 1983. Evolution of mesic and xeric habits in *Tillandsia* and *Vriesea* (Bromeliaceae). *Syst. Bot.* 8:233–242.

Greenwood, R. M., and I. A. E. Atkinson. 1977. Evolution of divaricating plants in New Zealand in relation to browsing. *Proc. N. Z. Ecol. Soc.* 24:21–33.

Grime, J. P. 1977. Evidence for the existence of three primary strategies in plants and its relevance to ecological and evolutionary theory. *Am. Natur.* 111:1169–1194.

Hagen, A. F., and F. N. Anderson. 1967. Nutrient imbalance and leaf pubescence in corn as factors influencing leaf injury by the adult western corn rootworm. *J. Econ. Entomol.* 60:1071–1073.

Hall, H. K., M. H. Quazi, and R. M. Skirvin. 1986. Isolation of a pure thornless loganberry by meristem tip culture. *Euphytica* 35:1039–1044.

Hargreaves, J. A., and J. P. R. Keon. 1986. Cell wall modifications associated with the resistance of cereals to fungal pathogens. In J. Bailey (ed.), *Biology and Molecular Biology of Plant–Pathogen Interactions*. NATO ASI Series, Vol H1. Berlin: Springer, pp. 133–140.

Harvell, C. D. 1986. The ecology and evolution of inducible defenses in a marine bryozoan: cues, costs, and consequences. *Am. Natur.* 128:810–823.

Harvey, T. L., and T. J. Martin. 1980. Effects of wheat pubescence on infestations of wheat curl mite and incidence of wheat streak mosaic. *J. Econ. Entomol.* 73:225–227.

Havel, J. 1986. Predator-induced defenses: a review. In W. C. Kerfoot and A. Sih (eds.), *Predation: Direct and Indirect Impacts of Aquatic Communities*. Hanover, New Hampshire: University Press of New England, pp. 263–278.

Huntley, B. J. 1982. Southern African savannas. In B. J. Huntley and B. H. Walker (eds.), *Ecology of Tropical Savannas*. New York: Springer.

Janick, J. 1986. *Horticultural Science*. New York: Freeman.

Janzen, D. H. 1984. Tropical black-water rivers, animals and mast fruiting by the Dipterocarpaceae. *Biotropica* 6:69–103.

Janzen, D. H. 1986. Chihuahuan desert nopaleras: defaunated big mammal vegetation. *Annu. Rev. Ecol. Syst.* 17:595–636.

Jennings, D. L. 1986. Breeding for spinelessness in blackberries and blackberry–raspberry hybrids: a review. *Acta Horticulturae* 123:59–66.

Johnson, H. B. 1975. Plant pubescence: an ecological perspective. *Bot. Rev.* 41:233–258.

King, P. 1986. Plant identikit: common plants of Central Australia. Conservation Commission Northern Territory, Australia.

Kupper, W., and P. Roshardt. 1960. *Cacti*. Edinburgh: Nelson.

Leuthold, W. 1970. Preliminary observations of food habits of gerenuk in Tsavo National Park, Kenya. *E. Afr. Wildl. J.* 8:73–84.

Levin, D. A. 1973. The role of trichomes in plant defense. *Quart. Rev. Biol.* 48:3–15.

Maier, F. E. 1982. Effects of physical defenses on vine and epiphyte growth in palms. *Tropical Ecol.* 23:212–217.

Mauseth, J. D., and W. Halperin. 1975. Hormonal control of organogenesis in *Opuntia polyacantha*. *Am. J. Bot.* 62:869–877.

McNaughton, S. J., and J. L. Tarrants. 1983. Grass leaf silicification: natural selection for an inducible defense against herbivores. *Proc. Nat. Acad. Sci. U.S.A.* 80:790–791.

Meijden, E. van der, M. Wijn, and H. J. Verkaar. 1988. Defense and regrowth, alternative plant strategies in the struggle against herbivores. *Oikos* 51:35–363.

Myers, J. H. 1987. Nutrient availability and the deployment of mechanical defenses in grazed plants: a new experimental approach to the optimal defense theory. *Oikos* 49:350–351.

Myers, J. H. 1988. The induced defense hypothesis: does it apply to the population dynamics of insects? In K. Spencer (ed.), *Chemical Mediation of Coevolution*. New York: Academic, pp. 530–557.

Nobel, P. S. 1978. Surface temperatures of cacti-influences of environmental and morphological factors. *Ecology* 59:986–996.

Nobel, P. S. 1980. Influences of minimum stem temperature on ranges of cacti in southwestern United States and central Chile. *Oecologia (Berlin)* 47:10–15.

Norris, D. M., and M. Kogan. 1980. Biochemical and morphological bases of resistance. In F. G. Maxwell and P. R. Jennings (eds.), *Breeding Plant Resistance to Insects*. New York: Wiley, pp. 23–61.

Okuda, T. 1987. The phenotypic variation of *Glochidion obovatum* Sieb. et Zucc. in relation to the deer browsing. *Hikobia* 10:13–19.

Owen-Smith, N. 1982. Factors influencing the consumption of plant products by large herbivores. In B. J. Huntley and B. H. Walker (eds.), *Ecology of Tropical Savannas*. New York: Springer, pp. 359–404.

Paiva, M., and J. Janick. 1980. Relationship between leaf pubescence and resistance to European red mite in apple. *HortScience* 15:511–512.

Palmer, A. R. 1985. The adaptive value of shell variation in *Thais lamellosa*: effect of thick shells on vulnerability to prey and to preferences by crabs. *Veliger* 27:349–356.

Pellew, R. A. 1984. Food consumption and energy budgets of giraffe. *J. Appl. Ecol.* 21:141–159.

Peters, K. M., and R. E. Berry. 1980. Effect of hop leaf morphology on two spotted spider mite. *J. Econ. Entomol.* 73:235–238.

Poincelot, R. P. 1980. *Horticulture: Principles and Practical Applications*. Englewood Cliffs, NJ: Prentice Hall.

Pollard, A. J. 1986. Variation in *Dnidoscolus texanus* in relation to herbivory. *Oecologia* 70:411–413.

Pollard, A. J., and D. Briggs. 1984. Geneocological studies of *Urtica dioica* L. III. Stinging hairs and plant–herbivore interactions. *New Phytol.* 97:507–522.

Poole, A. L., and N. M. Adams. 1980. *Trees and Shrubs of New Zealand*. P. D. Hasselberg. Christchurch.

Potter, D. A. and T. W. Kimmerer. 1988. Do holly leaf spines really deter herbivory? *Oecologia* 75:216–221.

Provenza, F. D., and J. C. Malechek. 1984. Diet selection by domestic goats in relation to blackbrush twig chemistry. *J. Appl. Ecol.* 21:831–841.

Richards, P. W. 1952. *The Tropical Rain Forest*. Cambridge, UK: Cambridge University Press.

Seif El Din, A., and M. Obeid. 1971. Ecological studies of the vegetation on the Sudan. IV. The effect of simulated grazing on the growth of *Acacia senegal* (L.) Willd. seedlings. *J. Appl. Ecol.* 8:211–216.

Smith, D. H., T. Ninan, E. Rathke, and C. E. Cress. 1971. Weight gain of cereal leaf beetle larvae on normal and induced leaf pubescence. *Crop Sci.* 11:639–641.

Supnick, M. 1983. On the function of leaf spines in *Ilex opaca*. *Bull. Torrey Bot. Club* 110:228–230.

Watson, M. A., and B. B. Casper. 1984. Morphogenetic constraints on patterns of carbon distribution in plants. *Annu. Rev. Ecol. Syst.* 15:233–258.

Webster, J. A. 1975. Association of plant hairs and insect resistance. *Agric. Res. Service, USDA Misc. Publ. No. 1297*.

Westwood, M. N. 1978. *Temperate-Zone Pomology*. San Francisco: Freeman.

White, P. S. 1988. Prickle distribution in *Aralia spinosa* (Araliaceae). *Am. J. Bot.* 75:282–285.

Wilson, C. L., and W. E. Loomis. 1962. *Botany*. New York: Holt, Rinehart, & Winston.

Young, T. P. 1987. Increased thorn length in *Acacia depranolobium*: an induced response to browsing. *Oecologia* 71:436–438.

PART III

AGRICULTURAL IMPLICATIONS OF INDUCIBLE PHYTOCHEMICALS

CHAPTER 15

INDUCIBLE DEFENSES IN SOYBEAN AGAINST HERBIVOROUS INSECTS

MARCOS KOGAN and DANIEL C. FISCHER

Office of Agricultural Entomology, University of Illinois and Illinois Natural History Survey, Champaign, IL 61820

1 Introduction
2 Soybean phytoalexins
   2.1 Biosynthesis of isoflavonoids and relationships to other plant phenolics
   2.2 Soybean resistance to fungal pathogens
   2.3 *Phytophthora* resistance and phytoalexin accumulation in soybean
   2.4 Elicitors of phyloalexins
3 Effects of soybean phytoalexins on herbivore behavior and nutritional physiology
   3.1 Specific phytoalexins
   3.2 Phenolic acids and their derivatives
4 Does insect feeding induce resistance in soybean?
   4.1 Induction by previous herbivory: greenhouse plants
   4.2 Induction by previous herbivory: field plants
   4.3 Postingestive effects of induced resistance
5 Chemical nature of allomonal compounds induced by herbivory
6 Enzymatic inhibition
7 Ecological effects of inducible defenses and their potential use in integrated pest management
   7.1 Potential applications in soybean integrated pest management
   References

1 INTRODUCTION

Soybean (*Glycine max*) is adapted to grow under temperate zone conditions. Within a latitudinal band in which the crop cycle is completed in 80–120 days, a combination of ecological factors and plant defense mechanisms greatly dampen the impact of potentially damaging herbivore pests. By contrast, soybean grown under subtropical or tropical conditions is increasingly vulnerable to injury (Kogan and Turnipseed 1987), probably because eco-

347

logical conditions favor the pests and in some way suppress the expression or the magnitude of the plant's defense mechanisms. Throughout its range of cultivation, however, multiple mechanisms of ecological and genetic resistance seem to operate in soybean at various stages of growth. Rapid seed germination and plant maturation, for example, allow soybean to escape seed and seedling pests early in the season and pod- and seed-feeding pests late in the season. Pubescence is an effective defense against leafhoppers and other small sucking arthropods (Turnipseed 1977), and most commercially grown varieties are profusely pubescent. Finally, soybean is remarkably tolerant to stand thinning, defoliation, and loss of blossoms and pods. Compensatory growth of neighboring plants makes up for whole plant losses within and among rows, and supplementary blossom development and seed expansion compensate for blossom and pod losses, depending on the stage of growth when such losses occur (Turnipseed and Kogan 1987).

Soybean chemistry, however, is the most likely source of defensive strategies against herbivory as well as other biotic stresses. Seeds are protected by powerful allomones, including saponins and proteinase inhibitors, that render the seeds virtually immune to insect attack (Sirisingh and Kogan 1982), and leaves are rich in allomones that affect both the metabolism (antibiosis) and behavior (antixenosis) of phytophagous arthropods.

Marked differences in the level of resistance to herbivores are found among soybean cultivars and lines (Kogan 1972). Highest levels of resistance to foliage-feeding arthropods have been consistently recorded under diverse testing conditions for the oriental genotypes "Kosamame" (PI 171451), "Miyako White" (PI 227687), and "Sodendaizu" (PI 229358) (Van Duyn et al. 1971, Clark et al. 1972, Kogan 1972, Luedders and Dickerson 1977, Rossetto et al. 1977, Layton et al. 1987, Talekar et al. 1988). Some of the inherent differences among genotypes are the result of constitutive phytochemical resistance (Smith 1985, Chiang et al. 1986, Kogan 1986, 1989, Norris et al. 1988). Although characterization of the constitutive factors is incomplete, progress has been made in the comparative phytochemistry of normally susceptible and resistant genotypes (Tester 1977, Dreyer et al. 1979, Grunwald and Kogan 1981, Smith and Fischer 1983, Binder and Waiss 1984, Caballero et al. 1986, Liu et al. 1988, Norris et al. 1988).

Leaf volatiles may be involved in the initial phases of the host selection process as attractants in susceptible genotypes, or as repellents in resistant genotypes (Liu et al. 1988). It seems that primary metabolites, mainly low concentrations of reducing sugars, and secondary metabolites, mainly isoflavonoids, are in part involved in some aspects of resistance. The elusiveness of the explanation of this phenomenon, however, is illustrated by the fact that unusually high (6.6–17.7 mg/g) concentrations of pinitol (1-D-3-0-methylchiroinositol) are found in many soybean varieties (Phillips et al. 1982). Dreyer et al. (1979) found higher concentrations of pinitol in the resistant PI 229358 than in the susceptible cultivar "Davis." In their experiments, larvae of the corn earworm, *Heliothis zea*, gained less weight when fed a standard medium containing 0.8% pinitol than those fed the medium alone. Because these results could not be replicated elsewhere and because pinitol is so prevalent both in susceptible and in resistant genotypes, the role of pinitol in soybean resistance to insects remains unclear (Gardner et al. 1984).

On the other hand, mounting evidence suggests that soybean responds to physical and biotic stresses through the enhanced synthesis of compounds previously present or through the de novo synthesis of "defensive chemicals." These defensive chemicals fit the definition of phytoalexins: "antimicrobial components of a usually related group of compounds synthesized by a plant in response to infection or stress" (Deverall 1982). The possible role of phytoalexins in soybean resistance to plant pathogens has been often noted (Keen and Kennedy 1974, Yoshikawa et al. 1978, Ingham et al. 1981, Mansfield 1982). Only when Hart et al. (1983) tested the behavioral response of the Mexican bean beetle, *Epilachna varivestis*, to phytoalexin-rich soybean cotyledon tissue was a potential antiherbivory role of soybean phytoalexins suspected. Since that work, much research on the various aspects of induced resistance in soybean to herbivores has been initiated. In this chapter we review findings since Hart et al. (1983) and address five fundamental questions:

1. What is the effect of soybean phytoalexins on herbivore behavior and nutritional physiology?
2. Does insect feeding induce resistance in soybean?
3. What is the chemical nature of the allomonal compounds induced by herbivory?
4. Is resistance in elicited plants suppressed by specific inhibition of enzymes associated with phytoalexin biosynthesis?
5. What is the ecological role of induced resistance in soybean and how can induced resistance be used to control arthropod pests?

As background for addressing these questions, we summarize current information on soybean phytoalexins, drawing mainly from the rich phytopathological literature.

2 SOYBEAN PHYTOALEXINS

Known soybean phytoalexins are isoflavonoids, most of which are derived from isoflavone or pterocarpan skeletons (Ingham 1982) (Fig. 1). The isoflavonoid phytoalexins that have been elicited either from soybean plants or from soybean tissue cultures are shown in Figure 2.

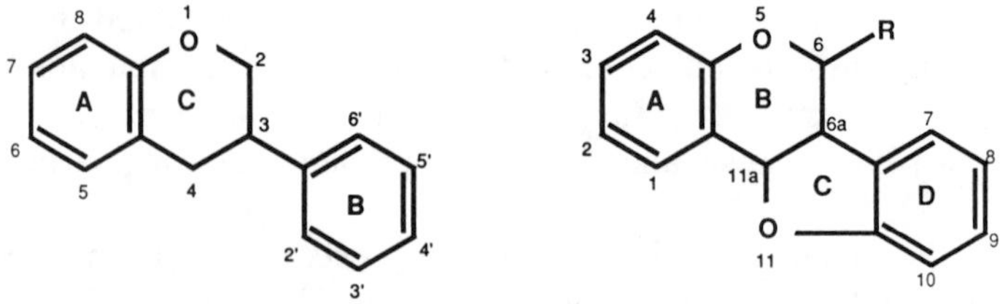

Isoflavones

Pterocarpans (R=H)

Coumestans (R=O)

Figure 1 Basic skeleton of soybean phytoalexins with ring numbering system. (Adapted from Ingham 1982.)

2.1 Biosynthesis of Isoflavonoids and Relationships to Other Plant Phenolics

The pathways in the biosynthesis of soybean phytoalexins are intimately connected with the production of benzoic and cinnamic acid and their derivatives, some of which seem to be present constitutively in soybean plants (Hardin 1979, Porter et al. 1986). Because an antiherbivory role may exist for those lower molecular weight phenolic acid derivatives (Fischer et al. 1990b) in addition to the purported defensive role of the phytoalexins, these biochemical relationships must be clarified. Excellent summaries of the current understanding of the biosynthesis of soybean phytoalexins are presented by Hahlbrock and Grisebach (1979), Hanson and Havir (1979), Harborne (1980), Stossel (1982), Darvill and Albersheim (1984), and Ebel (1986). Precursors of isoflavonoids are synthesized from the shikimate pathway, the major metabolic route leading to the formation of aromatic compounds in microorganisms and plants (Floss 1980).

Although the pathway is complex and gives rise to many aromatic compounds, of special interest are the steps that lead to the synthesis of phenylalanine and tyrosine (Fig. 3). These two amino acids are precursors in the synthesis of the phenylpropanoids from which are derived most of the compounds generically known as phenolics, including hydroxycoumarins, lignans, and lignins, as well as the flavonoids and isoflavonoids. Hydroxybenzoic acids, their phenols, and some quinone derivatives may come directly from shikimic acid or from phenylalanine via *p*-hydroxycinnamic acid (see Harborne 1980) (Fig. 4). The core sequence in the system is the deamination of L-phenylalanine and L-tyrosine to *trans*-cinnamate and *trans*-coumarate, respectively (Hanson and Havir 1979). The enzymes responsible for the reactions are phenylalanine ammonia lyase (PAL) and tyrosine ammonia lyase (TAL). Increases in PAL and TAL concentrations in plant tissues are indicative of the metabolic activities that lead to accumulations of various phenolic compounds, many of which are involved in the induced resistance of plants against pathogens and, in all likelihood, against herbivores also.

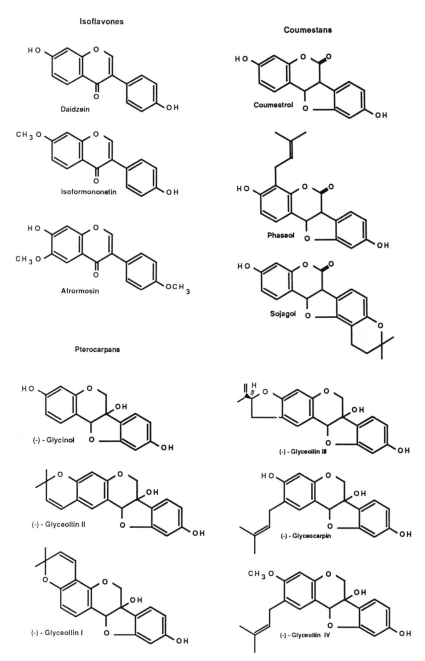

Figure 2   Structures of soybean isoflavonoids induced from various tissues. (Based on Ingham (1982), except afrormosin and phaseol, which are based on Caballero et al. (1986).)

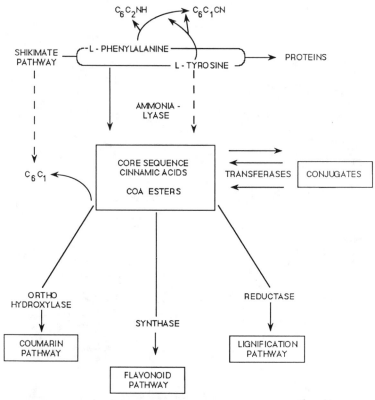

Figure 3 General scheme illustrating the biochemical pathways leading to the biosynthesis of phenolic products from shikimic acid.

2.2 Soybean Resistance to Fungal Pathogens

The soybean disease *Phytophthora* root and stem rot, caused by the fungus *Phytophthora megasperma* f.sp. *glycinea*, may reduce yields by more than 50% in susceptible cultivars. The extent of losses depends on soil type, rainfall, tillage practices, and soybean genotype (Sinclair 1982). There are some 16 races of *P. megasperma* f.sp. *glycinea* (Keen and Yoshikawa 1983) and at least five genes and four alleles are responsible for the resistance that has been incorporated into most of the soybean cultivars grown in North America (Sinclair 1982).

The soybean–*Phytophthora* system has been extensively used in studies of the role of phytoalexins in disease resistance (Paxton 1983). Inoculation through wounds usually yields consistent results, but a method to inoculate intact hypocotyls with zoosporidia has also been developed and used in recent studies of soybean phytoalexins (Mansfield 1982). Mycelium or zoospore inoculation of susceptible hypocotyls results in spreading lesions within 24 h. Resistance is expressed by the appearance of localized dark

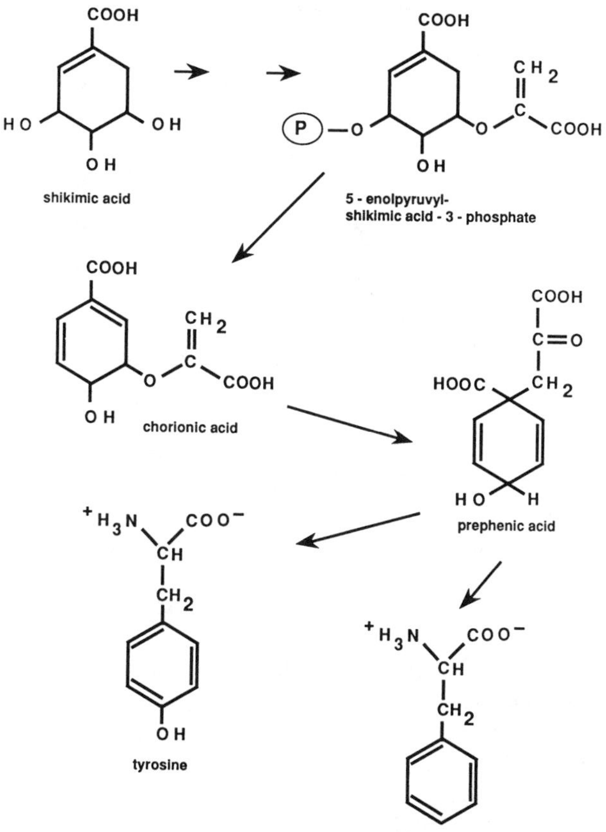

Figure 4   Steps in the biosynthesis of phenylalanine and tyrosine from shikimic acid. (Modified from Floss 1980.)

brown lesions caused by the necrosis of tissues. Fungal growth is restricted within these lesions, and resistance, therefore, has been described as a hypersensitive reaction.

2.3 *Phytophthora* Resistance and Phytoalexin Accumulation in Soybean

Yoshikawa et al. (1978) have demonstrated that the accumulation of glyceollins (soybean phytoalexin isomers) in tissues is directly correlated with the cessation of growth in soybean hypocotyls of *P. megasperma* f.sp. *glycinea*. Hypocotyls of "Harosoy" (susceptible) and "Harosoy 63" (resistant) soybean were inoculated with race 1 of the fungus, and the spread of lesions and the accumulation of glyceollin were monitored. In other experiments

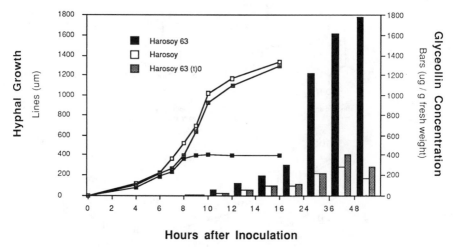

Hours after Inoculation

Figure 5 Accumulation of glyceollin in soybean leaves and corresponding growth of hyphae of *Phytophthora megasperma* f.sp. *glycinea*. Harosoy 63 (t) is Harosoy 63 treated with blasticidin S as an inhibitor of protein synthesis. (Redrawn from Yoshikawa et al. 1978.)

plants also were treated with blasticidin S, a protein synthesis inhibitor that renders the hypocotyls of Harosoy 63 susceptible to *Phytophthora*. Time-course experiments showed that glyceollins accumulated at greater rates and reached higher levels in Harosoy 63 than in Harosoy. ED_{50} (effective dosage) values of glyceollin mixtures in vitro against *P. megasperma* f.sp. *glycinea* have ranged from 25 to 100 µg/mL (Bhattacharyya and Ward 1986). The glyceollin levels recorded in inoculated Harosoy 63 did not reach the ED_{50} of about 100 µg/mL until 16 h after inoculation, but glyceollins accumulated rapidly in direct correlation with the restriction of fungal growth (Yoshikawa et al. 1978) (Fig. 5). Glyceollin concentrations reached ED_{90} values (ca. 200 µg/mL) 8 h after inoculation in tissues where many invading hyphae were present, but glyceollin was not detected in tissues not yet invaded by the hyphae. The cessation of fungal growth always coincided with the time of greatest accumulation of glyceollin in Harosoy 63. Although glyceollin concentration exceeded the ED_{90} in the susceptible Harosoy or blasticidin-treated Harosoy 63 after 24 h, these levels were not reached in time to prevent fungal growth and plant death.

Detailed experiments conducted by Bhattacharyya and Ward (1986) showed that different tissues of the soybean plant have diverse patterns of accumulation of glyceollin isomers I, II, and III. In the hypocotyls of Harosoy 63, glyceollin I reached a peak concentration at 48 h and declined thereafter, whereas glyceollin II and III concentrations continued to increase. Concentrations of all three isomers in leaves declined after reaching a peak. In the leaves of Harosoy, highest concentrations occurred at 24 h in the lesions, and production of glyceollin III continued at a fairly constant rate around the edges of the expanding lesions. Concentrations of the glyceollins in the border tissue in Harosoy (susceptible) equaled or exceeded those detected in Harosoy 63; at 24 h, concentrations within the lesions did not differ substantially between the two isolines. The authors concluded that initial patterns of glyceollin accumulation in leaves were similar in both Harosoy 63 and Harosoy, and that much of the additional accumulation in the former occurred after the lesions had ceased to spread. Phytoalexin accumulation and reversal of resistance were obtained by mechanical transfer of phytoalexins (Chamberlain and Paxton 1968) or by the inhibition of PAL activity by means of specific inhibitors (Moesta and Grisebach 1982, see below).

Although the experimental evidence connecting phytoalexin accumulation and resistance offers a compelling argument for a cause/effect relationship (Keen and Yoshikawa 1983), the argument is not without contradictions. Ersek and Király (1986) argued that, although significant, phytoalexins probably are not the sole factors that condition resistance to pathogens. Instead, phytoalexin accumulation may merely be a consequence of resistance effects that limit the growth of the pathogen and concurrently elicit phytoalexin accumulation. In fact, Keen et al. (1983) have identified a β-1,3-endoglucanase from wounded soybean cotyledons. The enzyme reportedly attacks the cell walls of *P. megasperma* f.sp. *glycinea*, releasing glucomannans that in turn act as elicitors of phytoalexin synthesis in a race-specific manner.

Although our understanding of phytoalexins in disease resistance is incomplete, investigators seem to agree that phytoalexins play a key role. They not only are one of several requirements for resistance against incompatible races of cultivar-specific pathogens, but are probably very important in preventing infections by organisms that are not normally pathogenic on the plant (Darvill and Albersheim 1984). The association of isoflavonoid and isoflavone phytoalexins with resistance to other pathogens, for example, the bacteria *Xanthomonas campestris* pv. *glycinea* (Fett 1984) and *Pseudomonas syringae* pv. *glycinea* has been also established (Fett and Zacharius 1983, Long et al. 1985).

2.4 Elicitors of Phytoalexins

The synthesis of phytoalexins in soybean is elicited by a variety of abiotic and biotic agents. Abiotic chemical elicitors include salts of heavy metals such as mercury and copper; respiratory inhibitors such as sodium iodoacetate, sodium fluoride, potassium cyanide, and 2,4-dinitrophenol; surfactants such as Triton X-100 or Nonidet P40; fungicides such as maneb (manganese ethylenebisdithiocarbamate) and benomyl [methyl-1-(butylcarbamoyl)-2-benzimidazolecarbamate] (Bailey 1982); and herbicides such as Acifluorfen {5-[2-chloro-4-(trifluoromethyl) phenoxy]-2-nitrobenzoate} (Cosio et al. 1985). Among the abiotic physical elicitors that have been

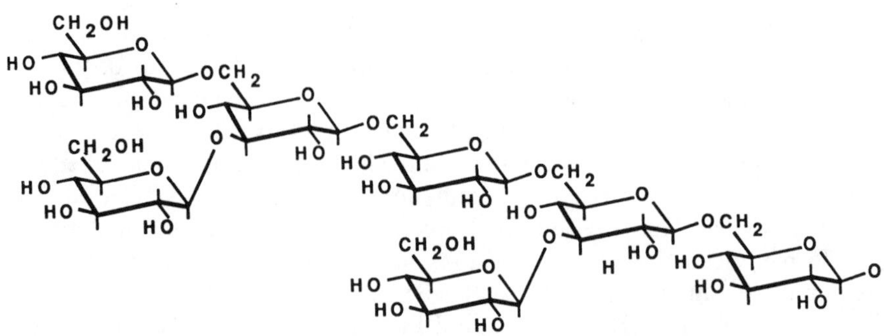

Figure 6   Structure of the hepta-*b*-glucoside alditol as the smallest elicitor-active oligosaccharide from *P. megasperma* f.sp. *glycinea* cell walls. (Based on Darvill and Albersheim 1984.)

recorded are the mechanical wounding of cotyledons, partial freezing, and UV irradiation.

Elicitors of biotic origin have been derived from plant cell walls and from cell walls of challenging microorganisms. Extensive work with *P. megasperma* f.sp. *glycinea* has resulted in the isolation of a very active fraction characterized as a glucan, and specifically a hepta-*b*-glucoside alditol, with the structure shown in Figure 6 (after Darvill and Albersheim 1984). This glucan was the smallest elicitor-active oligosaccharide obtained by partial acid hydrolysis of the cell walls of *P. megasperma* f.sp. *glycinea*. Other experiments have shown that the hepta-*b*-glucoside was a broad spectrum elicitor, neither race- nor species-specific (Darvill and Albersheim 1984). Structural definition of elicitors may lead to the identification of receptor sites in the plant cell that are part of the early sequence of events leading to phytoalexin synthesis and accumulation.

In addition to these exogenous elicitors, plant cell wall fractions also seem to be involved in the elicitation process. These endogenous elicitors probably act in concert with the exogenous elicitors. The dual elicitor system, it has been hypothesized (Bailey 1982), would prevent plants injured by mechanical factors alone, for example, wind or hail, from activating chemical defenses if microbial invasion was not also imminent. A polygalacturonide has been suspected as a possible endogenous elicitor from soybean (Darvill and Albersheim 1984). A scheme for the chain of events in the elicitation of phytoalexin synthesis and accumulation has been proposed by Bailey (1982) and Darvill and Albersheim (1984), based on this dual-elicitor system (Fig. 7). Such a seemingly effective inducible defense system against pathogens would have even greater adaptive value if equally effective against herbivores. It is thus reasonable to ask whether these same phytoalexins and related compounds also have antiherbivory properties.

Elicitor (E) inactive in
uninfected cells

Elicitor released after
injury due to infection

De novo synthesis of
phytoalexins

Phytoalexins accumulate
in dying infected cell

Phytoalexins localized in
hypersensitive cell.
Fungal growth inhibited

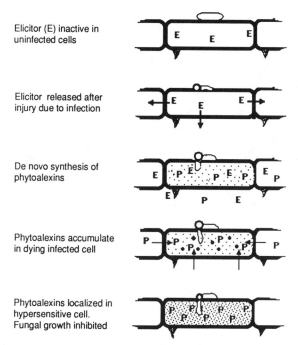

Figure 7   Scheme for the chain of events in the elicitation of phytoalexin synthesis. (Adapted from Bailey 1982.)

3   EFFECTS OF SOYBEAN PHYTOALEXINS ON HERBIVORE BEHAVIOR AND NUTRITIONAL PHYSIOLOGY

The effect of soybean phytoalexins on the feeding behavior of adult and larval Mexican bean beetle, *Epilachna varivestis* (Coleoptera:Coccinellidae), an oligophagous species that is a common pest of legume crops in North America, has been tested (Hart et al. 1983, Kogan and Paxton 1983). In preliminary experiments, detached cotyledons from the soybean cultivar "Clark 63" were inoculated with an extract of *P. megasperma* f.sp. *glycinea*. The inoculated cotyledons were incubated for 48 h to elicit phytoalexin production as described by Frank and Paxton (1971). Behavioral responses to the phytoalexins were analyzed using dual choice tests comparing treated (inoculated) versus untreated (control) cotyledons originating from the same seeds. Results from these preliminary tests suggested that the feeding of Mexican bean beetles was deterred on the elicited cotyledons.

Since the elicitor itself could have had a deterrent effect on beetle feeding, further tests were conducted using ultraviolet (UV) light as the elicitor and following a procedure adapted from Bridge and Klarman (1973). UV-treated cotyledons were incubated for 2 days and phytoalexin production was vis-

Figure 8   Soybean cotyledons showing markings of Mexican bean beetle feeding. *Left:* Phytoalexin free cotyledon with normal feeding markings; *right:* phytoalexin-rich cotyledon with only a few probing mandibular markings. (From Hart et al. 1983.)

ually determined by the typical dark-reddish coloration that accompanies the accumulation of phenolics on the treated cotyledon surface. Control cotyledons were also incubated for 2 days; however, they received the UV light treatment immediately prior to the experiments. Thus, control cotyledons were exposed to the same dosage of UV light as the treated cotyledons but did not have time to concentrate phytoalexins prior to submittal to the beetles. The presence of glyceollins in the UV-treated and incubated cotyledons was confirmed by thin-layer chromatography. No glyceollin was detected in the control cotyledons. In experiments conducted with both larvae and adult beetles, feeding was significantly greater on the control (phytoalexin-free) than on treatment (phytoalexin-rich) cotyledons. It was evident that some component(s) of the phytoalexin complex deterred feeding by the Mexican bean beetle. Single-bite mandible markings were found on phytoalexin-rich cotyledons (Fig. 8), suggesting that deterrence was due to a distasteful property of the phytoalexin-rich tissue (Hart et al. 1983, Kogan and Paxton 1983).

3.1 Specific Phytoalexins

The effect of specific phytoalexins was tested using young soybean looper larvae, *Pseudoplusia includens* (Lepidoptera:Noctuidae), and an artificial medium containing varying doses of coumestrol or glyceollins (Hart et al. 1983, Kogan and Paxton 1983). Survival of larvae after 8 days was reduced by about 30% with coumestrol doses ranging from 0.1 to 1.0% (dry weight)

TABLE 1 Soybean Looper Survival and Weight Gain to 8 Days on Artificial Media + Coumestrol or Glyceollin

Concentration (%)	% Survival		% Weight Gain	
	Coumestrol	Glyceollin	Coumestrol	Glyceollin
0	96.0	83.2	100.0	100.0
0.1	64.0	74.0	44.1	76.4
0.5	68.0	75.5	54.1	139.2
1.0	68.0	58.0	96.5	127.7

in the medium (Table 1). Survival was reduced by 9% on medium containing 0.1% glyceollin and by 25% on medium containing 1.0% glyceollin. However, weight gain of the surviving larvae was not affected, suggesting that the polyphagous soybean looper population used in the experiment may have had considerable genetic variability in susceptibility to these two phytoalexins (Kogan and Paxton 1983).

Recently, Rose et al. (1989) demonstrated that soybean looper larvae fed a medium containing 0.01% coumestrol in dimethyl sulfoxide had a substantially increased activity of glutathione transferase. This activity was similar to the activity observed in larvae fed foliage of the resistant PI 227687 and was considerably higher than the activity in larvae fed the susceptible cultivar "Bragg." This study provides indirect evidence of the allomonal role of coumestrol against the looper. Large amounts of coumestrol have been detected, apparently as a constitutive product, in the leaves of PI 227687 (Smith 1985). The constitutive nature of some isoflavonoids of the resistant PI 227687 was confirmed also by Caballero et al. (1986), who identified coumestrol, phaseol, and afrormosin in methanol extract fractions of foliage.

The deterrent effects of glyceollins were tested in experiments conducted with an aerosol spraying technique in which the test compounds were dissolved in acetone (10.0 nm) (Fischer et al. 1990c). The procedure insured uniform distribution of the test compounds on the surface of the leaf. For tests with the Mexican bean beetle, leaflets from its preferred host, *Phaseolus vulgaris*, were used. The rationale for using common bean leaves instead of leaves from soybean was to offer the Mexican bean beetle a food plant that was readily accepted and did not naturally contain the chemicals being tested. The surface application of the compounds did provide an unnatural condition because phytoalexins occur intracellularly; however, the surface exposure of chemicals permitted a direct, rapid, and rigorous test for deterrency. If the Mexican bean beetle could be deterred from feeding on a preferred host (common bean) by a surface-applied chemical, chances are that the chemical would also have antifeedant effect within the cell environment of the less preferred host (soybean). Given the peculiar feeding process of the Mexican bean beetle (Kogan 1973) and the anatomy of chemosensors attached to the mouthparts, the chemosensilla are likely to be immediately exposed to the intracellular contents of the leaf upon the first few bites (Fischer and Kogan 1986).

The feeding preference tests used in these experiments were performed in arenas made from 18.0-cm-diam glass petri dishes containing a layer of hardened plaster of Paris. This layer is saturated with tap water and covered with filter paper prior to the tests. Leaf disks were cut from the sprayed area of a leaflet. Test disks were from leaflets sprayed with glyceollin and control disks derived from leaflets sprayed with pure acetone. The use of leaf disks facilitated quantification of results, but it has been criticized because of the assumption that the cutting itself could elicit an induced response in the disk tissue. Preliminary tests optimized the testing conditions so that all preference tests could be terminated within 6 h. Within this time, induction caused by cutting of the disks was minimal, if it occurred at all. Leaf disks 24 h old were compared with leaf disks from the same plants that had been punched immediately prior to the test; no differences in preference were found (D. Fischer, unpublished results).

In each replicate of the preference test, three treated disks and three control disks were alternated along the perimeter of the petri dish arena. Four adult female Mexican bean beetles (4–10 days old) were starved for about 12 h prior to use in the tests. They were then placed in the center of the arena and allowed to eat until approximately half of the leaf material was consumed. Starvation primed the beetles to start feeding immediately and thus reduced the length of the tests. After each test, the disks were removed and the remaining area was measured with a leaf area meter. The area eaten from each disk was the difference between the initial area (constant) and the area of the uneaten leaf disk. Analyses were based on the preference index $PI = 2T/(C + T)$, where C is the area consumed from control leaf disks and T is the area consumed from test disks. PI values range from 0 to 2 with $1 =$ no preference, $0 =$ maximum rejection of test disk, and $2 =$ maximum acceptance. The use of this index helps normalize the variance often associated with tests using percentages as a measure of results. This index was first proposed by Kogan and Goeden (1972) and was recently adopted by Croxford et al. (1989), who called it a palatability index. This same preference test was used in a series of experiments mentioned in subsequent sections of this chapter.

Glyceollin used in the tests was extracted from induced "Williams 82" soybean cotyledons removed from 5-day-old seedlings. Induction was elicited by removing a sliver of tissue from the abaxial surface of each cotyledon and applying a drop of autoclaved cultured cells of *P. megasperma* f.sp. *glycinea* to the wounded area. After a 48-h incubation period, the cotyledons were extracted with a buffered (pH 7.2) 50% ethanol solution. The extract was concentrated and finally purified by HPLC (the detailed extraction and purification procedure is found in Fischer et al. 1990c). The identity of the purified glyceollin was confirmed by UV spectroscopy.

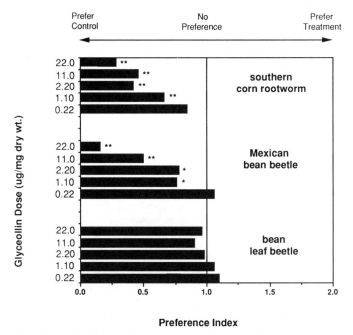

Figure 9 Results of feeding preference tests with glyceollin. Bars represent mean of eight preference tests at each dose. * $p < .05$, ** $p < .01$ (paired t test). (Reproduced from Fischer et al. 1990c.)

Glyceollin concentrations ranging from 0.22 to 22.0 mg/g dry wt of leaf tissue were tested using the aerosol spraying technique (Fischer et al. 1990c) and dual choice preference tests with adult female southern corn rootworm (*Diabrotica undecimpunctata howardi*), bean leaf beetle (*Ceromoma trifurcata*), and Mexican bean beetle. The physiological range of concentrations of phytoalexins induced by pathogens is 1–5 mg/g dry wt of leaf tissue (Long et al. 1985). Results confirmed that glyceollin deterred feeding by the Mexican bean beetle and the southern corn rootworm, but not by the bean leaf beetle (Fig. 9). An EC_{50} (defined as the concentration resulting in a preference index of 0.5) was calculated as a measure of the antixenotic effect of glyceollin (Fig. 10). The EC_{50} was 9.1 mg/g of glyceollin for the southern corn rootworm and 12.5 mg/g for the Mexican bean beetle. These concentrations are about twice the physiological concentrations resulting from pathogen induction; however, all doses above the lowest significantly deterred feeding by these two species. The bean leaf beetle was only mildly affected by the two highest dosages.

In summary, glyceollin deterred the polyphagous southern corn rootworm and the oligophagous Mexican bean beetle, which are generally poorly adapted to feeding on soybean, but it was not active against the bean leaf beetle, one of the native North American species best adapted to feeding

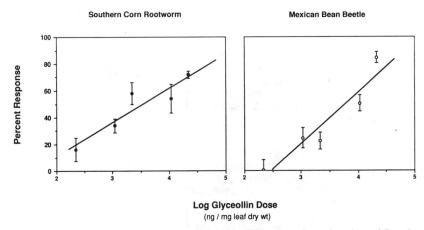

Figure 10 Glyceollin effect on feeding activity of Mexican bean beetle and Southern corn rootworm in preference tests (EC_{50}).

on soybean. Glyceollin, therefore, is potentially a factor discouraging herbivory in the soybean defense system, one that is particularly effective against a polyphagous species or against marginally adapted specialists.

3.2 Phenolic Acids and Their Derivatives

Phenolic acids occur in most plant tissues as glycosides or are condensed to form complex molecules (Robinson 1980). However, high levels (over 40% of total) of free ferulic acid have been detected in alfalfa and in spinach leaves (Huang et al. 1986).

The antiherbivory role of plant phenolics in many insect/plant systems has been well documented (Levin 1971, Fischer et al. 1990b). Phenolic acids and their methylated or hydroxylated derivatives have been reported from soybean (Hardin 1979, Hardin and Stutte 1980, Seo and Morr 1984, Porter et al. 1986). However, some of the analyses, were made after acid hydrolysis; whether the acids were free or esterified cannot be known.

Because of the biosynthetic relationships among soybean phenolics and isoflavonoid phytoalexins, a large series of compounds were tested using the aerosol spraying technique described above (Fischer et al. 1990b). Twenty-three compounds were tested including benzoic and cinnamic acids, simple phenolic acids (hydroxylated benzoic and cinnamic acids), and their methoxylated analogs (Fig. 11). Although not all are reported to occur naturally in soybean, many are (Hardin and Stutte 1980, Porter et al. 1986). Figure 12 shows the results of the preference tests with the hydroxylated derivatives. Seven compounds 2-hydroxybenzoic, 2,3-, 2,4-, 2,5-, and 2,6-dihydroxybenzoic, 2-hydroxycinnamic, and 3,5-methoxy-4-hydroxycinnamic acids, caused significant rejection of common bean leaves by Mexican

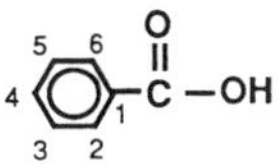 **1. BENZOIC ACID**	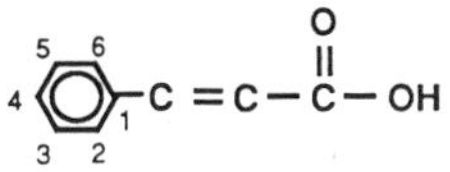 **4. CINNAMIC ACID**
2. Hydroxy Benzoic Acids 2-HBA (Salicylic Acid) 2,3-DHBA 2,4-DHBA 2,5-DHBA (Gentisic Acid) 2,6-DHBA 3,4-DHBA (Protocatechuic Acid) 3,5-DHBA 3,4,5-DHB (Gallic Acid) 4-HBA (p-Hydroxy Benzoic Acid)	**5. Hydroxy Cinnamic Acids** 2-HCA 3-HCA 4-HCA (p-Coumaric Acid) 3,4-DHCA (Caffeic Acid) **6. Methoxy Cinnamic Acids** 4-H-3,5-DMCA (Sinapic Acid) 4-MCA (p-Anisic Acid) 4-H-3-MCA (Ferulic Acid)
3. Methoxy Benzoic Acids 2-MBA 4-MBA 3-H-4-MBA 4-H-3,5-MBA (Syringic Acid) 4-H-3-MBA (Vanillic Acid)	

Figure 11   Phenolic compounds applied to snap bean leaves (BA, benzoic acid; CA, cinnamic acid; H, hydroxy; M, methoxy; D, di). (Reproduced with permission from Fischer et al. 1990b.)

bean beetles. Six of those compounds are hydroxylated acids with the hydroxyl adjacent to the carboxylic acid group on the benzene ring. Two of the compounds, gentisic acid (2,5-dihydroxybenzoic) and salicylic acid (2-hydroxybenzoic), have been detected in soybean leaves (Hardin 1979, Porter et al. 1986). The approximate concentration of total phenolic acids in soybean is 1.0 μg/g fresh wt; however, the behaviorally active dosage was considerably higher in these experiments. These results suggest that single phenolics may be only mildly active against the Mexican bean beetle. In the context of a multifaceted defense, however, they may play a role.

The postingestive effect of salicylic acid was investigated using soybean looper larvae (3rd and 4th instar). The larvae were reared for a full stage on an artificial medium containing 0, 10^2, 10^3, or 10^4 μg salicylic acid per gram artificial medium. Survival, developmental rate, weight gain, and consumption parameters were measured. Although results of our preliminary experiments were inconclusive, evidence indicated a considerable reduction in food consumed and a consequent reduction in normal rates of weight gain (Fig. 13). Efficiency of conversion of ingested food (ECI) was not significantly affected. These results suggest that the main effect of salicylic acid

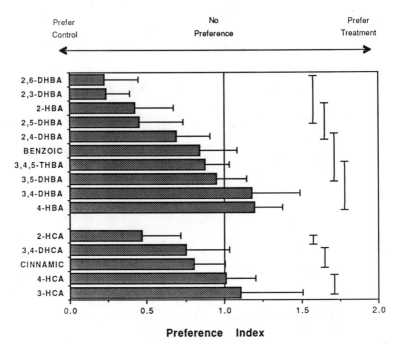

Figure 12   Preference indices for phenolic acids with hydroxyl substitutions only. Upper, dark bars indicate benzoic acid based compounds. Lower lighter bars indicate cinnamic acid based compounds. Vertical bars indicate groups of compounds whose PI's were not significantly different (ANOVA and Scheffe's f test, $p < .05$).

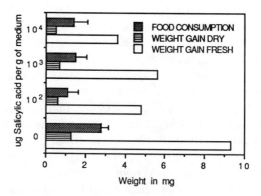

Figure 13   Weight gain and food consumption by soybean looper larvae on media containing various amounts of salicylic acid. (Based on unpublished results from C. Heister and M. Kogan, University of Illinois.)

was as a feeding deterrent. Once ingested, however, the acid had little or no antibiotic effect on larvae. Total mortality for all treatments did not significantly differ from that of the controls (C. Heister and M. Kogan, University of Illinois, unpublished results).

In summary, current evidence indicates that soybean phytoalexins and some simple phenolic acids are active feeding deterrents for the Mexican bean beetle. Low molecular weight phenolics with hydroxyls attached to the benzene ring in positions 2 or 6, that is, adjacent to the carboxyl group, were the most active, albeit at concentrations considerably higher than those occurring naturally in soybean leaves. The postingestive effect of glyceollin and salicylic acid suggests different modes of action. Glyceollin did not deter feeding by soybean looper larvae, but ECI was reduced; salicylic acid was a deterrent (low consumption rates), but ECI was not reduced. Coumestrol affected levels of glutathione transferase activity in soybean looper larvae, suggesting the mobilization of detoxification enzymes (Rose et al. 1989). The following question then is whether herbivory itself can elicit the processes that produce increased levels of resistance to subsequent herbivory.

4 DOES INSECT FEEDING INDUCE RESISTANCE IN SOYBEAN?

Wounding the resistant soybean PI 227687 causes increased levels of resistance to subsequent herbivory. Reynolds and Smith (1985) injured the leaves of PI 227687 by rolling a round steel file over them. The growth rates of 6th instar soybean looper larvae fed this wounded foliage were reduced by about 30–60%. Mexican bean beetles were deterred from feeding on PI 227687 when plants had been injured by Mexican bean beetle feeding 72 h prior to the tests (Chiang et al. 1986). Croxford et al. (1989) reported a reduction in palatability to *Spodoptera littoralis* of soybean leaf tissue that surrounded a hole punched in the center of a leaflet. The variety used, however, was not specified. Since PI 227687 also has powerful constitutive resistance (Smith 1985, Caballero et al. 1986, Chiang et al. 1987), we investigated the potential induced changes in a normally susceptible variety.

Since 1986 we have conducted a series of experiments using the variety Williams 82 (susceptible) with Mexican bean beetle adult females and soybean looper larvae. Tests were based principally on dual choice preference tests (see above) using leaf disks. Both greenhouse and field experiments were conducted and induction was elicited either by mechanical means or by soybean looper herbivory. The most salient aspects of this research are summarized in the following sections.

4.1 Induction by Previous Herbivory: Greenhouse Plants

Soybean plants (Williams 82) were grown in a greenhouse under metal halide supplemental light that provided a 16-h photophase. When plants had

6–8 completely open trifoliolates, the lower 3 or 4 trifoliolates were infested with 4th- or 5th-instar soybean looper larvae at a density of two larvae per trifoliolate. Larvae were allowed to feed for 3 or 4 days as necessary to produce the desired level of defoliation. After defoliation was completed, larvae were removed by hand and plants were left to recover for 2 weeks, at which time two new trifoliolates had usually opened completely.

Dual choice preference tests and weight gain tests were conducted using leaf tissue from the defoliated plants and tissue from similar plants grown under identical conditions but kept injury free. Paired comparisons were made using the most significant combinations. Injured and uninjured leaf tissue from induced (previously injured) plants were compared with leaf tissue from uninjured plants (control); leaves from the uninjured plants originated from nodes analogous to those that provided the induced tissue. Other comparisons included injury by mechanical means, removal of entire leaflets, and removal of large leaf segments.

Figure 14 summarizes the results of these experiments and describes the nature of the comparisons. Data were obtained from S. Amit and M. Kogan (University of Illinois, unpublished results), Iannone (1989), Lin et al. (1990), and Lin (1989).

In all tests involving direct previous injury by soybean looper feeding (except for one series reported by Amit and Kogan) (SBL-CxT in Fig. 14), the palatability of the plants to the Mexican bean beetle was significantly reduced for about 14–21 days after initial injury. These tests, covering a period of 2 years and over 100 replications, suggest that the susceptible Williams 82 becomes about 70–80% less palatable as measured by the preference index (PI) after 20–40% of the leaf area has been removed by the feeding of soybean looper. Within the same plant both injured and uninjured foliage reached similar levels of resistance to subsequent herbivory (see comparisons SBL-CxT, SBL-CxTi, SBL-CxTi7, SBL-C9xT9, and SBL-T7xTi7; Fig. 14). In this series of tests Ti denotes directly injured foliage, and T indicates uninjured foliage from the same or from another injured plant. The comparison T7xT7i is the most convincing demonstration of the systemic nature of the induced response because it shows no difference in palatability between directly wounded and nonwounded leaf tissue from the same plant. Mechanical injury of all kinds, although capable of inducing resistance to Mexican bean beetle feeding, is a less efficient inducer than insect feeding itself. Finally, of all methods of mechanically injuring the plant, the complete removal of entire leaflets at the petiole was the least effective in inducing a response (CUT-CxT, in Fig. 14). In another series of tests, Lin et al. (1990) showed that coating a mechanically (needle-punctured) injured leaflet with soybean looper regurgitate accentuated the induced response.

4.2 Induction by Previous Herbivory: Field Plants

Tests discussed in the previous sections were conducted with greenhouse plants, but the effect of previous herbivory on the acceptability of plants

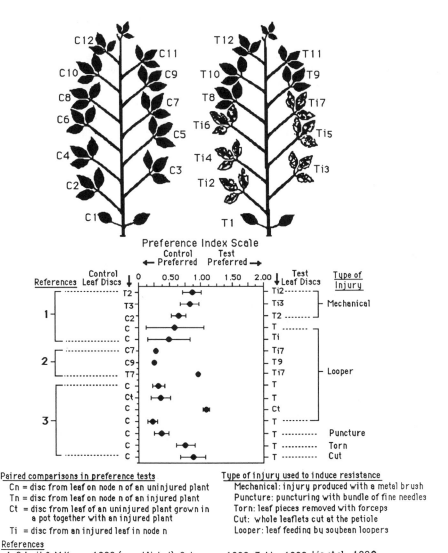

Figure 14 Scheme of dual choice preference tests comparing various treatments used to induce resistance in soybean (Williams 82). T, leaf tissue from plant previously injured (induced); C, leaf tissue from uninjured plant; C_T, leaf tissue from uninjured plant grown in same pot as previously injured plant; Ti, tissue from leaf fed upon by soybean looper larvae.

grown under normal field conditions is of even greater interest. A stand of Williams 82 was established and randomly selected 2 × 2-m plots were covered with a zippered cage at full bloom (R2 stage). Plants within the cages were infested with 4th-instar soybean looper larvae at rates of 2 or 4 per plant. Larvae were allowed to feed until pupation, at which time the reduction in leaf area was measured. Cages remained on the plots only during the period of feeding, and control plots (uninfested) were similarly caged. The acceptability of the injured plants to the Mexican bean beetle was measured by dual choice preference tests conducted at 3–4 day intervals over a period of 50 days beginning 8 days after infestation. The mean level of defoliation finally obtained was about 20% in plants infested at the 2 larvae rate and about 35% in plants infested at the 4-larvae rate. Plants from plots with a mean 35% level of defoliation reached a higher level of antixenosis and maintained this level longer (about 20 days) than plants with a 20% mean level of defoliation (Iannone 1989).

In summary, insect herbivory seems to induce a systemic response in soybean that increases the levels of antixenosis of plants to subsequent Mexican bean beetle feeding. The response is observed in greenhouse and in field-grown plants. The intensity of the response seems to increase with the severity of the injury. Insect-produced injury is more efficient than mechanically produced injury in the induction of resistance.

4.3 Postingestive Effects of Induced Resistance

Foliage from previously injured soybean plants has a modest effect on the growth and development of soybean looper larvae. The mean weight of larvae grown on previously injured plants was approximately 13% lower than that of larvae grown on uninjured plants (Amit and Kogan unpublished results). Iannone (1989) reported 19% lower pupal weights and an increase of 4 days in the duration of larval development when larvae were reared from the second stage on foliage from previously injured soybean plants. Mortality was not significantly increased but fecundity was somewhat depressed. Lin and Kogan (1990) used standard gravimetric techniques to assess the effect of induced resistance on the consumption and utilization of soybean foliage by the soybean looper and the Mexican bean beetle. Induced resistance significantly retarded development and growth of both species. Soybean looper larvae fed induced plants took 8.5% longer to complete larval development and Mexican bean beetles took 9.1% longer than when either species was reared on uninjured plants. Pupal weights were reduced 10% for the soybean looper and 16.5% for the Mexican bean beetle. However, total food consumption by either species was not affected by induced resistance. As a consequence, all nutritional indices (relative consumption rate, growth rate, efficiency of conversion of ingested and digested food, and approximate digestibility) diverged from the indices obtained when larvae fed on noninduced plants. The effects on the Mexican bean beetle were

TABLE 2 Biological and Nutritional Parameters of Soybean Loopers (SBL) (5th and 6th instars) and Mexican Bean Beetle (MBB) (3rd and 4th instars) Reared on Leaves from Previously Injured and Uninjured (control) Williams Soybeans Plants (adapted from Lin and Kogan, 1990)

	% Effect[b]	
Parameters[a]	SBL	MBB
Development time	+2.8	+9.9
Food consumed	No difference[c]	No difference
Pupal weight	−10.4	−16.5
WG	−5.9	−15.6
MBW	−5.4	−13.7
RCR	+4.3	+16.9
GR	−3.	−10.7
ECI	−7.1	−24.6
AD	No difference	+1.1
ECD	−5.7	−25.3

[a] WG, weight; MBW, mean body weight; RCR, relative consumption rate; GR, growth rate; ECI, efficiency of conversion of ingested food; AD, approximate digestibility; ECD, efficiency of conversion of digested food. All parameters computed following Kogan (1987) on indices defined in Waldbauer (1968).
[b] % Effect = [(parameter value of control − parameter value of treatment)/parameter value of control] × 100.
[c] No difference = no measurable difference between parameter values.

generally more accentuated than those obtained with the soybean looper (Table 2).

In conclusion, herbivory induces a plant response that affects both palatability of food (antixenosis) and growth and development of the insect (antibiosis). Both effects are more noticeable on the oligophagous Mexican bean beetle than on the polyphagous soybean looper.

5 CHEMICAL NATURE OF ALLOMONAL COMPOUNDS INDUCED BY HERBIVORY

Convincing evidence indicates that soybean phytoalexins and some phenolic acids have an allomonal effect on several species of phytophagous Coleoptera and Lepidoptera (see above). Furthermore, herbivory induces a plant response that results in increased levels of resistance to subsequent herbivory. Relationships between the two sets of evidence, however, are yet to be confirmed. H. Lin, M. Kogan, and D. C. Fischer (University of Illinois, unpublished results) made 80% ethanol extractions of soybean plants that had been exposed to soybean looper herbivory, and of comparable uninjured plants. Induction of resistance in the previously injured plants was confirmed by dual choice preference tests with the Mexican bean beetle and the soybean looper.

The extracts were analyzed by HPLC with UV detection (280 nm). The phytochemical profile of the extract at that wavelength revealed 12 peaks, all of which occurred in both tissues from previously injured (induced) and uninjured plants. However, four of the peaks were significantly higher in chromatograms of extracts from leaves of induced plants than those from leaves of noninduced plants. The net increases in combined areas of two peaks were correlated with the increase in induced resistance as measured by preference tests with the Mexican bean beetle. Increases in the combined areas of the four peaks were similarly correlated with soybean looper preferences. These correlations suggest that the higher peaks may relate to the presence of compounds involved in the deterrent effect of foliage from induced plants on the feeding of Mexican bean beetle and soybean looper. These peaks, however, may not correspond at all to deterrent compounds or to other resistance factors but instead may be inactive precursors of deterrent compounds produced by the biochemical machinery activated by induction. The identity of these compounds is currently under investigation and the compounds will be tested for specific activity as they become available.

Although de novo synthesis of known soybean phytoalexins did not appear to be elicited by herbivory, compounds with similar absorbance and chromatographic properties seem to be involved in the induced-resistance process. These experiments suggest an increase of previously existing compounds (constitutive) rather than de novo synthesis.

6   ENZYMATIC INHIBITION

According to Moesta and Grisebach (1982), an ideal system to test the role of phytoalexins in resistance would be a mutant in which phytoalexin synthesis is blocked at a single step. For want of such a mutant, they inhibited phytoalexin synthesis with the specific PAL inhibitor, L-2-aminooxy-3-phenylpropionic acid (L-AOPP), in soybean seedlings induced by infection with *P. megasperma* f.sp. *glycinea*. The results showed a good correlation between inhibition by L-AOPP of glyceollin accumulation in tissue at the infection site and the size of the infection area in the normally resistant Harosoy 63 soybean.

The hypothesis that the accumulation of glyceollin could be blocked by enzyme inhibition was tested in our laboratory using Williams 82 plants induced by previous herbivory (soybean looper feeding). Attempts to block the induced response of the soybean with the herbicide glyphosate, *N*-(phosphonomethyl)glycine (we actually used the commercial product Roundup, which is an isopropylamine salt of glyphosate), followed a procedure described by Simcox and Paxton (1984). Glyphosate is an inhibitor of 5-enolpyruvylshikimic acid-3-phosphate synthase that mediates the conversion of shikimate to chorismate. Cotyledons of mechanically injured Williams 82 plants were injected with 2 and 4 μL of 5 μg/mL of Roundup. Controls were glyphosate-injected uninjured plants and water-injected injured *and* uninjured plants. Injections were made concurrently with injury to plants made with bundled needles or a short-bristled metal brush. Mechanically produced wounds were coated with distilled water (controls) or a cell suspension of *P. megasperma* f.sp. *glycinea* (treatments). After a 48-h incubation period, leaves from treatment and control plants were compared using the dual choice preference tests with Mexican bean beetle adult females.

Results of several series of tests were inconclusive but tended to show that glyphosate injection did not alter acceptability of previously injured plants. In most tests, however, the level of resistance induced by mechanical injury + water or mechanical injury + elicitor was much less than the level of induction obtained in later experiments using soybean looper feeding (see above). These experiments suggest that if inhibition of glyceollin production by glyphosate occurred, it had little effect on the palatability of soybean to the Mexican bean beetle. Together with the chromatographic analyses discussed in the previous sections, results of the glyphosate experiments indicate that the antiherbivory effect of induction in soybean may involve mostly constitutive compounds that increase in concentration as a response to herbivory rather than the synthesis of novel compounds not found in the noninduced plants.

7 ECOLOGICAL EFFECTS OF INDUCIBLE DEFENSES AND THEIR POTENTIAL USE IN INTEGRATED PEST MANAGEMENT

Information on the impact of inducible defenses on the population dynamics of herbivores is scant. Edelstein-Keshet and Rausher (1989) proposed a generalized mathematical model for mobile, nonselective (polyphagous) herbivores with overlapping generations and concluded that inducible defenses may depress herbivore populations but only rarely can those defenses cause persistent population fluctuations. In our soybean system, we dealt with the nonselective and highly mobile (in the adult stage) soybean looper and the selective Mexican bean beetle. The level of deterrency of the Mexican bean beetle in induced plants was much higher than that achieved against the soybean looper. In addition, the postingestive effects of induced plants were more profound on the Mexican bean beetle than on the soybean looper. How these factors may impact on the population dynamics of either species will depend on the intensity of the induction process and the duration of the induced response in the plant.

Based on the "cost–risk–value" hypothesis (Rhoades 1979, Kogan 1986), inducible defenses are ideal for crop plants because the plants would not have to invest energy for defense unless an increase in their defensive posture were triggered by herbivore attacks or by pathogen infections. The antiherbivory and antimicrobial properties of soybean inducible defenses provide an efficient defense against multiple challenges with the advantage that a plant's overall fitness would not be reduced when challenges were absent. Furthermore, previous challenge by any single factor, that is, pathogen or herbivore, predisposes the plant to increased resistance to subsequent challenges. Finally, previous challenge at a stage of growth when injury is tolerated without reducing yield increases resistance to subsequent challenge at a stage when injury is more likely to reduce yield. For example, injury during the midvegetative stage, when effects on yield are less, would reduce injury at the midreproductive stage, when injury frequently results in lower seed yield.

The inducible response to herbivory may cause redistribution of a herbivore population within or among plants. For instance, stink bug nymphs tend to disperse along soybean rows as the nymphs grow older (Panizzi et al. 1980). This dispersal reduces competition among siblings and reduces concentration of resources at the disposal of predators. Both of these consequences are disadvantageous to the plants. On the other hand, dispersal tends to dilute the concentration of herbivores within the plant population, thus attenuating the impact of herbivory on any single plant. Finally, the pattern of colonization by herbivores of soybean fields in the Western Hemisphere may, at least hypothetically, be due to inducible defenses (Kogan 1991). Arthropod pests that have colonized soybean in the New World are characterized by high mobility, polyphagy, or preadaptive selectivity to other leguminous plants. Conspicuously absent are sedentary species or species that tend to form large colonies, such as aphids. Such species would be more vulnerable to an inducible plant defense system.

7.1 Potential Applications in Soybean Integrated Pest Management

It is apparent that there is considerable genetic diversity among soybean genotypes with regard to their capacity to respond to factors inducing resistance. Most current screening procedures in breeding programs assume static levels of constitutive resistance throughout plant development. New screening protocols that take into account inducible resistance must be devised. Such protocols should incorporate challenging the plants with elicitors of inducible resistance prior to the screening procedure. The primary objective of such a breeding program would be to identify plants with a rapid and intense response that persisted long enough to protect the plants during the most sensitive stages of plant growth.

From the standpoint of integrated pest management (IPM), understanding the influence of induced resistance in the dynamics of pest populations may lead to a reassessment of economic injury levels for early season pests. Early infestations that would render plants more resistant to subsequent infestations might prove beneficial to the crop.

Researchers have suggested that investigations of induced resistance may lead to a new line of pesticides that would behave as chemical elicitors of the defense response (Kogan and Paxton 1983, Fischer et al. 1990a). However, crop plants are always subject to environmental stresses that are natural elicitors, and it may be futile to develop a strategy based on a preventive mode with pesticide-like action. The main advantage of induced resistance is its mobilization only in the event of a challenge; if induction is triggered in a preventive way that advantage would be lost.

In conclusion, we believe that inducible defenses in soybean are effective against herbivorous insects and pathogens. The defensive capacity of the crop should be increased by classical breeding with the use of genes that regulate the intensity, response time, and duration of the induced response. After the mechanisms of chemical induction are known, the genes and gene products that mediate the biosynthesis of the active compounds can be identified. With that knowledge, current techniques of gene transfer may be used to produce new and more reactive genotypes. Concurrently with gene manipulations it will be necessary to assess the impact of induced resistance on population dynamics of pests with a realistic evaluation of the interactive effect of predators, parasitoids, and pathogens.

ACKNOWLEDGMENTS

We thank Jack Paxton and Charles Helm for a critical review of this paper, and Audrey Hodgins for editorial review. Jenny Kogan, coordinator, Soybean Insect Research Information Center, and Carla Heister, Librarian, Illinois Natural History Survey, helped with literature surveys. Research reported here was supported in part by grants from the American Soybean Association, the USDA-CSRS-CGO, Grant 86-CRCR-1-2174, and Illinois Agricultural Experiment Station Project 12-324.

REFERENCES

Bailey, J. A. 1982. Mechanisms of phytoalexin accumulation. In J. A. Bailey and J. W. Mansfield (eds), *Phytoalexins*. New York: Wiley, pp. 289–318.

Bhattacharyya, M. K., and E. W. B. Ward. 1986. Resistance, susceptibility and accumulation of glyceollins I–III in soybean organs inoculated with *Phytophthora megasperma* f. sp. *glycinea*. *Physiol. Mol. Plant Pathol.* 29:227–237.

Binder, R. G., and A. C. Waiss, Jr. 1984. Effects of soybean leaf extracts on growth and mortality of bollworm (Lepidoptera:Noctuidae) larvae. *J. Econ. Entomol.* 77:1585–1588.

Bridge, M., and W. L. Klarman. 1973. Soybean phytoalexin, hydroxyphaseollin, induced by ultraviolet irradiation. *Phytopathology* 63:606–609.

Caballero, P., C. M. Smith, F. R. Fronczek, and N. H. Fischer. 1986. Isoflavones from an insect-resistant variety of soybean and the molecular structure of afrormosin. *J. Nat. Prod.* 49:1126–1129.

Chamberlain, D. W., and J. D. Paxton. 1968. Protection of soybean plants by phytoalexins. *Phytopathology* 58:1349–1350.

Chiang, H. S., D. M. Norris, A. Ciepela, A. Oosterwyk, P. Shapiro, and M. Jackson. 1986. Comparative constitutive resistance in soybean lines to Mexican bean beetle. *Entomol. Exp. Applic.* 42:19–26.

Chiang, H. S., D. M. Norris, A. Ciepela, P. Shapiro, and A. Oosterwyk. 1987. Inducible versus constitutive PI 227687 soybean resistance to Mexican bean beetle, *Epilachna varivestis. J. Chem. Ecol.* 13:741–749.

Clark, W. J. , F. A. Harris, F. G. Maxwell, and E. E. Hartwig. 1972. Resistance of certain soybean cultivars to bean leaf beetle, striped blister beetle and bollworm. *J. Econ. Entomol.* 65:1669–1672.

Cosio, E. G., G. Weissenbock, and J. W. McClure. 1985. Acifluorfen-induced isoflavonoids and enzymes of their biosynthesis in mature soybean leaves. *Plant Physiol.* 78:14–19.

Croxford, A. C., P. J. Edwards, and S. D. Wratten. 1989. Temporal and spatial variation in palatability of soybean and cotton leaves following wounding. *Oecologia* 79:520–525.

Darvill, A. G., and P. Albersheim. 1984. Phytoalexins and their elicitors: a defense against microbial infection in plants. *Annu. Rev. Plant Physiol.* 35:243–275.

Deverall, B. J. 1982. Introduction. In J. A. Bailey and J. W. Mansfield (eds.), *Phytoalexins.* New York: Wiley, pp. 1–20.

Dreyer, D. L., R. G. Binder, B. G. Chan, A. C. Waiss, Jr., E. E. Hartwig, and G. L. Beland. 1979. Pinitol, a larval growth inhibitor for *Heliothis zea* in soybeans. *Experientia* 35:1182–1183.

Ebel, J. 1986. Phytoalexin synthesis: the biochemical analysis of the induction process. *Annu. Rev. Phytopathol.* 24:235–264.

Edelstein-Keshet, L., and M. D. Rausher. 1989. The effects of inducible plant defenses on herbivore populations. 1. Mobile herbivores in continuous time. *Am. Natur.* 133:787–810.

Ersek, T., and Z. Kiraly. 1986. Phytoalexins: warding-off compounds in plants? *Physiol. Plant.* 68:343–346.

Fett, W. F. 1984. Accumulation of isoflavonoids and isoflavone glucosides after inoculation of soybean leaves with *Xanthomonas campestris* pv. *glycines* and pv. *campestris* and a study of their role in resistance. *Physiol. Plant. Pathol.* 24:303–320.

Fett, W. F., and R. M. Zacharius. 1983. Bacterial growth and phytoalexin elicitation in soybean cell suspension cultures inoculated with *Pseudomonas* pathovars. *Physiol. Plant Pathol.* 22:151–172.

Fischer, D. C., and M. Kogan. 1986. Chemoreceptors of adult Mexican bean beetles: external morphology and role in food preference. *Entomol. Exp. Applic.* 40:3–12.

Fischer, D. C., M. Kogan, and P. Greany. 1990a. Inducers of plant resistance to insects. In E. Hodgson and R. J. Kuhr (eds.), *Safer Insecticides: Development and Use*. New York: Dekker, pp. 255–278.

Fischer, D. C., M. Kogan, and J. Paxton. 1990b. Deterrency of Mexican bean beetle (Coleoptera:Coccinellidae) feeding by free phenolic acids. *J. Entomol. Sci.* 25:230–238.

Fischer, D. C., M. Kogan, and J. Paxton. 1990c. Effect of glyceollin, a soybean phytoalexin, on feeding by three phytophagous beetles (Coleoptera:Coccinellidae and Chryrsomelidae): dose vs. response. *Environ. Entomol.* 19:1278–1282.

Floss, H. G. 1980. The shikimate pathway. In T. Swain, J. B. Harborne and C. F. Van Sumere (eds.), *Recent Advances in Phytochemistry*, Vol. 12, *Biochemistry of Plant Phenolics*. New York: Plenum, pp. 59–89.

Frank, J. A., and J. D. Paxton. 1971. An inducer of soybean phytoalexin and its role in the resistance of soybeans to *Phytophthora* rot. *Phytopathology* 61:954–958.

Gardner, W. A., D. V. Phillips, and A. E. Smith. 1984. Effect of pinitol on the growth of *Heliothis zea* and *Trichoplusia ni* larvae. *J. Agric. Entomol.* 1:101–105.

Grunwald, C., and M. Kogan. 1981. Sterols of soybeans differing in insect resistance and maturity group. *Phytochemistry* 20:765–768.

Hahlbrock, K., and H. Grisebach. 1979. Enzymatic controls in the biosynthesis of lignin and flavonoids. *Annu. Rev. Plant Physiol.* 30:105–130.

Hanson, K. R., and E. A. Havir. 1979. An introduction to the enzymology of phenyl propanoid biosynthesis. In T. Swain, J. B. Harborne, and C. F. Van Sumere (eds.), *Recent Advances in Phytochemistry*, Vol 12, *Biochemistry of Plant Phenolics*. New York: Plenum, pp. 91–137.

Harborne, J. B. 1980. Plant phenolics. In E. A Bell and B. V. Charlwood (eds.), *Secondary Plant Products*. Berlin: Springer, pp. 330–402.

Hardin, J. M. T. 1979. Phenolic acids of soybeans resistant and nonresistant to leaf-feeding larvae. M. S. Thesis, University of Arkansas.

Hardin, J. M., and C. A. Stutte. 1980. Analyses of phenolic and flavonoid compounds by high pressure liquid chromatography. *Anal. Biochem.* 102:171–175.

Hart, S. V., M. Kogan, and J. Paxton. 1983. Effect of soybean phytoalexins on the herbivorous insects Mexican bean beetle and soybean looper. *J. Chem. Ecol.* 9:657–672.

Huang, H. M., G. L. Johanning, and B. L. Odell. 1986. Phenolic-acid content of food plants and possible nutritional implications. *J. Agric. Food Chem.* 34:48–51.

Iannone, N. 1989. Soybean resistance induced by previous herbivory: extent of the response, levels of antibiosis, field evidence and effect on yield. M. S. Thesis, University of Illinois, Urbana-Champaign.

Ingham, J. L. 1982. Phytoalexins from the Leguminosae. In J. A. Bailey and J. W. Mansfield (eds.), *Phytoalexins*. New York: Wiley, pp. 21–80.

Ingham, J. L., N. T. Keen, L. J. Mulheirn, and R. L. Lyre. 1981. Inducibly-formed isoflavonoids from leaves of soybean (*Glycine max*). *Phytochemistry* 20:795–798.

Keen, N. T., and B. W. Kennedy. 1974. Hydroxyphaseollin and related isoflavanoids in the hypersensitive reaction of soybean to *Pseudomonas glycinea*. *Physiol. Plant Pathol.* 6:173–185.

Keen, N. T., and M. Yoshikawa. 1983. Physiology of disease and the nature of resistance to *Phytophthora*. In D. C. Erwin, S. Bartnicki-Garcia, and P. H. Tsao (eds.). *Phytophthora: Its Biology, Taxonomy, Ecology and Pathology.* St. Paul, MN: Am. Phytopathol. Soc., pp. 279–287.

Keen, N. T., M. Yoshikawa, and M. C. Wang. 1983. Phytoalexin elicitor activity of carbohydrates from *Phytophthora megasperma* f.sp. *glycinea* and other sources. *Plant Physiol.* 71:466–471.

Kogan, M. 1972. Intake and utilization of natural diets by the Mexican bean beetle, *Epilachna varivestis*: a multivariate analysis. In J. G. Rodriguez (ed.), *Insect and Mite Nutrition: Significance and Implications in Ecology and Pest Management.* Amsterdam: North Holland, pp. 107–126.

Kogan, M. 1973. Automatic recordings of masticatory motions of leaf-chewing insects. *Ann. Entomol. Soc. Am.* 66:66–69.

Kogan, M. 1986. Natural chemicals in plant resistance to insects. *Iowa State J. Res.* 60:501–527.

Kogan, M. 1987. Bioassays for measuring quality of insect food. In J. Miller and T. Miller (eds.), *Insect–Plant Interactions.* New York: Springer, pp. 155–190.

Kogan, M. 1989. Plant resistance in soybean insect control. In A. J. Pascali (ed.), *Proceedings of the World Soybean Research Conference-IV.* Buenos Aires, Argentina, pp. 1519–1525.

Kogan, M. 1991. Contemporary adaptations of herbivores to introduced legume crops. In P. W. Price et al. (eds.), *Plant Animals Interactions: Evolutionary Ecology in Tropical and Temperate Regions.* New York: Wiley.

Kogan, M., and R. Goeden. 1972. The host-plant range of *Lema trilineata daturaphila* (Coleoptera: Chrysomelidae). *Ann. Entomol. Soc. Am.* 63:1175–1180.

Kogan, M., and J. Paxton. 1983. Natural inducers of plant resistance to insects. In P. A. Hedin (ed.), *ACS Symposium Series*, Vol. 208, *Plant Resistance to Insects.* Washington, DC: American Chemical Society, pp. 153–171.

Kogan, M., and S. G. Turnipseed. 1987. Ecology and management of soybean arthropods. *Annu. Rev. Entomol.* 32:507–538.

Layton, M. B., D. J. Boethel, and C. M. Smith. 1987. Resistance to adult bean leaf beetle and banded cucumber beetle (Coleoptera:Chrysomelidae) in soybean. *J. Econ. Entomol.* 80:151–155.

Levin, D. A. 1971. Plant phenolics: an ecological perspective. *Am. Natur.* 105:157–181.

Lin, H. 1989. Induced resistance in soybean: studies with the Mexican bean beetle and the soybean looper. PhD. Thesis, University of Illinois, Urbana-Champaign.

Lin, H., and M. Kogan. 1990. Influence of induced resistance in soybean on the development and nutrition of the soybean looper and the Mexican bean beetle. *Entomol. Exp. Applic.* 55:131–138.

Lin, H., M. Kogan, and D. Fischer. 1990. Induced resistance in soybean to the Mexican bean beetle (Coleoptera:Coccinellidae): comparison of inducing factors. *Environ. Entomol.* 19:1852–1857.

Liu, S. H, D. M.Norris, and E. Marti. 1988. Behavioral response of female adult

Trichoplusia ni to volatiles from soybean versus a preferred host, lima bean. *Entomol. Exp. Applic.* 49:99–109.

Long, M., P. Barton-Willis, B. J. Starkawicz, D. Dahlbeck, and N. T. Keen. 1985. Further studies on the relationship between glyceollin acummulation and the resistance of soybean leaves to *Pseudomonas syringae* pv. *glycinea. Phytopathology* 75:235–239.

Luedders, V. D., and W. A. Dickerson. 1977. Resistance of selected soybean genotypes and segregating populations to cabbage looper feeding. *Crop Sci.* 17:395–397.

Mansfield, J. W. 1982. The role of phytoalexins in disease resistance. In J. A. Bailey and J. W. Mansfield (eds.), *Phytoalexins.* New York: Wiley, pp. 252–288.

Moesta, P., and H. Grisebach. 1982. L-2-Aminooxy-3-phenylpropionic acid inhibits phytoalexin accumulation in soybean with concommitant loss of resistance against *Phytophthora megasperma* f.sp. *glycinea. Physiol. Plant Pathol.* 21:65–70.

Norris, D. M, H. S.Chiang, A. Ciepiela, Z. R. Khan, H. Sharma, F. Neupane, N. Weiss, and S. Liu. 1988. Soybean allelochemicals affecting insect orientation, feeding, growth, development and reproductive processes. In F. Senhal, A. Zabza, and D. L. Denlinger (eds.), *Endocrinological Frontiers in Physiological Insect Ecology.* Wroclaw, Poland: Wroclaw Technical University Press, pp. 27–31.

Panizzi, A. R., M. H. M. Galileo, H. A. O. Gastal, J. F. F. Toledo, and C. H. Wild. 1980. Dispersal of *Nezara viridula* and *Piezodorus guildinii* nymphs in soybeans. *Eviron. Entomol.* 9:293–297.

Paxton, J. D. 1983. *Phytophthora* root and stem rot of soybean. In J. A. Gallow (ed.), *Biochemical Plant Pathology.* London: Wiley, pp. 19–27.

Phillips, D. V., D. F. Dougherty, and A. E. Smith. 1982. Cyclitols in soybean. *J. Agric. Food Chem.* 30:456–458.

Porter, P. M., W. L. Banwart, and J. J. Hassett. 1986. Phenolic acids and flavonoids in soybean (*Glycine max*) cultivar Forrest root and leaf extracts. *Environ. Exp. Bot.* 26:65–74.

Reynolds, G. W., and C. M. Smith. 1985. Effects of leaf position, leaf wounding, and plant age of two soybean genotypes on soybean looper (Lepidoptera:Noctuidae) growth. *Environ. Entomol.* 14:475–478.

Rhoades, D. F. 1979. Evolution of plant chemical defense against herbivores. In G. A. Rosenthal and D. H. Janzen (eds.), *Herbivores: Their Interactions with Secondary Plant Metabolites.* New York: Academic, pp. 3–52.

Robinson, T. 1980. *The Organic Constituents of Higher Plants.* Amherst, MA: Cordus.

Rose, R. L., T. C. Sparks, and C. M. Smith. 1989. The influence of resistant soybean (PI 227687) foliage and coumestrol on the metabolism of xenobiotics by the soybean looper, *Pseudoplusia includens* (Walker). *Pest. Biochem. Physiol.* 34:17–26.

Rossetto, D., A. S. Costa, M. A. Miranda, V. Nagai, and E. Abramides. 1977. Diferenças na oviposição de *Bemisia tabaci* em variedades de soja. *An. Soc. Entomol. Brasil* 6:256–263.

Seo, A., and C. V. Morr. 1984. Improved high performance liquid chromatographic analysis of phenolic acids and isoflavonoids from soybean protein products. *Agric. Food Chem.* 32:530–533.

Simcox, K. D., and J. D. Paxton. 1984. Induced pathogenicity in glyphosate-treated soybeans. *Phytochemist ry* 76:1271.

Sinclair, J. B. 1982. *Compendium of Soybean Diseases*. St. Paul, MN: Amer. Phytopathol. Soc.

Sirisingh, S., and M. Kogan. 1982. Insects affecting soybeans in storage. In J. B. Sinclair and J. A. Jackobs (eds.), *Soybean Seed Quality and Stand Establishment*. Proceedings of a Conference for Scientists of Asia. Colombo, Sri Lanka, Jan. 25–31, 1981. College of Agriculture, University of Illinois at Urbana-Champaign, INTSOY Series, pp.77–82.

Smith, C. M. 1985. Expression, mechanisms and chemistry of resistance in soybean, *Glycine max* L. (Merrill) to soybean looper, *Pseudoplusia includens* (Walker). *Insect Sci. Appl.* 6:243–248.

Smith, C. M., and N. H. Fischer. 1983. Chemical factors of an insect-resistant soybean genotype affecting growth and survival of the soybean looper. *Entomol. Exp. Applic.* 33:343–345.

Stossel, A. 1982. Biosynthesis of phytoalexins. In J. A. Bailey and J. W. Mansfield (eds.), *Phytoalexins*. New York: Wiley, pp. 133–180.

Talekar, N. S., H. R. Lee, and Suharsono. 1988. Resistance of soybean to four defoliator species in Taiwan. *J. Econ. Entomol.* 81:1469–1473.

Tester, C. F. 1977. Constituents of soybean cultivars differing in insect resistance. *Phytochemistry* 16:1899–1901.

Turnipseed, S. G. 1977. Influence of trichome variations on populations of small phytophagous insects in soybean. *Environ. Entomol.* 6:815–817.

Turnipseed, S., and M. Kogan. 1987. Integrated control of insect pests. In J. R. Wilcox (ed.), *Soybeans: Improvement, Production, and Uses*, 2d ed. *Agronomy Ser. 16*, Madison, WI: Am. Soc. Agronomy, pp. 779–817.

Van Duyn, J. W., S. G. Turnipseed, and J. D. Maxwell. 1971. Resistance in soybeans to the Mexican bean beetle. I. Sources of resistance. *Crop Sci.* 11:572–573.

Waldbauer, G. 1968. The consumption, digestion and utilization of food by insects. *Adv. Insect Physiol.* 5:229–288.

Yoshikawa, M., K. Yamauchi, and H. Masago. 1978. Glyceollin: its role in restricting fungal growth in resistant soybean hypocotyls infected with *Phytophthora megasperma* var. *sojae*. *Physiol. Plant Pathol.* 12:73–82.

CHAPTER 16

BACTERIALLY INDUCED CHANGES IN THE COTTON PLANT–BOLL WEEVIL PARADIGM

J. H. BENEDICT
Texas Agricultural Experiment Station, Texas A&M University, Corpus Christi, TX

J. F. CHANG
Phero Tech, Inc., Vancouver, British Columbia, Canada

1   Introduction
2   Background
3   Boll weevil responses to bacterially treated cotton
    3.1   Field free-choice studies of plant damage and yield
    3.2   Boll weevil preference, survival, and reproductive biology
    3.3   Boll weevil pheromone production
4   Plant semiochemical responses to bacterial treatment
    4.1   Airborne monoterpenes from flower buds and leaves
    4.2   Terpene aldehydes and tannins in flower bud and leaf tissues
5   Conclusions and prospects
    References

1   INTRODUCTION

Phytopathologists recognize that pathogen attack on a plant can induce resistance to further attack (Agrios 1988). The commonly recognized mechanism for this resistance is a series of physiochemical changes in the plant, of which the rapid production and accumulation of secondary substances is primary (Bell 1981, Kuć 1981, 1987). In some plant species certain of the newly synthesized secondary substances and constitutive substances are active against both pathogens and arthropods (McIntyre et al. 1981, Hare 1983, Bell 1986, Karban et al. 1987, Hammond and Hardy 1988). Similarly, certain forms of wounding by arthropods can induce plants to resist pathogen and arthropod attack (Karban et al. 1987, Karban, this volume, Chapter 17).

379

For example, an aphid infestation on one leaf induced resistance to the plant pathogen *Colletotrichum lagenarium* in a different leaf (Kuc 1981). These observations indicate that in plants there may exist a common stimulus for inducing resistance and a common mechanism to resist pathogen and insect attack (Ryan 1983a).

A second form of induced resistance is immunization (Kuć 1987). It has been known for some time that resistance to pathogens can be induced in plants by an immunizing agent (Matta 1971). Kuć (1981, 1982, 1987) has demonstrated that certain economic plant species can be artificially immunized against future attack from virulent plant pathogens. This is accomplished using a topical spray or injection of an avirulent pathogen or other elicitor. Immunization-induced resistance in plants is systemic and may endure from several weeks to the life of the plant. Plants that are immunized against pathogens do not have elevated levels of secondary substances compared to nonimmunized plants (Kuć 1987). This is a clear distinction from the elevated levels of secondary substances commonly associated with plant wounding by insects or pathogens (Ryan 1983a,b, Kuc and Rush 1985). Immunization of plants against future attack by pathogens is thought to involve an elicitor that systemically sensitizes the plant to respond more rapidly to pathogen attack than do nonimmunized plants (Kuć 1982). There is evidence (Kuć 1981, Ryan 1983b) that suggests that the elicitors of immunization-induced resistance are not unique constituents common only to fungi, bacteria, and viruses, but rather the increased resistance is due to a common stimulus, that is, an elicitor that can exert low level, persistent metabolic stress on the plant.

A lesser known mechanism of plant defense is possessed by certain plants that grow in association with mutualistic microorganisms (Dickinson and Preece 1976, Windels and Lindow 1985). These organisms may actually defend the plant by competition with or exploitation of pathogens (Blakeman and Fokkema 1982). Most plants in natural and agricultural systems support microorganisms that behave as mutualistic or symbiotic endo- and exoflora in, on, or near plant surfaces (Dickinson and Preece 1976, Blakeman and Fokkema 1982).

In agriculture, pathogen injury to crop plants can be reduced and yields increased through a process known as bacterization (Brown 1974, Burr and Caesar 1984). Bacterization is the artificial application of certain mutualistic bacteria to seeds, tubers, roots, or aboveground plant parts. It results in a stimulation of plant growth in typical field soil compared to untreated plants. The best known of these beneficial bacteria are the plant-growth promoting rhizobacteria (PGPR). Beneficial bacteria are associated with the phylloplane (aboveground plant surfaces) as well as the rhizoplane (below-ground plant surfaces). Researchers studying PGPR are uncertain as to how PGPR bacteria promote plant growth. They may promote plant growth by producing antibiotics that inhibit growth and reproduction of pathogens (i.e., antagonism) or by acting as biological control agents against plant pathogens in the root area (Burr and Caesar 1984, Suslow and Schroth 1982a,b). In addition to these mechanisms, certain mutualistic PGPR, such as *Bacillus megaterium*, have been discovered that produce plant growth hormone-like substances that are similar to gibberellin in their effect on aerial plant parts (Brown 1974).

We have studied the yield and secondary substance responses of the cotton plant, *Gossypium hirsutum*, to sprays of mutualistic bacteria. We have also studied the behavior, survival, and reproductive biology of a cotton herbivore, the boll weevil, *Anthonomus grandis*, to these bacterially treated plants. The purpose of this chapter is to briefly highlight our significant findings, and propose mechanisms of interaction for this microbe–plant–insect paradigm. We also speculate on the utilization of these microbes as inducers of resistance in production agriculture and attempt to show cause and effect relationships.

2 BACKGROUND

During the past 10 years we have cooperated with the Texas A&M University Multi-Adversity Resistance Breeding Program (El-Zik and Thaxton 1989) in evaluating, advancing, and releasing several cotton varieties with improved resistance to insects and pathogens (Benedict 1980, Masud et al. 1981, Bird et al. 1983a, Zummo et al. 1983, 1984b). These varieties have improved resistance to both plant pathogens and insect pests, suggesting a common mechanism for resistance (Bird et al. 1979). This cross-resistance is particularly noteworthy because the breeding system employed direct selection for improvement in only a few seed and seedling traits (Bird et al. 1983b). The selection traits were:

1. Resistance of the seed coat to deterioration under low temperature and high humidity
2. The rate of germination of these stressed seeds at cold temperature
3. Resistance to seedling pathogens
4. Resistance to races of the bacterial blight pathogen, *Xanthomonas campestris* pv. *malvacearum*.

Analysis of exudates from seed and seedling roots revealed that the electrical resistance of exudates, relative amounts of exudates, and concentrations of sodium, calcium, potassium, magnesium, and carbohydrates were correlated with the levels of resistance to pathogens, earliness of fruit production, and total yield (Bird et al. 1979, Bush 1979, El-Zik and Thaxton 1989). Bird and coworkers (Bush 1979, Bird et al. 1980, 1981, El-Zik and Thaxton 1989) also found that differences in rhizoplane–rhizosphere actinomycete and bacteria species composition of 18-, 30-, and 55-day-old plants were correlated to characteristics and quantity of seed and root exudates. Exudate components and the microflora composition were also significantly correlated with resistance to *Phymatotrichum* root rot, *Verticillium* wilt, certain seedling diseases, earliness of fruit set, and final yield. On these bases Bird et al. (1979) theorized that cotton has the genetic potential to control quality and quantity of exudates from the seed coat, roots, and above-ground tissues. Moreover, these exudates provide a nutritional base that can be selectively favorable for mutualistic actinomycetes and bacteria, and unfavorable for cotton pathogens, depending upon cotton genotype.

Bird et al. (1979) also found that some cotton genotypes that harbored beneficial root microflora appeared to be more resistant to yield loss from cotton fleahopper, *Pseudomoscelis seriatus*, bollworm, *Heliothis zea*, and boll weevil. They found that three bacteria from leaf and flower bud tissue were consistently associated with plant genotypes resistant to insects (Bird et al. 1979, 1981). The bacteria were identified as *Bacillus megaterium*, *B. cereus*, and *B. licheniformis* (Howell et al. 1987). *Bacillus megaterium* is a known soil inhabitant and PGPR organism (Brown 1974, Burr and Caesar 1984). Also, *B. megaterium* and *B. cereus* are known antagonists and competitors of some pathogens on aerial plant parts (Brown, 1974, Blakeman and Brodie 1976, Doherty and Preece 1978, Burr and Caesar 1984). Bird et al. (1979, 1980, 1981), El-Zik et al. (1985), and Cook et al. (1987) have clearly demonstrated in bacterization and competition studies with cotton that these bacteria play a role in resistance to cotton pathogens, although the actual mechanism is uncertain.

In 1978, investigations were initiated to determine if spray applications of *B. cereus* and *B. megaterium* to aerial parts of cotton could increase the level of resistance to boll weevil in different cotton varieties growing under normal farming practices (Benedict et al. 1979). These two microbial mutualists were isolated from the leaves and flower buds of the cotton genotype, CAMD-E. In these 1978 studies we clearly showed that bacterial sprays could significantly alter boll weevil damage to flower buds and yields compared to unsprayed cotton (Benedict et al. 1979).

3 BOLL WEEVIL RESPONSES TO BACTERIALLY TREATED COTTON

3.1 Field Free-Choice Studies of Plant Damage and Yield

Cotton plants grown under normal field conditions and management practices were sprayed with a mixture of *B. cereus* and *B. megaterium* in a 2:1 ratio, respectively. Field plots were sprayed with low rates of approximately 10×10^7 total colony forming units (cfu) of bacteria in 5–10 mL of distilled water per foot of crop row, and high rates of about 20×10^7 cfu of bacteria in 5–10 mL of water per foot. In tests where no boll weevils were present

TABLE 1   Cotton Yields and Numbers of Flower Buds Damaged by Boll Weevils from Cotton in Combined Field Tests with or without Boll Weevils Present, and with or without Topical Sprays of Mutualistic Bacteria, 1978–1987, Corpus Christi, Texas

	Mean Fiber Yields (lb/acre)[a]		
Treatments	Fields without Boll Weevil	Fields with Boll Weevil	Mean Buds Damaged[b]
Untreated	501	500	15.5
Treated at low rate	490	512	12.5
Treated at high rate	509	570	11.2

[a] Data were combined for four tests (over 3 years) without boll weevil present to reduce yields, and three tests (over 2 years) with boll weevil present. No significant differences were found when data from individual tests were combined ($p \leq .05$, Duncan's Multiple Test) due to differences between tests.
[b] Mean buds damaged represents the average flower bud damage by boll weevil feeding and oviposition.

we found no increase in yields with bacterial treatments (Table 1) (Benedict et al. 1980) and concluded that the bacteria did not promote yield increases as has been found in other crops (Burr and Caesar 1984).

Applications of these microorganisms were made at different stages of plant growth, rates, and frequencies. Bacteria applied to plants at the first flower bud growth stage (ca. 60–75 days after planting) enhanced yield (maximum of 40%) and reduced boll weevil damage to flower buds the most (maximum of 65%, $p \leq .05$, in 3 tests) (Benedict et al. 1979, 1980, Benedict and Bird 1981). However, when data from comparable tests were combined, no significant differences were found in yields or boll weevil damage to flower buds, although there was a trend toward increased yield and reduced damage (Table 1). Yield changes for bacterially treated cotton in individual tests (total of five tests) ranged from -13% to $+40\%$, and boll weevil flower bud damage ranged from $+54\%$ and -65% compared to the unsprayed control. Yield and/or boll weevil damage were significantly altered by bacterial application in four of the five tests. In some tests we observed that shortly after bacterial treatments were applied (10–14 days later) boll weevils would damage untreated plants more frequently than treated plants. However, about 30–40 days after treatment, the trend would change and boll weevils would show no preference for either bacterially treated or untreated plants. In some cases, weevils preferred treated cotton more than untreated.

Several mechanisms can be proposed to explain why applications of bacteria alter plant preference and suitability to boll weevil, or any insect herbivore. One mechanism could be that, when consumed, the microbial mutualists induce changes in the insect. For example, the consumed microbes may produce a toxin, or exploit boll weevil gut flora in such a way as to reduce boll weevil fitness and/or plant preference. A second mechanism could be that microbes induce changes in the plant's secondary and/or primary metabolism in ways that alter boll weevil fitness and/or preference.

3.2   Boll Weevil Preference, Survival, and Reproductive Biology

The first question addressed was whether the two *Bacillus* spp. used in the topical applications to plants were actually toxic to boll weevils. To examine this question, an artificial cottonseed diet (Gast and Davich 1966) was prepared containing both microbes in a ratio of 2:1 as in field sprays. The live bacteria were incorporated (9×10^7 total cfu of bacteria/mL of diet) in the diet just as the agar began to gel (ca. 35 °C). Freshly laid boll weevil eggs were placed on the diet and allowed to develop to adulthood. The bacteria did not adversely affect boll weevil larval development. In fact, weevils reared on diet containing bacteria produced significantly heavier adults (13 mg per adult for the control diet vs. 16 mg for the diet with bacteria).

Next we asked how survival of adult boll weevils would be affected when beetles were fed flower buds freshly harvested from bacterially treated plants grown under normal field conditions. Mortality was high for adults fed flower buds from bacterially treated plants (Table 2). The trend toward increased adult mortality was also shown in a preference experiment of free-choice/no-choice feeding and oviposition behavior (Table 3). Adult boll weevils survived significantly longer under free-choice conditions where they could feed on flower buds from treated and untreated plants. Furthermore, boll weevils fed more on flower buds from untreated plants under free-choice conditions than under no-choice conditions. Fecundity was not significantly affected by any treatment, although there was a trend ($p \leq .15$, Student's t test) for reduced oviposition in flower buds from treated plants. These data suggest that bacterial treatment may have

TABLE 2   Percent Mortality of Boll Weevils Fed for 12 and 14 Days on Flower Buds from SP-37H Plants Treated with Two Mutualistic Bacteria[a]

	Percent Mortality[b]	
Treatment[c]	12 Days	14 Days
Untreated	11.6 b	19.2 b
Treated with bacteria at low rate	42.8 a	48.8 a
Treated with bacteria at high rate	50.3 a	54.7 a

[a] Flower buds were collected from cotton grown in field plots beginning 5 days after bacterial treatment, Corpus Christi, Texas, 1987.
[b] Means followed by the same letter are not significantly different ($p \leq .05$, Duncan's New Multiple Range Test).
[c] Bacteria were sprayed on 27 May at the first-$\frac{1}{3}$-grown flower bud stage, 62 days after planting. The low rate received 31.1×10^7 cfu of bacteria and the high rate received 43.2×10^7 cfu of bacteria in 8 mL of distilled water per row foot.

TABLE 3   Adult Boll Weevil Fecundity, Total Feeding Punctures, and Survival[a] in a No-Choice/Free-Choice Test in the Laboratory with Bacterially Treated and Untreated SP-37 Cotton, Corpus Christi, Texas, 1979

	Free Choice[b]			No Choice[c]		
Treatment	Fecundity[d] per Female	Total Feeding Punctures	Days Male and Female Survival	Fecundity[d] per Female	Total Feeding Punctures	Days Male and Female Survival
Untreated	26.4 a	62.1 a		31.9 a	39.1 a	12.0 a
			15.1[e]			
Treated	23.8 a	44.7 b		18.7 a	39.3 a	10.5 a

[a] Means within columns followed with a different letter are significantly different ($p \leq .05$; Student's t test).
[b] Free-choice was 10 replicates, each replicate consisting of 8 flower buds of each treatment (16 total), and one pair of boll weevils (male and female) 6 days old at the start of the test, in a 9-in. glass petri dish.
[c] No-choice was 10 replicates of each treatment, each replicate consisting of 16 flower buds, and one pair of boll weevils (male and female) 6 days old at start of the test, in a 9-in. glass petri dish.
[d] Fecundity was measured as the total number of eggs inserted in buds of each treatment per female lifetime.
[e] The days of survival for boll weevils under free-choice conditions were significantly greater than the days of survival for boll weevils feeding on the treated and untreated flower buds under no-choice conditions ($p \leq .05$, Duncan's New Multiple Range Test).

1. Induced plant biosynthesis of a new allelochemical(s)
2. Enhanced biosynthesis of a constitutive allelochemical(s)
3. Caused the microbial flora to synthesize a toxic factor
4. Reduced plant nutrition in treated flower buds compared to untreated buds
5. Produced a combination of these effects.

We then compared boll weevil egg hatch and larval survival in treated and untreated flower buds. The boll weevil female bores a short tunnel through the calyx and corolla into the anthers of a developing flower bud where she deposits a single egg and seals the opening with a frass plug (Sterling and Adkisson 1971, Cross 1973). Throughout its development, the larva feeds on the pollen in a single bud. Infested flower buds usually abscise from the plant and fall to the ground at 7–10 days after oviposition. To duplicate this sequence of events in the field, tunnels were mechanically bored into flower buds of bacterially treated and untreated cotton and implanted with newly laid boll weevil eggs. Significantly less larval mortality occurred in flower buds from bacterially treated buds than buds from untreated plants (Table 4). These results (Tables 2–4) suggest that buds from treated plants produced a phytochemical that was toxic to boll weevil adults on or in the calyx or corolla, but not in the anthers, or was present throughout the flower bud.

Feeding and oviposition results in Tables 2 and 3 indicate flower buds from untreated plants were more attractive for feeding and oviposition. To

TABLE 4 Days[a] from Boll Weevil Egg Implantation to Flower Bud Abscission and Adult Emergence, Weight of Emerging Adults, and Percent Mortality for Boll Weevil Life Stages, from Eggs Artificially Implanted in $\frac{1}{3}$-Grown Flower Buds of Bacterially Treated and Untreated SP-37 Cotton in the Field, Corpus Christi, Texas, 1979

Treatment	Days from Egg Implantation to Bud Abscision	Days from Egg Implantation to Adult Emergence	% Mortality of Starting Egg Cohort			% Emerging as Adults	Weight (mg) of Emerging Adult	
			(n) Eggs	Larvae	Pupae		Female	Male
Untreated	10.1 a	16 a	(66) 52 a	16 a	3.0 a	25 a	10 a	10 a
Treated	8.7 a	16 a	(64) 55 a	5 b	0.1 b	35 a	10 a	10 a

[a] Means within columns followed by different letters are significantly different from each other ($p \leq .05$, Student's t test).

measure boll weevil olfactory responses to the flower buds of bacterially treated and untreated plants, a multiple-choice, pedestrian olfactometer was utilized (Chang et al. 1986). This was a circular, central chamber attached to four equally spaced peripheral chambers. The air flow was through the four smaller peripheral chambers containing the flower buds, into the middle of the larger central chamber, and out of the system. Boll weevils walked from the middle of the central chamber to the flower buds of their choice by following the odor trail. The percentage of weevils responding to the odor of buds tended to be higher for buds from the high rate of treatment (Table 5). For the genotype SP-37H tested in 1987, we found significantly fewer boll weevils responding to buds from the low treatment rate compared to buds from the high rate of bacterial treatment, suggesting that bacterial dose and cotton genotype can affect olfactory attractiveness of cotton flower buds to boll weevils.

3.3 Boll Weevil Pheromone Production

Male boll weevils produce a volatile, four-component pheromone, grandlure, which acts as a sex attractant to females and an aggregant to other males

TABLE 5 Percentage of Boll Weevils Responding in a Multiple-Choice, Pedestrian Olfactometer to Flower Buds from Two Varieties of Bacterially Treated Cotton[a]

Treatment	Percent Responding[b]	
	SP-37H	SP-37
Control (empty)	0.0 c	0.0 c
Untreated	36.4 a	28.0 b
Treated at low rate	14.7 b	33.4 ab
Treated at high rate	49.0 a	41.1 a

[a] The varieties were tested separately in the olfactometer, Corpus Christi, Texas.
[b] Each replication was composed of 20 weevils and three flower buds of each treatment, placed in a multiple-choice olfactomer; the reponse was observed after 20 min. The SP-37H data are from 1987 and the SP-37 from 1985. Means within columns followed by the same letter are not significantly different ($p \leq .05$, Duncan's New Multiple Range Test).

in plant colonization (Hardee et al. 1972, Hedin 1977, Tumlinson et al. 1969). The four components are two terpenoid aldehydes and two terpenoid alcohols which are produced and released in association with frass as the males feed on pollen in flower buds. Apparently, many dietary constituents can serve as precursors for the biosynthesis of pheromone (Hardee 1970, Thompson and Mitlin 1979). When flower buds from different cotton genotypes grown under field conditions were fed to male boll weevils, the pheromone component ratios and amounts of each component were significantly altered depending upon cotton genotype (Chang et al. 1988c, 1989). The cause of these differences is not known. There is some evidence that volatile terpenes can be inhaled by male boll weevils and serve as precursors for pheromone synthesis (Hedin 1977). In terms of mate location, the influence of diet on pheromone quality may be very important (Hendry 1976). Our studies (Chang 1986, Chang et al. 1987, Benedict and Chang 1988) and those of others (Bull et al. 1973, Hardee et al. 1972) clearly demonstrate that the ratios and doses of each pheromone component in the blend influence boll weevil responses. Furthermore, we have found that component ratios and doses of plant monoterpene volatiles significantly influence boll weevil attraction to grandlure (Chang et al. 1987). Certain monoterpenes were found under some conditions to synergize male and female weevil attraction to a point source in field populations.

These findings led us to investigate the response of pheromone component ratios and rates of production when weevils were fed on flower buds from bacterially treated plants. Utilizing methods developed by Chang et al. (1989), we studied pheromone produced by boll weevils 6–14 days after adult emergence. We selected this range because boll weevils do not produce detectable amounts of pheromone until day 6 of adulthood (Chang et al. 1989).

We found that pheromone production was significantly increased by feeding on buds from bacterially treated CAMD-E plants (Table 6) (Chang et al. 1989). Differences were also found in component ratios and amounts of pheromone for weevils fed on buds from untreated and treated plants. In another

TABLE 6   Boll Weevil Pheromone Produced (μg/weevil)[a] on Days 6 to 14 after Adult Emergence When Feeding on Flower Buds from Bacterially Treated and Untreated CAMD-E Plants, College Station, Texas, 1985 (modified after Chang et al. 1988a)

	Days after Adult Emergence					
Treatment	6	8	10	12	14	Total[b]
Untreated	261 a	940 b	1663 a	3269 b	2413 b	15329 b
Treated	389 a	1589 a	2519 a	5079 a	2858 a	21692 a

[a] Means within columns followed by different letters are significantly different ($p \leq .05$, Student's t test).

[b] The total represents the sum for days 6, 8, 10, 11, 12, 13, and 14 after adult emergence.

TABLE 7 Boll Weevil Pheromone Production (μg/weevil)[a] on Flower Buds of SP-37H, 26 June, 30 Days after the First Treatment with Bacteria, Corpus Christi, Texas, 1987 (modified from Chang et al. 1988a)

Treatment	Pheromone Components[b] (μg/weevil)				
	I	II	III	IV	Total
Untreated	247 a	216 a	64 a	64 a	590 a
Treated	117 b	107 b	29 a	31 a	293 b

[a] Means within columns followed by different letters are significantly different ($p \le .10$, Student's t test).

[b] Pheromone components and the recovery efficiency (%) from the air surrounding feeding boll weevils were I, (+)-*cis*-2-isopropenyl-1-methylcyclobutaneethanol (85.5%); II, *cis*-3,3-dimethyl-Δ1-α-cyclohexaneethanol (82.0%); III, *cis*-3,3-dimethyl-Δ1-β-cyclohexaneacetaldehyde (90.7%); IV, *trans*-3,3-dimethyl-Δ1-β-cyclohexaneacetaldehyde (89.4%).

study, weevils produced significantly more pheromone when fed on SP-37H buds from untreated plants than when fed on treated buds (Table 7). We do not know the reason for these differences in pheromone production when boll weevils consume flower buds of one bacterially treated genotype compared to another.

4 PLANT SEMIOCHEMICAL RESPONSES TO BACTERIAL TREATMENT

Cotton is rich in secondary plant metabolites (Table 8) (Bell 1984, 1986), some of which are known to be biologically active against boll weevil (Hedin et al. 1975, 1976, 1977, Dickins 1984, Chang 1986, Chang et al. 1986, 1988a,b,c). Certain monoterpenes and the sesquiterpene, bisabolol, have been identified as semiochemicals, kairomones that are active olfactory attractants to adult boll weevil. A large portion of the odor of cotton is produced by a blend of monoterpenes that varies with cotton genotype, physical environment, plant age, and other factors (Chang et al. 1988b). Differences in boll weevil responses to bacterially treated cotton caused us to ask how bacterial treatment affects the monoterpene odor surrounding cotton plants. The effect of bacterial treatment on the concentrations of nonvolatile terpene aldehydes and condensed tannins in tissues was also investigated. These substances may mediate weevil feeding and oviposition behavior.

4.1 Airborne Monoterpenes from Flower Buds and Leaves

Volatile monoterpenes captured from the air surrounding flower buds and leaves of bacterially treated and untreated cotton were found to differ depending upon cotton genotype, tissue, plant age, terpene compound, and

TABLE 8   Major Secondary Plant Products from Cotton, *Gossypium* spp., with Known or Suspected Activity Against Insect Pests[a]

Phenolic acids	*Terpenes*
Benzoic acids	Monoterpenes
Gentisic acids	Limonene
Salicylic acid	Myrcene
Syringic acid	Ocimene
Vanillic acid	Pinene
Cinnamic acids	Sesquiterpenes
Caffeic acid	Bisabolene
Chlorogenic acid	Caryophyllene
Ferulic acid	Humulene
Sinapic acid	Sesquiterpenoid naphthols
Flavonoids	Caldalenes
Flavonols	Lacinilenes
Gossypetin glycoside	Terpenoid aldehydes
Herbacetin glycoside	Gossypol
Kaempferol glycoside	Heliocides
Quercetin glycoside	Hemigossypol
Flavones	*Fatty acids*
Apigenin glycoside	Cyclopropenes
Anthocyanidins	Malvalic acid
Cyanidin glycoside	Sterculic acid
Delphinidin	Cyclopropanes
Flavanols	Dehydromalvalic acid
Catechin	Dehydrosterculic acid
Gallocatechin	

[a] Many compounds also have activity against nematodes and/or plant pathogens.

rate of bacteria applied (Table 9) (Chang et al. 1988a). Individual monoterpenes and total terpenes from the air space surrounding flower buds (14 days after treatment) were significantly increased when treated with a high rate of bacteria, but only myrcene and ocimene were significantly increased at the low rate. Leaves showed a somewhat different response in that the low rate of treatment tended to reduce the concentrations of terpenes compared to untreated leaves, and significantly lowered all monoterpene concentrations compared to leaves receiving the high rate of bacteria treatment (Table 9) (Chang et al. 1988a). These data demonstrated a nonlinear rate response.

In studies of the seasonal patterns of terpene emissions into the atmosphere surrounding excised flower buds from bacterially treated plants, the relative production of total terpenes for the genotype SP-37 and CAMD-E changed with time and genotype (Figs. 1 and 2). These cottons were grown in the field under the same conditions and treated simultaneously with the same concentrations of mutualistic bacteria. These studies and our previous work on seasonal terpene production in cotton (Hedin 1976, Zummo et al.

TABLE 9 **Concentration of Volatile Terpenes Collected from the Atmostphere Surrounding SP-37H Flower Buds and Leaves from Bacterially Treated and Untreated Plants, at 14 days after Bacterial Treatment, Corpus Christi, Texas, 1987 (modified from Chang et al. 1988a)**

Treatment/plant part[a]	Concentration of Monoterpenes (ng/bud or ng/leaf)[a]					
	α-Pinene	β-Pinene	β-Myrcene	*d*-Limonene	β-Ocimene	Total
Untreated						
Buds	53.1 bc	39.4 bc	56.3 cd	3.8 c	23.7 d	176.4 bcd
Leaves	33.5 cd	25.0 cd	54.2 cd	5.3 bc	39.6 bc	157.6 cd
Treated at low rate						
Buds	76.3 b	57.7 b	89.5 ab	5.8 abc	38.8 bc	268.0 b
Leaves	21.2 d	15.3 d	33.8 d	4.9 bc	25.2 cd	134.9 d
Treated at high rate						
Buds	111.3 a	84.7 a	118.4 a	7.5 ab	47.4 b	373.5 a
Leaves	49.3 c	38.5 bc	81.9 bc	8.6 a	72.7 a	251.0 bc

[a] Means within columns followed by the same letters are not significantly different ($p \leq .05$, Duncan's New Multiple Range Test). Monoterpene values adjusted for Porapak Q recovery efficiency. Recovery efficiency was 99.6% for α-pinene, 90.2% for β-pinene, 89.9% for β-myrcene, 91.5% for *d*-limonene, and 82.5% for β-ocimene.

[b] The first application was made on 27 May 87 at first ⅓-grown flower bud stage of growth. The low rate of treatment received 31.1×10^7 cfu, *B. cereus* and *B. magaterium* (ratio of 2:1), in 8 mL of distilled water per foot of row at the first application. The high rate of treatment received 43.2×10^7 cfu in 8 mL of distilled water per foot of row at the first application. The second application was made on 15 June 87 at the first bloom stage of growth. The low rate received 65×10^7 cfu and the high rate received 98.6×10^7 cfu per foot of row at the second application.

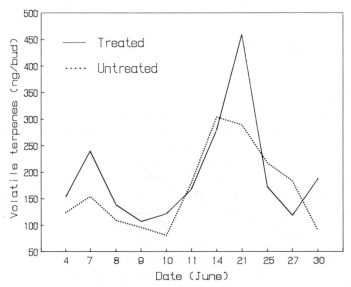

Figure 1 Volatile terpenes collected from the atmosphere surrounding excised flower buds from bacterially treated and untreated SP-37 cotton plants. Volatiles collected from early to late bloom stage of growth, Corpus Christi, Texas, 1985. (Modified from Chang et al. 1988a.)

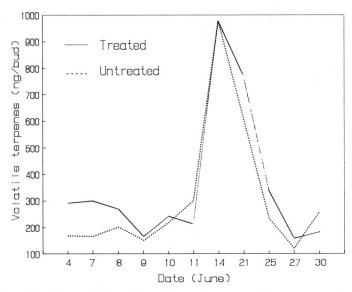

Figure 2   Volatile terpenes collected from the atmosphere surrounding excised flower buds from bacterially treated and untreated CAMD-E cotton plants. Volatiles collected from early to late bloom stage of growth, Corpus Christi, Texas, 1985. (Modified from Chang et al. 1988a.)

1984b, Chang et al. 1988b) demonstrated that the type and quantity of terpene production was dynamic and related to genetic, abiotic, and plant developmental factors.

Dickins (1984) found that the boll weevil readily senses volatile terpene compounds. In our work with terpene attractants to boll weevil in the field, we have found that the number of boll weevils responding to a point source is mediated, in part, by the varied quantities and types of these terpenes (Chang 1986, Chang et al. 1987, Benedict and Chang 1988). Organisms capable of perceiving terpene odor profiles are receiving information on the status of host–plant physiology and development (i.e., host–plant suitability) which may be significant to their fitness.

4.2   Terpene Aldehydes and Tannins in Flower Bud and Leaf Tissues

The bacterially induced changes in cotton volatile terpenes discussed above suggest that bacterial treatments result in a stress-induced response by the plant similar to that triggered by pathogen infection (Bell 1981, Essenberg et al. 1982, Altman et al. 1985, Hammond and Hardy 1988). The *B. cereus* used in these inoculations is a minor or facultative pathogen on roots of some plant species under certain conditions, and a beneficial symbiont under other conditions (Burr and Caesar 1984). Minor pathogens are unusual in that their infection of a host does not result in typical disease symptoms, such as necrotic tissue and cellular lysis. Instead, infection often results in unusual disease symptoms or no apparent symptoms at all. For example, *B. cereus* causes frenching disease of tobacco (Steinberg 1952). The symptoms of frenching disease also can be produced by the "toxin" of *B. cereus* or with isoleucine or leucine (Steinberg 1952). This microorganism is also regularly found on leaves of leek, in association with uredospores of the pathogenic organism, *Puccinia allii* (Blakeman 1985). A topical spray containing *B. cereus*, when applied to leaves of Douglas fir, *Pseudotsuga menziesii*, controlled needle rust, *Melampsora medusae* (Blakeman 1985). Spurr and Knudson (1985) also recorded *B. cereus*, isolated from crop leaves, as effective in controlling *Alternaria* leaf spot. Thus, *B. cereus* has varied roles on different plant species and under different environmental conditions. The microorganism, *B. megaterium*, is a biological control agent on rice where it is a dominant microorganism on the phylloplane. It is known to prevent brown spot of rice when sprayed on aerial parts (Islam and Nandi 1985).

Under natural field conditions microorganisms are transported to aerial plant surfaces by rain, airborne pollen, dust, debris, and insect activities (Warren 1976, Blakeman 1985). The sources of nutrients on aerial surfaces to support epiphytic and saprophytic microorganisms are quite diverse, but the main sources appear to be plant cellular exudates and leakages that seep onto plant surfaces (Godfrey 1976, Blakeman 1985). There is also a leaching process in which cellular components migrate onto aerial plant surfaces through leaching when the surface is wet with dew or rainwater. These exudates, leakages, and leachates can be chemically complex, containing simple sugars, amino acids, minerals, phenols, and terpenoids (Blakeman and Brodie 1976, Blakeman 1985). Thus, a wide range of organic and inorganic compounds and minerals are available on plant surfaces. Phytoalexins can be induced by pathogen attack (Kuć and Rush 1985, Harborne 1986), attempted penetration of plant surfaces by nonpathogenic microorganisms (Hislop 1976), and chemical elicitors (Kuć and Rush 1985) such as plant cell wall fragments (Ayers et al. 1985) and agrochemicals (Hislop 1976, Zummo et al. 1984a).

Since monoterpenes are induced by bacterial applications, we investigated the response of other phytoalexins, that is, tannins and nonvolatile terpenes, to cotton pathogens (Bell et al. 1986). These compounds are also biologically active as repellents, attractants, and toxicants of insects (Bell and Stipanovic 1978, Kogan and Paxton 1983, Stipanovic et al. 1986, Metcalf 1987).

Relative condensed tannins were measured during field studies using the butanol–HCl method (Zummo et al. 1983) in five separate tests (1980–1985). The relative astringency of tannin was measured using a protein precipitation method (Zummo et al. 1984b). All tests showed results similar to those in

TABLE 10   Relative Concentration of Tannins (percent absorbency, %A) in Fresh Leaf Tissue of Three Bacterially Treated and Untreated Cotton Genotypes in the Field, Corpus Christi, Texas, 1980

Genotype	Treatment[a]	Tannin %A[b]
SP-37	Untreated	0.5956 b
SP-37	Treated	0.8731 a
CAMD-E	Untreated	0.8903 a
CAMD-E	Treated	0.8798 a
Blightmaster	Untreated	0.8398 a
Blightmaster	Treated	0.9190 a
Mean, all genotypes	Untreated	0.7753 b
Mean, all genotypes	Treated	0.8906 a

[a] Bacteria were sprayed on "treated" plots 3 June 80 at the rate of 5×10^6 cfu in 5 mL of distilled water per row foot. The bacteria were a mixture of *B. cereus* and *B. megaterium* in the ratio of $2:1$, respectively.
[b] Means within columns followed by the same letter are not significantly different ($p \leq .05$, Duncan's New Multiple Range Test), except the "Means, all genotypes," which were analyzed using Student's *t* test ($p \leq .05$).
[c] Relative concentrations of tannin were determined using the first fully expanded terminal leaf, sampled on 18 June for SP-37 and CAMD-E, and 23 June for Blightmaster.

Table 10. Relative tannin concentration as well as astringency was significantly increased with bacterial treatment. In one test *B. megaterium* alone was sprayed on SP-37 cotton, at a low and high rate, and a mixture of *B. megaterium* and *B. cereus* was sprayed at a low and high rate. The relative tannin concentration was significantly reduced in both treatments of *B. megaterium* alone compared to the untreated check, while the mixtures significantly increased tannins compared to *B. megaterium* alone. These results suggest that either *B. cereus* or the mixture induce phytoalexin production.

The major active terpene aldehydes were measured by HPLC in a 1987 field study with SP-37H cotton. As with the monoterpenes (Table 9), responses differed by treatment depending upon the terpenoid compound, plant tissue, and rate of bacterial application (Chang et al. 1988a). Total terpene aldehydes were significantly increased in leaves receiving the high rate of bacterial treatment, and in flower buds at the low rate of treatment (Table 11). The major terpenoid aldehyde increase in flower buds was for gossypol, which is known to be an active allelochemical against *Heliothis* spp. (Stipanovic et al. 1986). In leaves, the major increase in terpene aldehydes at the high rate was for leaves receiving the high rate of bacterial treatment, and for flower buds receiving the low rate of treatment (Table 11) (Chang et al. 1988a). In leaves the major increase in terpene aldehydes was hemigossypol quinone and gossypol at the high rate of treatment. Hemigossypol quinone has also been associated with insect and plant pathogen resistance in cotton (Bell et al. 1986, Stipanovic et al. 1986).

TABLE 11   Concentrations of Terpene Aldehydes Determined by HPLC Analyses of Bacterially Treated and Untreated Flower Buds and Leaves of SP-37H Cotton, 14 days after Bacterial Treatment, Corpus Christi, Texas, 1987 (modified from Chang et al. 1988a)

Treatment	Terpene Concentrations (ppm)[a]						
	HGQ	G	H_1	H_2	H_3	H_4	Total
Untreated							
Leaves	7263b	736d	4036a	2292a	739a	1423a	17389b
Flower buds	678c	2086b	1356b	941b	251b	337b	5649c
Treated at low rate							
Leaves	7334b	775d	5084a	2146a	667a	1513a	17518b
Flower buds	733c	2465a	1395b	1002b	261b	297b	6153c
Treated at high rate							
Leaves	11048a	1153c	5135a	2198a	674a	1399a	21607a
Flower buds	784c	2404a	1133b	950b	247b	270b	5787c

[a] Terpenes are HGQ, hemigossypol quinone; G, gossypol; H_{1-4}, heliocides 1–4. Means within columns followed by the same letters are not significantly different ($p \leq .05$, Duncan's New Multiple Range Test).

5   CONCLUSIONS AND PROSPECTS

From these studies we can conclude much about the bacteria–plant–boll weevil interaction. One of the most interesting conclusions is that very subtle changes in host–plant chemistry can influence boll weevil host selection, colonization, survival, reproductive biology, and pheromone quantity and quality. Such evidence also suggests that subtle changes in secondary plant chemistry may influence reproductive biology and population dynamics of other organisms associated with cotton, including insect parasitoids that respond to monoterpene synomones (Price et al. 1980, Elzen et al. 1984).

Our work indicates that bacterial inoculation of cotton plants with certain microbial flora from the phylloplane and associated tissues of CAMD-E cotton does induce phytochemical changes (both volatile and nonvolatile compounds) as well as changes in boll weevil preference and population growth parameters. Furthermore, we found that the bacterially induced changes in plant secondary chemistry are related, in a poorly understood manner, to rate of application, cotton genotype, and plant age at application. It is also notable that plant responses are systemic and extend more than 30 days after application. We found that numbers and ratios of colony forming microorganisms on the leaf and flower bud surfaces were altered following applications of the microbes *B. cereus* and *B. megaterium*. However, results were variable, suggesting that environmental factors, plant age, and/or genotype play a role in regulating bacterial populations on and/or in aboveground tissues.

In 1969 Fraenkel stated that "host selection is the very heart of agricultural entomology, and that secondary plant substances are the clues to the [management of pest] problems." Yet in 1987 Metcalf noted that, "plant odorants and their kairomonal action are scarcely mentioned in a recent review (i.e., Prestwich 1985) on molecular communication of insects." Hedin (1976) and coworkers (Hedin et al. 1975) have recorded approximately 70 compounds collected from the air surrounding cotton. With such a large number of compounds and the dynamic complexities of their expression, as demonstrated by our studies of bacterially induced resistance, it is apparent that a major reason for the lack of knowledge on the kairomonal action of secondary plant substances is due, in part, to the difficulty in obtaining cause and effect relationships between volatile chemistry and insect response. Moreover, the semiochemically mediated responses of insects appear to be regulated by a number of chemical cues that must be present in the correct ratios and concentrations rather than by the concentration of a single plant compound. This is particularly true of the boll weevil where the plant odor is most attractive when mingled with boll weevil pheromone that is released from frass deposited on cotton plant surfaces as male weevils feed on pollen. Methodologies and research expertise are presently being developed to unravel these enigmas in managed and natural ecosystems.

The possibility that bacterially induced resistance could be utilized as a practical boll weevil management tactic is slight. We base this on the fact that the level of the bacterially induced antibiosis and nonpreference to boll weevil is relatively low and unpredictable. Furthermore, in large commercial cotton fields we suspect these slight differences in secondary chemistry induced by bacteria treatment would have little influence over rates of insect colonization, insect population dynamics, or plant damage.

These studies have raised many questions concerning the influence of plant mutualistic microorganisms on relationships between plants and herbivores. We suggest that mutualistic microorganisms on the phylloplane and in the tissues of many plant species play a role in mediating the ecological relationships between microbes, insects, and plants in natural and artificial communities.

ACKNOWLEDGMENTS

We are especially grateful to Luther Bird for his insight and guidance, and Mary Howell and Jim Stack for culturing the microbial symbionts used in these studies. We thank Michael Raupp and George Teetes for their valuable reviews of the early versions of this chapter. We also thank Kristine Schmidt, Carolyn Villanueva, Mike Treacy, Roger Anderson, James Segers, and Darryl George for their patient assistance throughout these studies. We acknowledge the following for financial support: Cotton Inc. under Grant 81-496; the EPA and USDA under Grants CR806277030 and 71-59-2481-1-2-030-1, respectively; the Natural Fibers and Food Protein Commission of Texas; and the Texas Agricultural Experiment Station under Expanded Research for Cotton Improvement. This manuscript was approved for publication by the Texas Agricultural Experiment Station as Technical Article 23969.

REFERENCES

Agrios, G. N. 1988. *Plant Pathology*. New York: Academic.

Altman, D. W., R. D. Stipanovic, D. M. Mitten, and P. Heinstein. 1985. Interaction of cotton tissue culture cells and *Verticillium dahliae. In Vitro Cell. Dev. Biol.* 21:659–664.

Ayers, A. R., J. J. Goodell, and P. L. DeAngelis. 1985. Plant detection of pathogens. In G. A. Cooper-Driver, T. Swain and E. E. Conn (eds.), *Recent Advances in Phytochemistry*, Vol. 19, *Chemically Mediated Interactions between Plants and Other Organisms*. New York: Plenum, pp. 1–21.

Bell, A. A. 1981. Biochemical mechanism of disease resistance. *Annu. Rev. Plant Physiol.* 32:21–81.

Bell, A. A. 1984. Morphology, chemistry, and genetics of *Gossypium* adaptations to pests. In B. N. Timmermann, C. Steelink, and F. A. Loewus (eds.), *Phytochemical Adaptations to Stress*. New York: Plenum, pp. 197–230.

Bell, A. A. 1986. Physiology of secondary products in cotton. In J. R. Mauney and J. McD. Stewart (eds.), *Cotton Physiology*. Memphis, TN: The Cotton Foundation, pp. 597–621.

Bell, A. A., and R. D. Stipanovic. 1978. Biochemistry of disease and pest resistance in cotton. *Mycopathology* 65:91–106.

Bell, A. A., M. E. Mace, and R. D. Stipanovic. 1986. Biochemistry of cotton (*Gossypium*) resistance to pathogens. In M. B. Green and P. A. Hedin (eds.). *Natural Resistance of Plants to Pests: Roles of Allelochemicals, ACS Symposium Series 296*, Washington, DC: American Chemical Society, pp. 36–54.

Benedict, J. H. 1980. Progress in breeding for insect resistance in cotton. *Texas Agric. Exp. Sta. Tech. Rep.* 80:1–5.

Benedict, J. H., and L. S. Bird. 1981. Relationship of microorganisms within the plant and resistance to insects and diseases. In J. M. Brown (ed.), *Proceedings Beltwide Cotton Production Research Conferences*. Memphis, TN: National Cotton Council of America, pp. 149–150.

Benedict, J. H., and J. F. Chang. 1988. Use of pheromone traps in the management of overwintered boll weevils on the lower Gulf Coast of Texas. In J. M. Brown (ed.), *Proceedings Beltwide Cotton Production Research Conferences*. Memphis, TN: National Cotton Council of America, pp. 205–206.

Benedict, J. H., L. S. Bird, and C. Liverman. 1979. Bacterial flora of MAR cottons as a boll weevil resisting character. In J. M. Brown (ed.), *Proceedings Beltwide Cotton Production Research Conferences*. Memphis, TN: National Cotton Council of America, pp. 228–230.

Benedict, J. H., L. S. Bird, and D. M. George. 1980. Boll weevil and bollworm antibiosis and antixeniosis studies of MAR cotton. In J. M. Brown (ed.), *Pro-

ceedings Beltwide Cotton Production Research Conferences. Memphis, TN: National Cotton Council of America, p. 286.

Bird, L. S., C. Liverman, R. G. Percy, and D. L. Bush. 1979. The mechanisms of multi-adversity resistance in cotton: theory and results. In J. M. Brown (ed.), *Proceedings Beltwide Cotton Production Research Conferences.* Memphis, TN: National Cotton Council of America, pp. 226–228.

Bird, L. S., C. Liverman, P. Thaxton, and R. G. Percy. 1980. Evidence that microorganisms in and on tissues have a role in a mechanism of multi-adversity resistance in cotton. In J. M. Brown (ed.), *Proceedings Beltwide Cotton Production Research Conferences.* Memphis, TN: National Cotton Council of America, pp. 283–285.

Bird, L. S., C. Liverman, G. R. Lazo, P. Thaxton, R. G. Percy, and J. H. Benedict. 1981. Do plants have genetic systems controlling selective colonization of tissues by symbiotic organisms which function in mechanisms of resistance to adversities? In J. M. Brown (ed.), *Proceedings Beltwide Cotton Production Research Conferences.* Memphis, TN: National Cotton Council of America, pp. 34–36.

Bird, L. S., K. M. El-Zik, R. G. Percy, P. Thaxton, J. H. Benedict, L. Reyes, R. A. Creelman, L. E. Clark, C. M. Heald, and A. J. Kappelman, Jr. 1983a. Improved multi-adversity resistant (MAR) cotton from the MAR-4 hybrid pool. *Texas Agric. Exp. Sta. Prog. Rep. 4128.*

Bird, L. S., K. M. El-Zik, R. G. Percy, P. Thaxton, J. H. Benedict, L. Reyes, R. A. Creelman, L. E. Clark, C. M. Heald, and A. J. Kappelman, Jr. 1983b. Improved fiber and other traits in new multi-adversity resistant cottons. In J. M. Brown (ed.), *Proceedings Beltwide Cotton Production Research Conferences.* Memphis, TN: National Cotton Council of America, pp. 103–105.

Blakeman, J. P. 1985. Ecological succession of leaf surface microorganisms in relation to biological control. In C. E. Windels and S. E. Lindow (eds.), *Biological Control on the Phylloplane.* St. Paul, MN: The American Phytopathological Society, pp. 6–30.

Blakeman, J. P., and I. D. S. Brodie. 1976. Inhibition of pathogens by epiphytic bacteria on aerial plant surfaces. In C. H. Dickinson and T. F. Preece (eds.), *Microbiology of Aerial Plant Surfaces.* New York: Academic, pp. 549–557.

Blakeman, J. P., and N. J. Fokkema. 1982. Potential for biological control of plant diseases on the phylloplane. *Annu. Rev. Phytopathol.* 20:167–192.

Brown, M. E. 1974. Seed and root bacterization. *Annu. Rev. Phytopathol.* 12:181–197.

Bull, D. L., J. R. Coppedge, D. D. Hardee, D. R. Rummel, G. H. McKibben, and V. S. House. 1973. Formulations for controlling the release of synthetic pheromone (grandlure) of the boll weevil. 3. Laboratory and field evaluations of 3 slow release preparations. *J. Environ. Entomol.* 2:905–909.

Burr, T. J., and A. Caesar. 1984. Beneficial plant bacteria. *CRC Crit. Rev. Plant Sci.* 2:1–20.

Bush, D. L. 1979. Variation in root leachate and rhizosphere–rhizoplane microflora amoung cultivars representing different levels of multi-adversity resistance in cotton. Ph.D. Dissertation, Texas A&M University, College Station, TX. 187 pp.

Chang, J. F. 1986. Influence of cotton cultivars on boll weevil (Coleoptera: Curculionidae) behavior and pheromone production. Ph.D. Dissertation, Texas A&M University, College Station, TX.

Chang, J. F., J. H. Benedict, T. L. Payne, and B. J. Camp. 1986. Methods for collection and identification of volatile terpenes from cotton and evaluation of their attractiveness to boll weevils. *Southwestern Entomol.* 11:233–241.

Chang, J. F., J. H. Benedict, T. L. Payne, and B. J. Camp. 1987. Attractiveness of cotton volatiles and grandlure to boll weevils in the field. In T. C. Nelson (ed.). *Proceedings Beltwide Cotton Production Research Conferences.* Memphis, TN: National Cotton Council of America, pp. 102–104.

Chang, J. F., J. H. Benedict, B. J. Camp, L. S. Bird, and R. D. Stipanovic. 1988a. Bacterially-induced changes in plant terpene chemistry and boll weevil phero- mone. In J. M. Brown (ed.), *Proceedings Beltwide Cotton Production Research Conferences.* Memphis, TN: National Cotton Council of America, pp. 282–285.

Chang, J. F., J. H. Benedict, T. L. Payne, and B. J. Camp. 1988b. Volative mon- oterpenes collected from the air surrounding flower buds of seven cotton geno- types. *Crop Sci.* 28:685–688.

Chang, J. F., J. H. Benedict, T. L. Payne, and B. J. Camp. 1988c. Pheromone production of boll weevils in response to seven cotton genotypes grown in two environments. *Environ. Entomol.* 17:921–925.

Chang, J. F., J. H. Benedict, T. L. Payne, B. J. Camp, and S. B. Vinson. 1989. Collection of pheromone from the atmosphere surrounding boll weevils, *An- thonomus grandis. J. Chem. Ecol.* 15:767–777.

Cook, C. G., K. M. El-Zik, L. S. Bird, and M. L. Howell. 1987. Effect of treatment with *Bacillus* species on cotton root traits, yield and *Phymatotrichium* root rot. In T. C. Nelson (ed.), *Proceedings Beltwide Cotton Production Research Con- ferences.* Memphis, TN: National Cotton Council of America, pp. 43–45.

Cross, W. H. 1973. Biology, control, and eradication of the boll weevil. *Annu. Rev. Entomol.* 18:17–46.

Dickins, J. C. 1984. Olfaction in the boll weevil, *Anthonomus grandis* Boh. (Cole- optera: Curculionidae): electroantennogram studies. *J. Chem. Ecol.* 10:1759– 1785.

Dickinson, C. H., and T. F. Preece. 1976. *Microbiology of Aerial Plant Surfaces.* New York: Academic.

Doherty, M. A., and T. F. Preece. 1978. *Bacillus cereus* prevents germination of uredospores of *Puccinia allii* and the development of rust disease of leek, *Allium porrum*, in controlled environments. *Physiol. Plant Pathol.* 12:123–132.

Elzen, G. W., J. H. Williams, and S. B. Vinson. 1984. Isolation and identification of cotton synomones mediating searching behavior by a parasitoid. *J. Chem. Ecol.* 10:1251–1264.

El-Zik, K. M., L. S. Bird, M. Howell, and P. M. Thaxton. 1985. Symbiotic organisms associated with plant parts of multi-adversity resistance (MAR) and non-MAR cottons. *Phytopathology* 75:1344.

El-Zik, K. M., and P. M. Thaxton. 1989. Genetic improvement for resistance to pests and stresses in cotton. In R. E. Frisbie, K. M. El-Zik, and L. T. Wilson (eds.), *Integrated Pest Management Systems and Cotton Production.* New York: Wiley, pp. 191–224.

Essenberg, M., M. D. Doherty, B. K. Hamilton, V. T. Henning, E. C. Cover, S. J. McFaul, and W. M. Johnson. 1982. Identification and effects on *Xanthomonas campestris* pv. *malvacearum* of two phytoalexins from leaves and cotyledons of resistant cotton. *Phytopathology* 72:1349–1356.

Fraenkel, G. 1969. Evaluation of our thought on secondary substances. *Entomol. Exp. Applic.* 12:473.

Gast, R. T., and T. B. Davich. 1966. Boll weevils. In C. N. Smith (ed.), *Insect Colonization and Mass Production.* New York: Academic, pp. 405–518.

Godfrey, B. E. S. 1976. Leachates from aerial parts of plants and their relation to plant surface microbial populations. In C. H. Dickinson and T. F. Preece (eds.), *Microbiology of Aerial Plant Surfaces.* New York: Academic pp. 433–449.

Hammond, A. M., and T. N. Hardy. 1988. Quality of diseased plants as hosts for insects. In E. A. Heinrichs (ed.), *Plant Stress–Insect Interactions.* New York: Wiley, pp. 381–432.

Harborne, J. B. 1986. The role of phytoalexins in natural plant resistance. In M. B. Green and P. A. Hedin (eds), *Natural Resistance of Plants to Pests: Roles of Allelochemical, American Chemical Society Symposium Series 296.* Washington, DC: American Chemical Society, pp. 22–35.

Hardee, D. D. 1970. Pheromone production by male boll weevils as affected by food and host factors. *Contributions from Boyce Thompson Institute* 24:315–322.

Hardee, D. D., G. H. McKibben, R. C. Gueldner, E. B. Mitchell, J. H. Tumlinson, and W. H. Cross. 1972. Boll weevils in nature respond to grandlure, a synthetic pheromone. *J. Econ. Entomol.* 65:97–100.

Hare, J. D. 1983. Manipulation of host suitability for herbivore pest management. In R. F. Denno and M. S. McClure (eds.), *Variable Plants and Herbivores in Natural and Managed Systems.* New York: Academic, pp. 655–680.

Hedin, P. A. 1976. Seasonal variations in the emissions of volatiles by cotton plants growing in the field. *Environ. Entomol.* 5:1234–1238.

Hedin, P. A. 1977. A study of factors that control biosynthesis of the compounds which comprise the boll weevil pheromone. *J. Chem. Ecol.* 3:279–289.

Hedin, P. A., A. C. Thompson, and R. C. Gueldner. 1975. Survey of the air space volatiles of the cotton plant. *Phytochemistry* 14:2088–2090.

Hedin, P. A., A. C. Thompson, and R. C. Gueldner. 1976. Cotton plant and insect constituents that control boll weevil behavior and development. *Recent Advances in Phytochemistry* 10:271–350.

Hedin, P. A., J. N. Jenkins, and F. G. Maxwell. 1977. Behavioral and developmental factors affecting host plant resistance to insects. In P. A. Hedin (ed.), *Host Plant Resistance to Pests, American Chemical Society Sympoisum Series 62,* Washington, DC: American Chemical Society, pp. 221–275.

Hendry, L. B. 1976. Insect pheromones: diet related? *Science* 192:143–145.

Hislop, E. C. 1976. Some effects of fungicides and other agrochemicals on the microbiology of the aerial surfaces of plants. In C. H. Dickinson and T. F. Preece (eds.), *Microbiology of Aerial Plant Surfaces.* New York: Academic, pp. 41–74.

Howell, M. L., L. S. Bird, K. M. El-Zik, and P. M. Thaxton. 1987. Identification of three bacteria associated with plant tissues of TAMCOT CAMD-E. In T. C.

Nelson (ed.), *Proceedings Beltwide Cotton Production Research Conferences.* Memphis, TN: National Cotton Council of America, p. 35.

Islam, K. Z., and B. Nandi. 1985. Control of brown spot of rice by *Bacillus megaterium. J. Plant Dis. Protect.* 92:241–246.

Karban, R., R. Adamchak, and W. C. Schnathorst. 1987. Induced resistance and interspecific competition between spider mites and a vascular wilt fungus. *Science* 235:678–679.

Kogan, M., and J. Paxton. 1983. Natural inducers of plant resistance to insects. In P. A. Hedin (ed.), *Plant Resistance to Insects, American Chemical Society Symposium Series 208.* Washington, DC: American Chemical Society, pp. 153–171.

Kuć, J. 1981. Multiple mechanisms, reaction rates, and induced resistance in plants. In R. C. Staples (ed.), *Disease Control.* New York: Wiley, pp. 259–272.

Kuć, J. 1982. Induced immunity to plant disease. *BioScience* 32:854–860.

Kuć, J. 1987. Plant immunization and its applicability for disease control. In I. Chet (ed.), *Approaches to Plant Disease Control.* New York: Wiley, pp. 255–274.

Kuć, J., and J. S. Rush. 1985. Phytoalexins. *Arch. Biochem. Biophys.* 236:455–472.

Masud, S. M., R. D. Lacewell, C. R. Taylor, J. H. Benedict, and L. A. Lippke. 1981. Economic impact of integrated pest management strategies for cotton production in the Coastal Bend Region of Texas. *Southern J. Agric. Econ.* Dec.:47–52.

Matta, A. 1971. Microbial penetration and immunization of uncongenial host plants. *Annu. Rev. Phytopathol.* 9:387–410.

McIntyre, J. L., J. A. Dodds, and J. D. Hare. 1981. Effects of localized infection of *Nicotiana tabacum* by tobacco mosaic virus on systemic resistance against diverse pathogens and an insect. *Phytopathology* 71:297–301.

Metcalf, R. L. 1987. Plant volatiles as insect attractants. *CRC Crit. Rev. Plant Sci.* 5:251–301.

Prestwich, G. D. 1985. Communication in insects. II. Molecular communication of insects. *Quart. Rev. Biol.* 60:437.

Price, P. W., C. E. Bouton, P. Gross, B. A. McPheron, J. N. Thompson, and A. E. Weiss. 1980. Interactions among three trophic levels: influence of plants on interactions between insect herbivores and natural enemies. *Annu. Rev. Ecol. Syst.* 11:41–65.

Ryan, C. A. 1983a. Insect-induced chemical signals regulating natural plant protection responses. In R. F. Denno and M. S. McClure (eds.), *Variable Plants and Herbivores in Natural and Managed Systems.* New York: Academic, pp. 43–60.

Ryan, C. A. 1983b. Wound-regulated synthesis and compartmentation of proteinase inhibitors in plant leaves. In O. Ciferri and L. Dure III (eds.), *Structure and Function of Plant Genomes.* New York: Plenum, pp. 337–345.

Spurr, H. W., and G. R. Knudsen. 1985. Biological control of leaf diseases with bacteria. In C. E. Windels and S. E. Lindow (eds.), *Biological Control on the Phylloplane.* St. Paul, MN: The American Phytopathological Society, pp. 45–62.

Steinberg, R. A. 1952. Frenching symptoms produced in *Nicotiana tabacum* and *Nicotiana rustica* with optical isomers of isoleucine and leucine and with *Bacillus cereus* toxin. *Plant Physiol.* 27:302.

Sterling, W. L., and P. L. Adkisson. 1971. Seasonal biology of the boll weevil in the High and Rolling Plains of Texas as compared with previous biological studies of this insect. *Texas Agric. Exp. Sta. Misc. Pub. 993.*

Stipanovic, R. D., J. H. Williams, and L. A. Smith. 1986. Cotton terpenoid inhibition of *Heliothis virescens* development. In M. B. Green and P. A. Hedin (eds.), *Natural Resistance of Plants to Pests: Roles of Allelochemicals, American Chemical Society Symposium Series 296.* Washington, DC: American Chemical Society, pp. 79–94.

Suslow, T. V., and M. N. Schroth. 1982a. Rhizobacteria of sugar beets: effects of seed application and root colonization on yield. *Phytopathology* 72:199–206.

Suslow, T. V., and M. N. Schroth. 1982b. Role of deleterious rhizobacteria as minor pathogens in reducing crop growth. *Phytopathology* 72:111–115.

Thompson, A. C., and N. Mitlin. 1979. Biosynthesis of the sex pheromone of the male boll weevil from monoterpene precursors. *Insect Biochem.* 9:293–294.

Tumlinson, J. H., R. C. Gueldner, D. D. Hardee, A. C. Thompson, P. A. Hedin, and J. P. Minyard. 1969. Sex pheromones produced by male boll weevil: isolation, identification and synthesis. *Science* 166:1010–1012.

Warren, R. C. 1976. Microbes associated with buds and leaves: some recent investigations on deciduous trees. In C. H. Dickinson and T. F. Preece (eds.), *Microbiology of Aerial Plant Surfaces.* New York: Academic, pp. 361–374.

Windels, C. E., and S. E. Lindow. 1985. *Biological Control on the Phylloplane.* St. Paul, MN: American Phytopathological Society.

Zummo, G. R., J. H. Benedict, and J. C. Segers. 1983. No-choice study of plant-insect interactions for *Heliothis zea* on selected cottons. *Environ. Entomol.* 12:1833–1836.

Zummo, G. R., J. H. Benedict, and J. C. Segers. 1984a. Effect of the plant growth regulator mepiquate chloride on host plant resistance in cotton to bollworm (Lepidoptera:Noctuidae). *J. Econ. Entomol.* 77:922–924.

Zummo, G. R., J. C. Segers, and J. H. Benedict. 1984b. Seasonal phenology of allelochemicals in cotton and resistance to bollworm. *Environ. Entomol.* 13:1287–1290.

CHAPTER 17

INDUCIBLE RESISTANCE IN AGRICULTURAL SYSTEMS

RICHARD KARBAN

Department of Entolomology, University of California, Davis, CA 95616

1   Introduction
2   Do plant changes affect herbivore populations?
    2.1   Scale
    2.2   Compensatory mortality
    2.3   Timing
    2.4   Extent of initial damage
    2.5   Environmental conditions
3   Induced resistance of cotton to spider mites
    3.1   Problems of scale
    3.2   Timing
    3.3   Extent of initial damage
    3.4   Environmental conditions
4   Practical consequences
5   Conclusion
    References

1   INTRODUCTION

Recently, entomologists have begun to explore the possibility of using induced plant resistance as a means of controlling pest populations (Kogan and Paxton 1983). Kogan and Paxton (1983) defined induced resistance as the "qualitative or quantitative enhancement of a plant's defense mechanisms against pests in response to extrinsic physical or chemical stimuli." Herbivores may be the stimuli, though other stresses or cultural manipulations may also induce resistance. This definition states that the induced plant must be better "defended" against pests. This does not necessarily imply that the plant benefits by induction in either ecological or evolutionary time. To the contrary, defense may be enhanced by deterioration of plant

403

metabolism; both plant and herbivore may experience reduced population growth and/or fitness.

Interest in the applications of induced resistance is not new; the entomological literature of the past four decades is filled with descriptions of plant conditioning influencing the performance of herbivorous insects and mites. Conditioned plant tissue is that which becomes less suitable to herbivores following initial exposure to herbivory. Most of the descriptions of plant conditioning are anecdotal accounts and the researchers involved made no attempt to utilize plant conditioning as a means of protecting crops from economically important pest damage. For example, walnut aphid, *Chromaphis juglandicola*, had reduced adult survival, nymphal survival, and fecundity when raised on conditioned leaves that had previously hosted aphids compared to control leaves (Davis 1950). Conditioned walnut foliage was associated with a reduced buildup of aphids throughout the growing season (Sluss 1967). However, no attempt has been made to incorporate these observations into a management program.

During the past decade entomologists have rediscovered this phenomenon and renamed it *induced resistance*. Current interest in the field was motivated largely by experiments that demonstrated that plants change chemically after herbivore attack. Much of the chemical profile of a plant is altered following herbivore damage, including secondary plant metabolites which are presumed to have deterrent or antibiotic effects on herbivores. For instance, damage to tomato or potato foliage by feeding of Colorado potato beetles (*Leptinotarsa decemlineata*) or by mechanical wounding resulted in increased foliar concentrations of proteinase inhibitor (Green and Ryan 1972). These chemical changes led Green and Ryan to hypothesize that induced increase in proteinase inhibitors functioned as a means of protection for the damaged plant.

Despite much careful work that has demonstrated induced chemical changes, the consequences of these induced changes on populations of herbivores are equivocal. For example, proteinase inhibitors, which increased in potato plants following damage, failed to produce a negative effect on Colorado potato beetles (Jeker et al. 1977). Larval growth of beet armyworm caterpillars (*Spodoptera exigua*) was reduced when they were fed induced tomato foliage compared to caterpillars fed controls (Broadway et al. 1986). However, under field conditions all tomato plants may be induced most or all of the time (Broadway and Duffey, personal communication).

Recently, many workers have found that herbivore performance is reduced when insects are fed tissue from damaged plants compared to undamaged controls. Reductions in survival, growth rate, adult size, fecundity, and sex ratio have been documented (reviewed by Rhoades 1979, Fowler and Lawton 1985, Karban 1987a, Karban and English-Loeb 1988). Several problems exist that make interpretation of these findings difficult. Several of these studies were not well designed and involve pseudoreplication (Fowler and Lawton 1985). Many of the documented cases of reductions in herbivore performance were small and, as such, may not cause significant changes in field populations of herbivores (Fowler and Lawton 1985). Many of the studies used artificial or mechanical damage to induce resistance rather than using actual herbivores. There is growing evidence that artificial defoliation does not adequately mimic defoliation caused by actual herbivory (Dyer and Bokhari 1976, Capinera and Roltsch 1980) and that this discrepancy can affect the interpretation of induced resistance (Haukioja and Neuvonen 1985). Despite these important problems, there is little doubt that prior host exposure to herbivores can reduce the subsequent performance of insects that feed on that host.

The extrapolation of estimates of herbivore performance to populations should not be assumed. Experimental bioassay results on insect life table parameters (survival, growth rate, size at maturity, etc.) do not necessarily translate into effects on population dynamics. Western tent caterpillars (*Malacosoma californicum pluviale*) that were fed alder foliage from trees that had been damaged developed more slowly and attained a smaller weight at metamorphosis than caterpillars from undamaged trees (Myers and Williams 1987). These differences in performance were insufficient to cause a decline in the tent caterpillar population, but they may have prevented further population growth.

In this chapter I will suggest several factors that should be considered if we are to learn whether induced chemical changes can play a major role in affecting populations of herbivores. I will then describe my attempts to address these factors by studying induced resistance of cotton against spider mites. Finally, I will speculate on practical consequences of induced resistance in agriculture and suggest various new directions for future research.

2 DO PLANT CHANGES AFFECT HERBIVORE POPULATIONS?

Induced resistance must have the capability of affecting populations of herbivores if it is to be considered an important ecological process in natural or agricultural systems. I envision five major steps or difficulties involved with translating the reductions in components of herbivore performance that have been documented to date into effects upon herbivore populations. These are problems of (1) scale, (2) compensatory mortality, (3) appropriate timing, (4) extent of initial damage, and (5) environmental conditions.

2.1 Scale

The diverse array of phenomena that are considered in this volume and elsewhere to be induced resistance against herbivores includes examples that occur at very different spatial scales. Differences in performance are compared for insects on the damaged or undamaged sides of single leaves (e.g., Bergelson et al. 1986), damaged or undamaged branchlets of the same tree (e.g., Raupp and Denno 1984), new foliage from previously damaged or undamaged plants (e.g., Karban and Carey 1984), and foliage from undamaged trees in close proximity or far from damaged trees (e.g., Rhoades 1983). Responses at these different spatial scales are likely to involve very different mechanisms and have disparate effects on herbivore populations.

Induced plant responses will affect insect populations only if they occur at a spatial scale that is appropriate. Many empirical studies of induced changes have examined phenomena at a scale at which they may be irrelevant to population dynamics of the herbivores. For instance, many investigators have compared the performance of herbivores on plants previously selected and damaged by insects to performance on plants not previously selected by insects. Since insects are biased in their initial selection of hosts, interpretation of results from studies relying on naturally selected damage is difficult (Neuvonen and Haukioja 1985). More rigorously designed experiments allow the investigator (rather than the insect) to randomly assign plants to different herbivore damage treatments. Herbivores are then forced to feed on plant tissue that was either induced or protected. However, these experiments must often be conducted on a small scale, which may potentially alter the generality of the results. For example, female tussock moth caterpillars (*Orgyia vetusta*) forced to consume lupine foliage previously damaged by woolly bears (*Platyprepia virginalis*) attained lower pupal weights and produced fewer eggs than tussock moths fed undamaged control foliage (Harrison and Karban 1986a). However, it is possible that ovipositing tussock moths or feeding caterpillars that are allowed to choose their food source would avoid previously damaged foliage so that the population would not be strongly affected by induced changes in a small fraction of the available host plants. In other words, processes that are important in small-scale, no-choice experiments may be irrelevant to natural populations.

Several lines of evidence may be collected to assess whether results from small-scale experiments apply at larger, more realistic spatial scales. Non-manipulative observations of populations of herbivores over larger spatial scales and periods of time can often provide insight. Two species suspected of negative interactions mediated through changes in host quality may be expected to co-occur less often than estimates of co-occurrence predicted by their independent frequencies. While such a pattern can neither prove nor disprove that the negative interaction is important, it can provide useful inference. Interactions that are found to be important by experimentation on a small spatial scale can be examined on a larger, more realistic scale, although this often necessitates a reduction in experimental replication.

2.2 Compensatory Mortality

Rarely is it possible to monitor herbivores through their entire life cycles and from one generation to the next. Most of the documented effects upon herbivore performance attributed to induced resistance involve differences in the larval stages. Differences in one developmental stage may be compensated for by factors in another stage such that the net effects on population dynamics are minimal (Karban 1984, Fowler and Lawton 1985). Induced resistance to lime aphids (*Eucallipterus tiliae*) can reduce survivorship but this effect may have no net impact on the population dynamics of this species because of more profound effects caused by weather and predation (Barlow and Dixon 1980). Similarly, if populations of a generalist herbivore are reduced on host species by induced phytochemicals, this may have little effect on local or regional population sizes because of more important effects that occur for individuals that associate with other host species. For example, induced gum production in response to oviposition by periodical cicadas (*Magicicada* spp.) reduced densities of nymphs beneath black cherry trees (Karban 1983), but gum defenses alone probably had no effect on either local or regional populations of cicadas because cherry is but one of many suitable hosts. To determine the effectiveness of induced responses we must monitor not only performance estimates of individual herbivores but also effects on herbivore populations, preferably over more than a single generation.

2.3 Timing

The amount of temporal separation between initial damage and the subsequent challenge may determine the magnitude of the effect of induced resistance upon herbivore populations. Infestation by beet flies (*Pegomya betae*) caused chemical changes in sugar beet plants that were associated with increased mortality of the flies (Rottger and Klingauf 1976). However, this response was strongest three weeks after infestation and was thereafter relaxed. Avocado brown mites (*Oligonychus punicae*) placed on damaged, bronzed avocado leaves produced significantly fewer eggs than females on less damaged foliage (McMurtry 1970). Leaves damaged three months earlier were more favorable for mites than leaves that had just received a comparable amount of damage, suggesting a relaxation of the response. Obviously, the timing of herbivore infestations relative to the times necessary for induction and relaxation are critical in determining the strength of the induced resistance.

Long-lived plants may retain induced resistance for extended periods of time. Induced resistance persists in birch trees following defoliation by *Epirrita autumnata* and on larch trees following defoliation by *Zeiraphera diniana* for at least 3–5 years (Tuomi et al. 1984, Baltensweiler 1985). Plants may acquire resistance to certain herbivores that lasts for the remainder of their lives, although the evidence for this phenomenon is equivocal (Karban 1987b).

2.4 Extent of Initial Damage

The extent of the initial damage upon the host may greatly affect the consequences of induced plant responses upon subsequent populations. For example, moderate levels of damage caused by western tent caterpillars altered foliage quality so that fall webworm larvae grew faster and attained heavier pupal weights on damaged versus control foliage (Williams and Myers 1984). However, heavily damaged foliage resulted in slower growth rates and smaller pupal sizes for fall webworms. Responses to varying amounts of initial damage may be nonlinear. Populations of spider mites (*Tetranychus urticae*) responded positively to increasing levels of damage to their bean plant hosts over some portion of the range of damage and negatively to increasing damage over other portions of the range of damage (English-Loeb, personal communication). Therefore, varying the extent of initial damage can completely reverse the effect of increased damage on herbivore populations.

As levels of initial damage are increased, the physiological functions of the plant may suffer. Even if damaging the plant results in reduced pest populations, such damage may be impractical for agricultural applications if it also reduces yield.

2.5   Environmental Conditions

The expression of constitutive (noninduced) host plant resistance varies markedly as environmental conditions change (Tingey and Singh 1980). The strength of induced resistance has also been shown to be affected by weather (Benz 1974, Baltensweiler et al. 1977), nutrient availability (Neuvonen and Haukioja 1984, Tuomi et al. 1984, Haukioja an Neuvonen 1985), and shading (Kimmerer, personal communication). Other sources of mortality may greatly alter the effects of induced responses on herbivore populations. The consequences of slowed insect growth caused by an induced plant response are likely to vary considerably depending upon the vulnerability of immature herbivores to predators, parasites, and other mortality factors (Schultz 1983).

Induced resistance can also be influenced by the genotype of the host plant and the herbivores. For example, induced responses of one soybean variety resulted in a 30% reduction of two spotted mite populations and a 50% reduction in fecundity (Hildebrand et al. 1986). However, a second variety showed essentially no induced response and no corresponding reduction in mites.

3   INDUCED RESISTANCE OF COTTON TO SPIDER MITES

My work with cotton and spider mites has focused on determining the effects of induced resistance on population dynamics of the mites. I will describe my basic experiment and then report results that attempt to address the five problems detailed above.

The basic experiment demonstrated that changes in cotton seedlings, in-

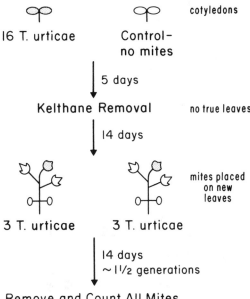

Figure 1 Experimental procedure for the basic experiment. Shaded leaves are sites of mite release. (From Karban 1987a; reprinted with permission.)

duced by prior experimental herbivory, reduced the population growth of spider mites. Cotton plants that were exposed briefly to mites supported smaller numbers of mites than unexposed controls when mites were reintroduced (Karban and Carey 1984).

Cotton seedlings (Acala SJ-2) were grown in plastic pots in environmentally controlled chambers. Plants were randomly assigned to an experimental group that was infested with *T. urticae* or to a control group that was not. The basic experiment is summarized in Figure 1. Five days after infestation, all mites were removed by treating plants of both groups with Kelthane. The plants were allowed to grow for 14 days, at which time they were challenged with three adult female *T. urticae* mites placed on the most recently expanded leaf of each plant. After an additional 14 days (one and a half generations for mites), each leaf was removed and mite populations were counted. Population growth of mites on plants that had been damaged by previous herbivory was about half the growth of mite populations on unexposed controls (Fig. 2; Karban and Carey 1984). Because actual numbers of mites were measured after more than one generation, this procedure eliminated the problem of compensatory mortality.

The induced resistance was systemic; leaves that were not present at the time of inoculation were characterized with resistance. The response was also not species-specific. Resistance against *T. urticae* was induced by damaging cotyledons with a second species of spider mite, *T. turkestani* Ugarov

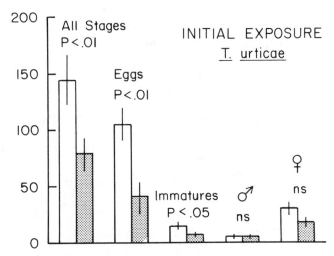

Figure 2 Population size of mites (*T. urticae*) on damaged plants (shaded bars) and undamaged controls (open bars). Means and standard errorbars are shown. (From Karban and Carey 1984; reprinted with permission. Copyright 1984 by the AAAS.)

and Nikolski (Karban and Carey 1984), by inoculating young stems with conidiaspores of the fungus *Verticillium dahliae* (Karban et al. 1987), or by damaging the cotyledons mechanically (Karban 1985). The induced response had deleterious effects on a variety of organisms that parasitize the plant. Seedlings whose cotyledons had been damaged briefly by *T. turkestani* were less suitable hosts for caterpillars of *S. exigua* (Karban 1988) and were less likely to show symptoms caused by verticillium wilt (Karban et al. 1987).

3.1 Problems of Scale

The results described in the basic experiment were all from laboratory experiments. It was conceivable that the effect on mite populations was a lab artifact and that all plants in the field either were not capable of induction or conversely were in a state of permanent induction. Clearly, field verification was necessary to resolve this question.

In the field, 30 plants were randomly assigned to each of four treatments: (1) exposed to caged mites during the cotyledon stage, (2) mechanically abraded during the cotyledon stage, (3) a cage control with no mites, and (4) a control with neither cage nor mites (Karban 1986). Seven to 11 days after establishing these treatments, mites were removed from all plants and natural population levels of mites were censused throughout the seasons of 1984 and 1985. In both years, plants of the control treatments were more likely to support mite populations early in the season than plants that had been exposed to mites or mechanically abraded (Karban 1986). However, late in the season, when populations of mites were growing rapidly, the early season differences disappeared. These results indicate that induced resistance is a real phenomenon under field conditions and not a laboratory artifact.

These results do not completely eliminate all problems of scale, however. The experimental treatments were conducted on single cotton plants. Mites may have preferentially left those plants that had become relatively more resistant in these experiments. Laboratory experiments indicate that adult female spider mites can assess whether plants have been previously damaged and preferentially move toward undamaged foliage (Harrison and Karban 1986b, Dicke 1986). Movements such as these may be responsible for the differences in mite populations observed in the field. These movements may be far less common when entire large fields, rather than single plants, vary in quality. However, if mites are unable to move to undamaged plants, the situation becomes similar to the no-choice lab experiment in which population growth of mites was reduced on induced plants.

3.2 Timing

The results described above from the basic experiment indicate that cotton plants can be rapidly induced by damage to the cotyledons and that resistance lasts for at least 14 days following damage. This raises questions about whether resistance can be induced by damaging the true leaves rather than the cotyledons. Will damage to the cotyledons induce resistance that lasts for periods of time longer than several weeks?

Rather than placing mites on the cotyledons as in Figure 1, I waited until the first true leaf had expanded before introducing 8 adult female *T. turkestani* mites to this first leaf of each experimental plant. Control plants received no mites. After 6 days, plants of both treatments were dipped in a Kelthane solution (200 ppm) which killed all mites. Fourteen days later 3 *T. urticae* females were placed on new leaves of each plant of both treatments. After an additional 14 days, all leaves were removed and mite populations were counted. Populations of *T. urticae* were markedly smaller on plants that had had previous mite herbivory on the first true leaf compared to undamaged controls (Fig. 3). This result indicates that there is nothing unique about the cotyledons; rather, resistance can be induced by allowing mites to feed on the first true leaves as well.

To determine if the induced resistance lasts longer than 14 days, I allowed the seedlings to grow older before challenging them with a second introduction of mites. In this experiment I had three treatments: (1) a group whose cotyledons were exposed to 8 *T. turkestani* females for 6 days, (2) a control group that was not damaged, and (3) a defoliation group from which I removed the cotyledons when the first true leaves were beginning to expand. Following treatment with Kelthane, the plants were allowed to grow for 60 days, at which time I placed 3 *T. urticae* adult females on each plant. This

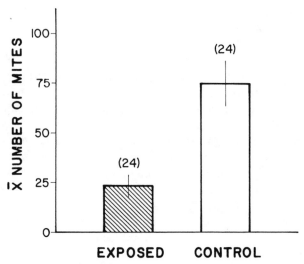

Figure 3   Population size of mites (*T. urticae*) on new growth of plants that had been damaged at the first true leaf stage (shaded bars) and on undamaged controls (open bars) ($t = 4.0$, df $= 46$, $p < .01$).

contrasts with 14 days between removal of mites and reintroduction in the basic experiment (Fig. 1).

After this longer time interval, plants that had been exposed to mites and plants that had been defoliated at the cotyledon stage supported smaller populations of mites than unexposed controls (Fig. 4). Unfortunately, the plants were physiologically no more advanced, despite being much older, than in the basic experiment. This is due, in part, to the suboptimal growing conditions associated with potted plants in a growth chamber. Although the plants were 3 months old they were generally at the 6-node stage.

In the field experiments described above, significant differences in the likelihood that a plant will support a large population of mites persisted for 3 months. Therefore, both lab and field results suggest that induced resistance persists for at least several months following the initial damage.

3.3   Extent of Initial Damage

The extent of the initial damage may determine the strength and nature of the induced plant response. To examine the relationship between initial damage and induced resistance I conducted experiments in which I varied the number of *T. turkestani* mites (from 0 to 16) that were placed on the cotyledons. Other procedures were identical to the basic experiment (Fig. 1). Populations of mites showed a monotonic, nonlinear decrease as the extent of the initial damage was increased (Karban 1987a). These results are similar to those reported for the response of citrus red mites (*Panonychus citri*) to

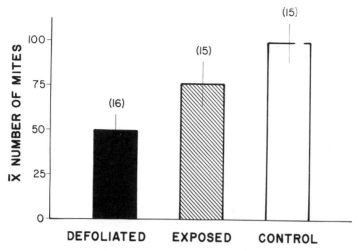

Figure 4 Population size of mites (*T. urticae*) on plants that were defoliated at the cotyledon stage (black bars), damaged by mites at the cotyledon stage (shaded bars), or undamaged controls (open bars) (one-way ANOVA $F_{2,43} = 5.5, p < .01$).

six levels of damage to orange foliage (Henderson and Holloway 1942). I found no evidence that a damage threshold must be exceeded before changes that affect mite population growth are induced.

Levels of damage that were sufficient to reduce population growth of spider mites did not cause changes in morphological characteristics of cotton plants in the lab (Karban 1987a) nor in the field (Karban 1986). Seed cotton yield was not significantly affected by these levels of damage (Karban 1986). In other studies cotton has been found to be relatively insensitive to early season damage (Wilson 1986, Kerby and Keeley 1987).

3.4 Environmental Conditions

I have now repeated the basic experiment of damaging cotyledons with *T. turkestani* and challenging with *T. urticae* 21 times. The experiments themselves have been conducted in environmental chambers under reasonably similar conditions (Karban 1987a). For some replicates, however, the effects of induced resistance were reduced or nonexistent (Karban 1985, 1987a). The growth rate of the *T. urticae* population on undamaged control plants varied by more than an order of magnitude between experiments. This growth rate on the control plants indicates the overall quality of environmental conditions for mites. When conditions for population growth are most favorable, the subtle plant changes caused by prior feeding had little or no effect. When conditions for population growth were otherwise marginal, the strength of induced resistance was greatest (Karban 1987a). Field results are consistent with this evaluation (Karban 1986). It is unclear at this time whether the conditions required for rapid population growth of mites are mediated through varying quality of the host plant or are direct environmental effects upon the mites themselves. These results suggest that the relatively slight changes in host quality caused by previous damage have a large impact on mite populations only when acting in concert with other factors that are slowing mite population growth.

4 PRACTICAL CONSEQUENCES

Immunization is the basis for preventive medicine involving animals. Recently, plants have been protected against bacteria, viruses, and fungi by the use of restricted inoculations of these plant parasites (Horsfall and Cowling 1980, Kuc 1982). For example, commercial tomato growers throughout much of Europe now routinely expose their seedlings to less virulent strains of mosaic viruses and achieve protection against economically damaging strains of these viruses (Fletcher 1978, Hamilton 1980). Despite these successes, entomologists have not applied similar techniques to protect plants against herbivorous insects and mites. It may be possible to use induced resistance as a practical management tool to reduce pest problems in the future (Kogan and Paxton 1983).

In cotton and other agricultural systems we are still far from being able to recommend the use of a specific management plan involving induced resistance. Much variability in responses needs to be explained. The mechanisms of induced responses remain largely unknown. Knowledge of protective mechanisms may help to clarify expectations about the protective effectiveness of induced resistance and greatly aid in developing applications of the induced phenomenon.

Damage by herbivores is not strictly required to induce resistance. In many instances, resistance against herbivores was induced by mechanical wounding (e.g. Raupp and Denno 1984, Karban 1985). Growth regulators and herbicides that are applied to the host plant also have been shown to affect insect performance (reviewed by Kogan and Paxton 1983). Such results suggest that induced resistance can be caused by a variety of factors that stress or wound the host plant. Therefore, it may be possible to culturally manipulate the host plant so as to induce resistance to pests of economic importance.

Often a host plant will be able to support high densities of one species without reductions in yield but much lower densities of a second species will cause economic damage. In cases such as these, it may be possible to inoculate the plant with the species that does not cause economic damage and to induce resistance against the more damaging species. This process is roughly analogous to a mammalian vaccination.

We are attempting to use a vaccination-like technique with two mite species that feed on grapes in California (English-Loeb and Karban 1988); Willamette mite, *Eotetranychus willamettei*, does not generally reduce yields, although Pacific mite, *T. pacificus*, is an important economic pest (Flaherty and Huffaker 1970, Flaherty et al. 1981). Willamette mite tends to build up earlier in the season than Pacific mite, and vines that support large Willamette mite populations are generally not damaged by Pacific mite (Flaherty and Hoy 1971, English-Loeb, personal communication). We are experimentally exposing vines to Willamette mite early in the season and comparing Pacific mite populations on exposed vines and unexposed controls.

Growers currently use introductions of less damaging and/or easily controlled strains of herbivores to build up larger populations of predators. For example, in Czechoslovakia, glasshouse growers introduced to their young cucumber plants populations of two-spotted mites from northern Russia that are very susceptible to miticides. Predaceous mites were introduced 10 days later. This technique provided good control of spider mites throughout the season, presumably because of high levels of predators (Z. Landa, personal communication). It is also conceivable that induced resistance was involved. In either case this example demonstrates that an introduction of a herbivore can be a feasible strategy.

Economics will probably determine whether induced resistance will become an important management tool for agriculture. Inducing resistance is almost certainly more labor intensive than aerial application of chemical pesticides. Factors such as value of the crop, economic losses to pests, costs and health restrictions on chemical pesticide use, and resistance of pest species to chemical pesticides will all affect whether growers will show interest in induced resistance techniques when and if they are developed. The more difficult the process of inducing resistance to pests, the less likely that growers will be willing to employ the techniques. Plants that are handled individually, for other reasons, may be the most cost-effective candidates for induced resistance. Spider mites on strawberries in California may be a good candidate for management that includes induced resistance. The crop is extremely valuable and damage caused by spider mites is very costly. Chemical control has become one of the major expenses associated with growing strawberries and mites are quickly developing resistance to the miticides that are currently registered. Commercial strawberry growers normally transplant cuttings that were started in nurseries. As such, each plant is already handled; additional handling to induce resistance should represent relatively less of an expense than would be the case for crops that are sown directly into the ground.

5 CONCLUSION

Numerous studies have found that damage caused by herbivory changes host plant quality such that herbivores that feed on damaged foliage perform less well than those feeding on undamaged control foliage. However, problems exist in extrapolating these reductions in insect performance to significant effects on herbivore populations. Nonetheless, induced resistance has been shown to reduce herbivore populations in at least some systems. It remains a challenge to entomologists to learn to manage induced resistance so that it becomes a useful pest control tool.

REFERENCES

Baltensweiler, W. 1985. On the extent and the mechanisms of the outbreaks of the larch bud moth (*Zeiraphera diniana* Gn., Lepidoptera, Tortricidae) and its impact on the subalpine larch-cembran pine forest ecosystem. In H. Turner and W. Tranquillini (eds.), *Establishment and Tending of Subalpine Forest: Research and Management*. Proc. 3rd IUFRO Workshop, pp. 215–219.

Baltensweiler, W., G. Benz, P. Bovey, and V. Delucchi. 1977. Dynamics of larch bud moth populations. *Annu. Rev. Entomol.* 22:79–100.

Barlow, N. D., and A. F. G. Dixon. 1980. *Simulation of Lime Aphid Population Dynamics*. Wageningen, Netherlands: Centre for Agricultural Publishing.

Benz, G. 1974. Negative rückkoppelung durch raum-und nahrungskonkurrenz sowie zyklische Veränderung der Nahrungsgrundlage als Regalprinzip in der Populations-dynamik des Grauen Lärchenwicklers, *Zeiraphera diniana* (Guenée). *Z. Angew. Entomol.* 76:196–228.

Bergelson, J., S. Fowler, and S. Hartley. 1986. The effects of foliage damage on casebearing moth larvae, *Coleoptera serratella*, feeding on birch. *Ecol. Entomol.* 11:241–250.

Broadway, R. M., S. S. Duffey, G. Pearce, and C. A. Ryan. 1986. Plant proteinase inhibitors: a defense against herbivorous insects? *Entomol. Exp. Applic.* 41:33–38.

Capinera, J. L., and W. J. Roltsch. 1980. Response of wheat seedlings to actual and simulated migratory grasshopper defoliation. *J. Econ. Entomol.* 73:258–270.

Davis, C. S. 1950. The biology of the walnut aphid (*Chromaphis juglandicola* (Kalt). Dissertation, University of California.

Dicke, M. 1986. Volatile spider-mite pheromone and host-plant kairomone, involved in spaced-out gregariousness in the spider mite *Tetranychus urticae*. *Physiol. Entomol.* 11:251.

Dyer, M. I., and U. G. Bokhari. 1976. Plant-animal interactions: studies of the effects of grasshopper grazing on blue grama grass. *Ecology* 57:762–772.

English-Loeb, G. M., and R. Karban. 1988. Negative interactions between Willamette mites and Pacific mites: possible management strategies for grapes. *Entomol. Exp. Appl.* 48:269–274.

Flaherty, D. L., and M. A. Hoy. 1971. Biological control of Pacific mites and Willamette mites in San Joaquin Valley vineyards. III. Role of Tydeid mites. *Res. Popul. Ecol.* 13:80–96.

Flaherty, D. L., and C. B. Huffaker. 1970. Biological control of Pacific mites and Willamette mites in San Joaquin Valley vineyards: Part I. The role of *Metaseiulus occidentalis*. *Hilgardia* 40:267–308.

Flaherty, D. L., M. A. Hoy, C. D. Lynn, and W. L. Peacock. 1981. Spider mites. In D. L. Flaherty, F. L. Jenson, A. N. Kasimatis, H. Kido, and W. J. Moller (eds.), *Grape Pest Management*. Berkeley, CA: Agricultural Sciences, pp. 111–125.

Fletcher, J. T. 1978. The use of avirulent virus strains to protect plants against the effects of virulent strains. *Ann. Appl. Biol.* 89:110–114.

Fowler, S. V., and J. H. Lawton. 1985. Rapidly induced defenses and talking trees: the devil's advocate position. *Am. Natur.* 126:181–195.

Green, T. R., and C. A. Ryan. 1972. Wound-induced proteinase inhibitor in plant leaves: a possible defense mechanism against insects. *Science* 175:776–777.

Hamilton, R. I. 1980. Defenses triggered by previous invaders: viruses. In J. Horsfall and E. B. Cowling (eds.), *Plant Disease*, Vol. 5. New York: Academic, pp. 279–303.

Harrison, S., and R. Karban. 1986a. Effects of an early-season folivorous moth on the success of a later-season species, mediated by a change in the quality of the shared host, *Lupinus arboreus Sims. Oecologia* 69:354–359.

Harrison, S., and R. Karban. 1986b. Behavioral response of spider mites (*Tetranychus urticae*) to induced resistance of cotton plants. *Ecol. Entomol.* 11:181–188.

Haukioja, E., and S. Neuvonen. 1985. Induced long-term resistance of birch foliage against defoliators: defensive or incidental? *Ecology* 66:1303–1308.

Henderson, C. F., and J. K. Holloway. 1942. Influence of leaf age and feeding injury on the citrus red mite. *J. Econ. Entomol.* 35:683–686.

Hildebrand, D. F., J. G. Rodriguez, G. C. Brown, K. T. Luu, and C. S. Volden. 1986. Peroxidative responses of leaves in two soybean genotypes injured by two-spotted spider mites (Acari: Tetranychidae). *J. Econ. Entomol.* 79:1459–1465.

Horsfall, J., and E. B. Cowling (eds.). 1980. *Plant Disease*, Vol. 5. New York: Academic.

Jeker, T., C. Mueller, and U. Volkart, cited in G. Benz. 1977. Insect induced resistance as a means of self defense. *Bull. SROP*.

Karban, R. 1983. Induced responses of cherry trees to periodical cicada oviposition. *Oecologia* 59:226–231.

Karban, R. 1984. Opposite density effects of nymphal and adult mortality for periodical cicadas. *Ecology* 65:1656–1661.

Karban, R. 1985. Resistance against spider mites in cotton induced by mechanical abrasion. *Entomol. Exp. Applic.* 37:137–141.

Karban, R. 1986. Induced resistance against spider mites in cotton: field verification. *Entomol. Exp. Applic.* 42:239–242.

Karban, R. 1987a. Environmental conditions affecting the strength of induced resistance against mites in cotton. *Oecologia* 73:414–419.

Karban, R. 1987b. Herbivory dependent on plant age: a hypothesis based on acquired resistance. *Oikos* 48:336–341.

Karban, R. 1988. Resistance to beet armyworms (*Spodoptera exigua*) induced by exposure to spider mites (*Tetranychus turkestani*) in cotton. *Am. Midl. Nat.* 119:77–82.

Karban, R., and J. R. Carey. 1984. Induced resistance of cotton seedlings to mites. *Science* 225:53–54.

Karban, R., and G. M. English-Loeb. 1988. Effects of herbivory and plant conditioning on the population dynamics of spider mites. In M. W. Sabelis and W. Helle (eds.), *Population Dynamics of Spider Mites and Predatory Mites.* Amsterdam: Elsevier.

Karban, R., R. Adamchak, and W. C. Schnathorst. 1987. Induced resistance and interspecific competition between spider mites and a vascular wilt fungus. *Science* 235:678–680.

Kerby, T. A., and M. Keeley. 1987. Cotton seedlings can withstand some early leaf loss. *Calif. Agric.* 41:18.

Kogan, M., and J. Paxton. 1983. Natural inducers of plant resistance to insects. In P. A. Hedin (ed.), *Plant Resistance to Insects.* Washington, DC: American Chemical Society, pp. 153–171.

Kuc, J. 1982. Induced immunity to plant disease. *BioScience* 32:854–860.

McMurtry, J. A. 1970. Some factors of foliage condition limiting population growth of *Oligonychus punicae* (Acarina: Tetranychidae). *Ann. Entomol. Soc. Am.* 63:406–412.

Myers, J. H., and K. S. Williams. 1987. Lack of short or long term inducible defenses in the red alder-western tent caterpillar system. *Oikos* 48:73–78.

Neuvonen, S., and E. Haukioja. 1984. Low nutritive quality as defense against herbivores: induced responses in birch. *Oecologia* 63:71–74.

Neuvonen, S., and E. Haukioja. 1985. How to study plant resistance? *Oecologia (Berlin)* 66:456–457.

Raupp, M. J., and R. F. Denno. 1984. The suitability of damaged willow leaves as food for the leaf beetle, *Plagiodera versicolora. Ecol. Entomol.* 9:443–448.

Rhoades, D. F. 1979. Evolution of plant chemical defense against herbivores. In G. A. Rosenthal and D. H. Janzen (eds.), *Herbivores: Their Interaction with Secondary Plant Metabolites.* New York: Academic, pp. 3–54.

Rhoades, D. F. 1983. Responses of alder and willow attack by tent caterpillars and webworms: evidence for pheromonal sensitivity of willows. In P. A. Hedin (ed.), *Plant Resistance to Insects.* Washington, DC: American Chemical Society, pp. 55–68.

Röttger, V. U., and F. Klingauf. 1976. Änderung im stoffwechsel von zuckerrubon blattern durch befall mit *Pegomya betae* Curt. (Muscidae: Anthomyidae). *Z. Angew. Entomol.* 83:220–227.

Schultz, J. C. 1983. Impact of variable plant defensive chemistry on susceptibility of insects to natural enemies. In P. A. Hedin (ed.), *Plant Resistance to Insects.* Washington, DC: American Chemical Society, pp. 37–54.

Sluss, R. R. 1967. Population dynamics of the walnut aphid, *Chromaphis juglandicola* (Kalt.) in northern California. *Ecology* 48:41–58.

Tingey, W. M., and S. R. Singh. 1980. Environmental factors influencing the magnitude and expression of resistance. In F. G. Maxwell and P. R. Jennings (eds.), *Breeding Plants Resistant to Insects.* New York: Wiley, pp. 86–113.

Tuomi, J., P. Niemela, E. Haukioja, S. Siren, and S. Neuvonen. 1984. Nutrient stress: an explanation for plant anti-herbivore responses to defoliation. *Oecologia* 61:208–210.

Williams, K. S., and J. H. Myers. 1984. Previous herbivore attack of red alder may improve food quality for fall webworm larvae. *Oecologia* 63:166–170.

Wilson, L. T. 1986. The compensatory response of cotton to leaf and fruit damage. *Proceedings Beltwide Cotton Production Research Conference.*

SPECIES INDEX

Abies grandis, 248, 250–252, 254, 255
Acacia, 329
  A. depranolobium, 331, 335
  A. raddiana, 331, 335
  A. senegal, 335
  A. victoriae, 328
Acalymma vittata, 21, 172
Acanthoscelides obtectus, 226, 229
Acer pseudoplatanus, 21, 97
Adzuki beans, 226
Aedes aegypti, 225
Aglais urticae, 27
Alaska feltleaf willow, *see Salix alaxensis*
Alaska paper birch, *see Betula resinifera*
  (= *papyrifera*) *humilis*
Alder, *see Alnus* spp.
Alfalfa, 26
Allium cepa, 176
Alnus spp.:
  A. crispa, 13, 140, 141
  A. glutinosa, 14, 206–211
  A. incana, 140
  A. rubra, 20, 21, 137
Alternaria spp., 392
Anthonomus grandis, 226, 381–388, 391, 394, 395
Aphids, 285, 299, 327, 372, 380
  Lime, *see Eucallipterus tiliae*
Apocheima pilosaria, 15, 35, 106, 108, 111, 115–117, 119, 120, 122–127, 286
Apple, 298, 327, 334

Aralia spinosa, 334
Armyworm, 206–209, 211
  beet, *see Spodoptera exigua*
  southern, 76
Aspen, *see Populus* spp.
Aulacophorina, 171
Autumnal moth, *see Epirrita* (= *Oporinia*)
  autumnata
Avocado, 407

Bacillus spp.:
  B. cereus, 382–394
  B. licheniformis, 382
  B. megaterium, 381–394
Bacterial blight, *see Xanthomonas campestris*
  malvacearum
Balsam poplar, *see Populus balsamifera*
Bark beetles, *see* Scolytidae
Barley, 10, 226
Bean, *see Phaseolis vulgaris*
Bean leaf beetle, *see Cerotoma trifurcata*
Beavers, 139
Bebb willow, *see Salix bebbiana*
Beet flies, *see Peganya betae*
Benincasa, 162
Betula spp., 183, 184, 407
  B. glandulosa, 140–142, 146, 148
  B. nana, 142
  B. papyrifera, 13, 32, 137, 140, 141, 146, 285
  B. pendula, 13, 14, 19, 32, 106, 107, 124,

Betula spp. (*Continued*)
   125, 140, 206, 207, 209, 211–213, 215, 281, 286, 287
   B. populifolia, 18, 19, 32, 286, 287
   B. pubescens, 14–19, 32, 106, 124, 137, 140, 147, 206, 207, 212, 219, 286, 287
      B. pubescens tortuosa, 93, 94, 97, 98, 277, 279, 281–283, 286, 287
      B. resinifera (= *papyrifera*) *humilis,* 137, 140, 141, 147, 287
   Birch, 4, 13–19, 29, 32, 33, 35, 106, 107, 117, 119, 122, 123, 125–128, 137, 140, 143, 146, 190, 206, 207, 209, 211, 212, 216, 217, 277, 280, 284, 285, 287, 288, 295–299, 303, 304, 407
   gray, *see Betula populifolia*
   mountain, *see Betula pubescens tortuosa*
   silver, *see Betula pendula*
   white, *see Betula papyrifera*
Bison, 327
Blackberry, *see Rubus vestitus*
Blackbrush, *see Coliogyne ramosissima*
Blister beetle, 226, 227
Bollworm, *see Heliothus zea*
Brephos parthenias, 16, 286
Bryozoa, 65, 335
Bull nettle, *see* Nettles

Cactaceae, 328
Cactus, 328, 330–333, 338
Calabash tree, 297
Callosobruchus spp.:
   C. chinensis, 229
   C. maculatus, 226, 230
Calosoma calidum, 225
Camels, *see Camelus dromedarius*
Camelus dromedarius, 329
Cameraria sp. nov., 309, 310, 312
Canthium spp., 328
Carabid beetles, 225
Carabus spp., 225
Carissa spp., 328
Carya spp., 8
Casebearer, 295
Cattle, 327, 331, 332
Cedrus spp., 248
Ceratocystis spp.:
   C. clavigera, 250, 251, 254, 255, 258
   C. ips, 250, 255, 264
   C. minor, 250, 255, 256
   C. montia, 251
   C. nigrocarpa, 264
   C. polonica, 250
Cereals, 10

Cerotoma trifurcata, 361
Cervus nippon, 334, 335
Cherry, *see Prunus serotina*
Choristoneura conflictana, 72, 75–77, 80, 81, 137, 147
Chromaphis juglandicola, 404
Chrysomelidae, 198
Cicadas, periodical, *see Magicicada* spp.
Cimex spp., 225
Citrullus spp., 162
Citrus, 334
Cnidoscolus texanus, 327
Coccinellidae, 161
Cockroach, 225
Coleophera serratella, 13, 106, 108, 121, 126, 127, 217, 285, 286
Coleopheridae, 108
Coleoptera, 227, 246
Coliogyne ramosissima, 331, 335
Colletotrichum lagenarium, 380
Colorado potato beetle, *see Leptinotarsa decemlineata*
Corn, *see Zea mays*
Corylus avellana, 212
Cotton, *see Gossypium hirsutum*
Cotton fleahopper, see *Pseudomoscelis seriatus*
Cottonwood, 12, 28, 30, 299
Cottonwood leaf rust, 12, 28
Cowpea, 226
Crataegus monogyna, 206–210
Cucumber, *see Cucumis sativus*
Cucumber beetles, 157, 171–175
Cucumis sativus, 162, 168, 175, 176
Cucurbita spp., 216, 295
   C. andreana, 157–159
   C. foetidissima, 172
   C. maxima, 165
   C. moschata, 21, 157, 167
   C. pepo, 158–160, 167, 168, 173, 217
   C. texana, 157, 158, 160, 165–167, 169–171
Cucurbitaceae, 156, 159, 161, 163, 168–171, 177
Culex nigripalpus, 225
Cylas formicarius elegantulus, 225

Datura stramonium, 52, 53, 62, 231
Deer, 329, 331, 334
   muntjack, *see Muntiacus reevesi*
   sika, *see Cervus nippon*
Dendroctonus spp., 246, 260
   D. brevicomis, 260
   D. frontalis, 247, 250, 251, 255
   D. micans, 259, 260

D. ponderosae, 247, 249–251, 254, 255, 258–260, 263, 264
D. pseudotsuga, 260
D. rufipennis, 260
D. terebrans, 259, 260
D. valens, 250, 255, 259, 260
Deporaus betulae, 287
Diabrotica spp., 156
D. undecimpunctata howardi, 160, 171, 172, 361, 362
Diabroticina, 171
Dineura virididosota, 16, 106, 286
Dioptid oak moth, 307
Diprionidae, 227
Douglas fir, *see Pseudotsuga menziesii*
Drepanosiphum platanoides, 21
Dwarf birch, *see Betula glandulosa*

Echinocystis spp., 162
Elatobium abietinum, 25
Epicauta pestifera, 226, 227
Epilachna spp., 155, 156, 162, 164
E. admirabilis, 161, 162
E. borealis, 24, 161–168, 170, 171, 173
E. cucurbitae, 162
E. dodecastigma, 162
E. paenulata, 162
E. pocohante, 162
E. septima, 161, 162, 177
E. tredecimnotata, 21, 155, 162, 164
E. tumida, 162
E. varivestis, 169, 226, 227, 231, 349, 357–363, 365, 369–371
E. vigintioctomaculata, 162
Epinotia spp., 309, 310
Epirrita (= *Oporinia*) spp.:
E. autumnata, 4, 15–18, 33, 106, 137, 147, 277, 279–285, 287, 407
E. dilutata, 108, 119, 120, 122, 123
Eriocrania spp., 108, 111, 121
Eriocraniidae, 108, 111
Eriogaster lanestris, 16, 286
Erranis defoliaria, 108, 119, 120, 279
Eucallipterus tiliae, 20, 295, 407
Euceraphis spp., 285
E. punctipennis, 19
Euproctis similis, 108, 119, 120
European pine sawfly, *see Neodiprion sertifer*

Fall webworm, *see Hyphantria cunea*
Feltleaf willow, *see Salix alaxensis*
Fir engraver, *see Scolytus ventralis*

Geometridae, 108, 279, 287, 297

Gerenuk, *see Litocranius walleri,* 329
Giraffa camelopardalis, 329
Giraffe, *see Giraffa camelopardalis*
Glochidion obovatum, 331, 334
Glycine max, 210, 226, 347–373, 408
Goats, 139, 146, 329, 331, 335
Goldenrod, 10
Gorillas, 327
Gossypium hirsutum, 23, 24, 210, 295, 296, 381–394, 408–414
Gracillaridae, 309
Grapes, *see Vitis* spp.
Grass, 331
Gynostemma pentaphyllum, 161, 162
Gypsy moth, *see Lymantria dispar*

Hares, 136, 139–142, 145–149, 164
mountain, *see Lepus timidus*
Hawthorn, 206–210, 217
Heliothos zea, 228, 348, 382, 393
Helix aspersa, 206
Hessian fly, *see Mayetiola destructor*
Holly, *see Ilex opaca*
Hymenoptera, 227
Hyphantria cunea, 20, 299, 329, 408

Ilex opaca, 329
Impala, 329
Ips sp., 246, 260
I. grandicollis, 260
I. pini, 250, 255, 260
I. typographus, 250, 260

Jimsonweed, *see Datura stramonium*

Kudu, 329

Lagenaria, 162
Larch, *see Larix* spp.
Large aspen tortrix, *see Choristoneura conflictana*
Larix spp., 32, 248, 304, 407
L. decidua, 137
L. laricina, 140
L. occidentalis, 254
Leek, 392
Legumes, 161, 163
Lema trilineata, 231
Lepidoptera, 183, 193, 218, 285, 309, 327, 358
Leptinotarsa decemlineata, 226, 228–231, 404
Leptographium terebrantis, 250, 255, 259, 264
Lepus spp.:
L. americanus, 13, 20, 32, 136, 139, 140–142
L. timidus, 139–141
Leucophaea maderae, 225

Lima bean, 225, 226, 231
Lime tree, 20, 295
Liriodendron spp., 8
Litocranius walleri, 329
Loblolly pine, *see Pinus taeda*
Lodgepole pine, *see Pinus contorta latifolia*
Luffa spp., 162
Lycopersicon esculentum, 12, 22, 62, 168, 216, 217, 223, 226, 228, 232, 234–236, 238, 404, 414
Lymantria dispar, 18, 19, 108, 286, 287, 298, 300
Lymantriidae, 108

Magicicada spp., 407
Malacosoma californicum, 20, 21, 137, 297, 299, 405, 408
Mammals, 135, 136, 139–142, 145–149
  feeding models of, 136
  population dynamics of, 136, 148, 149
Manduca sexta, 48, 53, 56, 57, 60, 225, 228, 233–238
Marginal blister beetle, *see Epicauta pestifera*
Marma bean, 226
Mayetiola destructor, 225
Maytenus spp., 328
Melampsora medusae, 392
Melittia snowii, 172
Meloidae, 227
Melothria spp., 162
Membranspora membranacea, 335
Mexican bean beetle, *see Epilachna varivestis*
Millet, 226
Mimosa pudica, 331
Mites, 187, 295, 296, 302, 327
  avocado brown, *see Oligenychus punicae*
  citrus red, *see Panonychus citri*
  spider, *see Tetranychus* spp.
Momordica, 161, 162
Moose, 14, 139, 140, 143, 147, 149, 285
Mosquitoes, 225
Mountain birch, *see Betula pubescens*
Mountain hare, *see Lepus timidus*
Mountain pine beetle, *see Dendroctonus ponderosae*
Muntiacus reevesi, 335
Muntjack, *see Muntiacus reevesi*

Needle rust, *see Melampsora medusae*
Nematodes, 389
Neodiprion sertifer, 226, 227
Nettles, 27, 297, 327, 331, 337
Neurobathra bohartiella, 307

Nicotiana spp., 47, 53, 55, 63, 64
  N. glauca, 55
  N. sylvestris, 48–52, 54, 58, 59, 62, 63, 159
Noctuidae, 108, 213, 358
Nymphalidae, 297

Oak, 14, 18, 19, 26, 295, 298, 300–313
  black, *see Quercus velutina*
  emory, *see Quercus emoryi*
  red, 300
  water, *see Quercus nigra*
Oligenychus punicae, 407
Operophtera spp., 279
Ophiostoma, see Ceratocystis
Oporinia, see Epirrita
Opuntia, spp., 331–333
  O. bigelovii, 330
  O. stricta, 332, 333
Orthosia stabilis, 213–215
Orygia spp.:
  O. antigua, 19, 207
  O. vetusta, 406
Ostrinia nubilalis, 228

Panolis flammea, 25, 191
Panonychus citri, 412, 413
Pegomya betae, 24, 407
Phaseolus vulgaris, 9, 338, 359, 408
Phrygonidia californica, 307
Phyllobius spp., 285
Phyllonorycter herrisella, 26
Phymatotrichum spp., 382
Phytophthora megasperma glycinea, 352–357, 360, 370, 371
Picea spp., 248
  P. abies, 25, 250
  P. stichensis, 25
Pine, 11, 25, 247, 250, 296, 297
  loblolly, *see Pinus taeda*
  lodgepole, *see Pinus contorta latifolia*
Pinus spp., 247, 250
  P. banksiana, 250, 255, 264
  P. contorta latifolia, 25, 53, 247, 250, 252, 254, 255, 258, 264, 265
  P. palustris, 250, 255
  P. pinaster, 250, 255
  P. resinosa, 250, 255, 264
  P. taeda, 250, 252, 256, 264
Pine beauty moth, *see Panolis flammea*
Plagiodera versicolora, 21, 184, 186–189, 190–196, 197
Platyprepia virginalis, 406
Plum, 334

Poplar, 14, 32, 184, 185, 197
  balsam, *see Populus balsamifera*
  quaking aspen, *see Populus tremuloides*
Populus spp., 8, 184, 197
  P. balsamifera, 13, 74, 140–142
  P. nigra, 14
  P. tremuloides, 13, 72, 73, 76, 77, 79–81, 137, 140, 185
Potato, 223, 226, 228, 230, 232, 404
Potentilla spp., 337, 338
Prickly wattle, *see Acacia victoriae*
Pristophora spp., 16, 286
Prunus serotina, 407
Psedomonas syringae glycinea, 355
Pseudomoscelis seriatus, 382
Pseudoplusia includens, 358, 359, 364, 367, 369–371
Pseudotsuga menziesii, 254, 392
Pteronidea spp., 16, 286
Puccinia allii, 392

Quaker moth, *see Orthosia stabilis*
Quaking aspen, *see Populus tremuloides*
Quercus spp., 8
  Q. agrifolia, 307
  Q. emoryi, 301–312
  Q. nigra, 306, 307
  Q. robur, 14, 26
  Q. velutina, 18, 19, 304

Rabbits, 327, 329
Ragwort, 26
Raphanus sativus, 7
Red flour beetle, *see Tribolium castaneum*
Rheumaptera hastata, 137, 147, 287
Rhodnius prolixus, 225
Rice, 226, 392
Rodents, microtine, 139
Rubus spp., 328, 331, 334, 335, 338
  R. trivialis, 331
  R. vestitus, 328–330, 335–337, 340, 341
Rye, 226

Salix spp., 9, 20, 21, 33, 198, 297
  S. alaxensis, 20, 140–142, 145, 146, 184
  S. alba, 21, 184–186, 189, 197
  S. arbusculoides, 140
  S. babylonica, 21, 184, 185, 187, 189, 192, 193, 196
  S. bebbiana, 140, 142, 184
  S. caprea, 140, 141
  S. glauca, 140
  S. lapponum, 93

S. nigricans, 140
S. pentandra, 140
S. phyllicifolia, 140
S. sitchensis, 184
Sawflies, 106, 285, 287, 296, 297
Schizopepon bryoniaefolius, 161, 162
Scolytidae, 246, 247, 249–255, 258–260, 263, 264, 267
Scolytus spp., 246, 260, 263, 266, 267
  S. ventralis, 250–255, 259, 260
Sechium spp., 162
Sheep, 146, 327, 329
Shrankia microphylla, 332
Slugs, 285
Smilax spp., 328
Snails, 14
Snowshoe hare, *see Lepus americanus*
Solanaceae, 47, 161, 163
Sorghum, 226
Southern pine beetle, *see Dendroctonus frontalis*
Soybean, *see Glycine max*
Spear-marked black moth, *see Rheumaptera hastata*
Spodoptera spp.:
  S. exigua, 228, 233, 404, 410
  S. littoralis, 12, 19, 22, 106, 108, 123–126, 206–209, 211, 232, 365
Spotted cucumber beetle, *see Diabrotica undecimpunctata howardi*
Spruce, 32
  Norway, 25
  Sitka, 25
Squash beetle, *see Epilachna* spp.
Squash vine borer, *see Melittia snowii*
Stable fly, *see Stomoxys calcitrans*
Stilbosis juvantis, 191, 306, 307
Stink bugs, 372
Stomoxys calcitrans, 225, 228
Strawberry, 415
Sugar beet, 24
Sycamore, 22, 29

Tent caterpillars, *see Malacosoma californicum*
Tetranychus spp., 57, 408–415
  T. urticae, 23, 24
Three-lined potato beetle, *see Lema trilineata*
Thuja plicata, 254
Tiger swallowtail, 75, 76
Tilia spp., 20
Tillandsia spp., 338
Tobacco hornworm, *see Manduca sexta*
Tomato, *see Lycopersicon esculentum*

Tomicinus piniperda, 250, 255
Tortricidae, 309
Tribolium spp.:
　T. castaneum, 226
　T. confusum, 229
Trichiosoma lucorum, 286
Trichosanthes, 162
Trichoplusia ni, 56, 57
Trichosporium symbioticum, 250,
　251–255
Triticum aestivum, 338
Tsuga spp., 248
Tussock moth, *see Orgyia vetusta*
Tyria jacobaeae, 26

Ungulates, 329, 331
Urtica spp., 338
　U. dioica, 27, 327

Verticicladiella spp., 250, 255
Verticillium dahliae, 410
Verticillium wilt, *see Verticillium dahliae*
Vitis spp., 414, 415

Vole, 26

Walnut, 297
Walnut aphid, *see Chromaphis juglandicola*
Weevils, 285, 287
　bean, *see Acanthoscelides obtectus*
　boll, *see Anthonomus grandis*
　cowpea, *see Callosobruchus maculatus*
　sweet potato, *see Cylas formicarius*
　　eleganthus
Wheat, 226
Willow, *see Salix* spp.:
　sitka, *see Salix sitchensis*
Winter moth, 298
Wolly bear, *see Platyprepia virginalis*

Xanthomonas campestris:
　p.v. *glycinea,* 355
　p.v. *malvacearum,* 381

Zea mays, 226, 338
Zeiraphera diniana, 137, 407
Zucchini, 24, 164

SUBJECT INDEX

Abscission, 304–306, 309, 314
Adaptations to phytochemical induction:
   avoidance, 248, 250–253, 257–260, 262
   cooperative behavior, 257, 258, 261–263
   physiological tolerance, 257–260
   trenching behavior, *see* Trenching behavior
   vein cutting, 257
Agriculture, 294, 372, 373, 403–416
Alkaloids, 326
   autotoxicity from, 58, 61
   biosynthetic enzymes for, 55
   effects of auxins on, 51, 52, 55, 58, 59
   effects on herbivores, 47, 48, 56, 60
   effects of plant architecture on, 55, 65
   effects of root stress on, 50, 51, 56
   effects of water on, 55
   evolution of, 64, 65
   glycoalkaloids, 62
   in nitrogen transport, 63, 65
   responses to herbivory, 48, 53, 54, 64
   in singlet oxygen quenching, 62, 63
   synthesis sites, 48–51, 63
   tropane alkaloids, 62
Allomones, 171, 348
Anabasine, 55
Antibiosis, 348, 369
Antixenosis, 348, 368, 369
Aphids:
   on soybean, 372
   walnut aphid, 404

Armyworms, 12, 19, 22, 106, 108, 123–126,
   206–211, 228, 232, 233, 365, 404, 410
Auxin, 333

B-1,3-endoglucanase, 355
Bacterial blight, 355, 381
Bacterization, 380, 382
Bark beetles:
   behavioral interactions with induced
     responses, 250–253, 257–263
   fungal symbionts, 247, 253–255
   life histories, 246
   pheromones, 247, 257–258, 261–263
   population dynamics and management,
     266–268
   susceptibility to induced responses,
     250–253
Behavior, effects of induction on herbivore:
   dispersion, 190–193, 198
   feeding choices, 106, 118–126, 128, 184,
     187–198, 206–211, 216, 217, 232, 233,
     285, 295–299, 307, 308, 327, 357–362,
     364, 365, 368, 386–388, 394, 395, 406, 411
   foraging patterns, 118, 125, 126, 206–220,
     248, 250–253, 257–260, 262, 278, 283, 308
   temporal activity, 193–196, 198
β-glucosidase, 161
Bollworm, 228, 348, 382, 393
Bowman-Birk inhibitor, 225, 229, 230
Bradford assay, 108, 111, 112

Brown spot, 392

Carbohydrate, 142, 144, 145, 184
Carbon, 57, 58, 61, 63, 85, 86, 88–93, 95,
  97–100
  carbon-based 'defenses', 86–101, 339
  carbon stress, 145
Carbon/nutrient balance hypothesis, 79, 89,
  90, 97–101, 339
Ceiling hypothesis, 219
Chitosan, subcortical induction of, 255–256
Colorado potato beetle, 226, 228–231
Compensatory growth, 88
Compensatory mortality, 406, 407, 409
Competition, 302, 308–310, 313, 314
Conifers, 246–268. *See also* Pine
Constitutive resistance, *see* Resistance to
  herbivores, constitutive
Cost–risk–value hypothesis, 371
Cotton, 23, 24, 210, 295, 296, 381–394, 408–414
Cottonwood, 12, 28, 30, 299
Cucumber, 415
Cucumber beetles, 156, 157, 160, 171–175
Cucurbitacins, 155
  aglycones, 161
  effects on herbivores, 156, 162–165,
    172–174, 177, 216
  elicitors of, 176
  evolutionary role of, 156
  glycosides, 161, 172
  induction of, 156–161, 167, 168, 173, 174
  structures of, 157
  within-plant distribution of, 156, 160, 172,
    174–176
  in wound repair, 176
Cucurbits, 155, 156, 167, 171, 175
Cultural control, 414
Cytokinin, 333

Damage:
  causing nutrient loss, 93, 135, 136, 145, 146
  causing phenological change, 304–306, 314
  effects of damage degree on induction,
    407, 408, 412, 413
  effects of damage timing on induction,
    407, 409, 411, 412, 415
  effects on phytochemical induction from
    abrasion, 410
  artificial injury, 48, 53, 54, 72, 74, 76,
    77, 81, 98, 99, 107–111, 113–124, 126,
    128, 157–159, 166, 167, 173, 183,
    185–189, 192, 206, 207, 209–213,
    232–237, 253, 254, 256, 280–283, 285,

286, 288, 295–299, 300, 301, 305, 331,
  333, 366, 367, 371, 405, 410, 414
defoliation:
  artificial, 74, 76–78, 97, 98, 137–139,
    143, 144, 148
  natural, 76, 81, 91–94, 136, 137, 144,
    147, 279, 281, 285, 287, 288
  insect herbivory, 48, 53, 54, 64, 106,
    107, 111–120, 122–124, 128, 167, 171,
    183–185, 187, 189–197, 232, 256,
    295–299, 300, 301, 304, 365–370, 371,
    379, 380, 407, 408, 410, 412, 413
  interactions with pathogens, 99, 248,
    253–255
  localized injury, 96, 97, 208–212, 264
  pruning:
    artificial, 138, 147
    vertebrate, 136, 138, 139, 143–145
  ultraviolet light, 357
  vertebrate browsing, 135, 136, 139, 143,
    145–149, 285, 328–332, 334–336
  histological changes from, 248, 255, 256,
    264
  image analysis of, 210–212
Deer, 329, 331, 334
  muntjack, 335
  sika, 334, 335
Delayed inducible resistance, *see* Induction,
  phytochemical, in plant defense,
  long-term (LTI)
Density dependence, 301, 302

E-64, 225, 228–232
Economic injury level, 373
Economics, 415
Elaterase, 161
Equilibrium hypothesis, 219
Ethylene, 99, 256

Fall webworm, 20, 299, 329, 408
Fiber, 137
Fir engraver, 250–255, 259, 260
Fitness, effects of phytochemical induction:
  on herbivores, 56, 72, 75–79, 106, 107,
    110–112, 118–127, 136, 139–142, 145–149,
    183, 184, 187–198, 250–253, 262, 263,
    278–283, 285–287, 293–299, 303, 306–308,
    314, 327, 333, 338, 358, 359, 363, 364
  on host plants, 85, 94–96, 246–248, 264
Fixed allocation model, 88–93, 100
  deciduous *vs.* evergreen trees, 88, 92
Flavonoids, *see* Phenolics, flavonoids
Flexible allocation model, 89–93, 97, 100

Folin Denis assay, 108–112, 114, 115, 120, 124
Frenching disease, 392
Fungi, 116, 117, 128, 129, 410
  interactions with bark beetles, 247, 250–255

Genotype, plant, 5, 7, 9–13, 28, 35, 294, 303, 408
  interactions with environment, 10
Gibberellin, 333
Glyceollin:
  EC_{50} for, 361
  EC_{90} for, 354
Glyphosate, 370
Gossypol, 393, 394
Greenic acid, 141
Growth patterns, plant:
  determinate, 8, 25, 31, 32
  fixed, 8, 32, 36
  free, 8, 31, 32, 36
  habit, 8, 28, 31, 36
  indeterminate, 8, 20, 28
  regulators of, 414
  semi-indeterminate, 20
Gum production, 407
Gypsy moth, 18, 19, 108, 286, 287, 298, 300

Hawthorn, 206–210, 217
Hemigossypol quinone, 393, 394
Hepta-*b*-glucoside alditol, 356
Herbicide, 414
Herbivores, effect of phytochemical induction:
  on communities, 125–127, 196, 197, 310–314
  on fitness, see Fitness
  on populations, 72, 77–79, 106, 107, 125–127, 218, 219, 266–268, 277–279, 283, 285, 287, 288, 293–299, 301–303, 307, 313, 314, 371–372, 394, 395, 404–414, 416
Hypersensitivity, 353

Immunization, 380, 414
Induction, physical:
  from artificial damage, 331, 333, 336
  effect of environment on, 337–339
  effect of fertilization on, 336
  effect of light intensity on, 333, 335–339, 340
  effect of plant age on, 334, 339–341
  hairs, 325–327, 337, 338, 340
  from herbivory, 304, 328–332, 334–336
  metabolic costs of, 85–87, 330, 331, 335
  in plant defense:
    active, 326, 337
    passive, 326, 337, 341
  prickles, 325, 327, 330, 336, 338, 340
  spines, 325–333, 337, 338, 340
  thorns, 325–331, 334, 335, 337–340
  trichomes, 326
Induction, phytochemical, 48, 57, 58, 64, 65, 71, 137, 155, 156–161, 168, 172, 293–314, 379
  cues triggering, 51–53, 64, 86, 98–100, 114, 115, 159, 168, 232, 233, 278, 281
  definitions of, 71, 72, 92, 403
  effects of damage on, see Damage
  effects of environment on, 55, 224, 228, 232–238, 263, 264, 408, 410, 411, 413, 414
  effects on herbivores, see Herbivores, effect of phytochemical induction
  effects of plant age on, 5, 7–10, 12, 13, 28–33, 35, 36, 108–110, 122, 136, 139–142, 145, 148, 149, 185, 186, 190–193, 195, 233, 235–238, 264, 294, 300, 303–306, 314, 389–391, 394
  effects of root stress on, 50, 51, 56
  elicitors of, 176, 248, 253–255, 355–357, 379, 380, 384, 392, 394, 395
  in plant defense, 86, 92, 205, 206, 216–220, 223, 224, 403
    active, 4, 28–31, 36, 64, 79, 85, 86, 101, 147–149, 283, 284, 293
    long term (LTI), 4, 72, 76–81, 86, 93, 96, 100, 136–147, 149, 184, 205, 219, 279, 281–285, 287, 288
    mechanics of, 48, 49, 74, 80, 136, 143–147, 149, 158, 160, 177, 256, 383–385
    metabolic costs of, 57–61, 63, 85–88, 95, 96, 256
    passive, 4, 28–30, 33, 34, 36, 49, 64, 79, 80, 147–149, 283, 284, 305, 306
    rates of, 155, 157–160, 206, 217, 254, 295–300, 303, 314, 407, 411
    relaxation of, 98, 206, 217, 256, 283, 302, 303, 314, 394, 407, 411, 412
    short term (STI), 4, 11, 36, 72, 74–76, 79–81, 86, 96, 99, 105–132, 184–187, 205, 279–281, 285, 207, 288
    specificity of, 34, 254
    systemic, 106, 111, 114, 115, 117, 119, 128, 195, 207, 208, 217, 219, 232, 281, 394, 409
    theoretical models of, 85, 86, 91–98, 100, 144, 147, 219, 280, 302
    variation in, 8, 26, 36, 106, 110, 111, 115, 117, 127, 259, 264

Inhibitors:
    Kunitz, 225, 228, 230–232
    phenylalanine ammonia lyase, 370
    potato chymotrypsin, 228, 230
    trypsin, 224, 228, 229, 230, 234
Integrated pest management (IPM), 372, 373
Integrated physiological unit (IPU), 95
Isocoumarins, 255

Kairomones, 171, 388, 395

Larch, 32, 137, 140, 248, 254, 304, 407
Leafminers, 108, 111–114, 117, 119–122, 126,
    191, 193, 295, 298, 302, 304, 306, 307,
    309–313
Leaf tiers, 309, 310
Lepidoptera, 183, 193, 218, 285, 309, 327, 358
Lignin, 184
Lipids, 256

Maternal care, 259
Mevalonic acid, 250
Mineral nutrients:
    availability, 335, 339, 340, 408
    nitrogen, 47–49, 53, 55–58, 62–65, 74, 77,
        79, 81, 87, 89, 93, 94, 97, 98, 126, 137–141,
        143, 185–189, 197, 281, 304, 339
    phosphorus, 74, 77, 87, 89, 147
    uptake, 9
Mites, 187, 295, 296, 302, 327
    citrus red mite, 412, 413
    spider mite, 23, 24, 57, 408–415
Modules, plant, 94–97, 100, 101
Monoterpenes, 53, 247, 249–255
Moose, 14, 139, 140, 143, 147, 149, 285
Mosaic virus, 414

Nitrogen, *see* Mineral nutrients
N-methylputrescine oxidase, 55
Nutrient stress hypothesis, 93, 143, 145, 147,
    283, 284

Oak, 14, 18, 19, 26, 295, 298, 300–313
Ornithine decarboxylase, 55
Ozone, 12, 28

Papyriferic acid, 141
Parasites, 125, 126, 408
Pepstatin, 225, 228, 229
Pesticide resistance, 415
Phenolics, 93, 94, 98, 106–111, 114–199, 122,
    127, 128, 137–139, 141, 143, 144, 184,
    185, 187, 197, 216, 250, 253, 281, 300–303,
    310, 334, 362–365, 392
    anthocyanins, 99

biosynthesis of, 107, 112, 114, 116, 117,
    123, 124, 126, 128
flavonoids, 389
glycosides, 73, 75, 77, 79–81
phenylpropanoids, 89, 97, 99
proanthocyanidin, 184
tannins, 73, 77, 79, 99, 184–186, 300–303,
    309–311, 335, 339, 392, 393
Phenotypic plasticity, 85, 89–92, 96, 100
Phenylalanine ammonia lyase (PAL), 99,
    114–117, 119, 123, 128, 250, 350, 355
    inhibitor of, *see* Inhibitors
Photosynthesis, 5, 8, 9, 31
Phytoalexins, 116, 128, 392, 393
    afrormosin, 351, 359
    biosynthesis of, 350, 352, 353
    coumestrol, 351, 358, 359, 365
    daidzein, 351
    definition of, 349
    elicitors of, 355, 356
    glyceocarpin, 351
    glycinol, 351
    isoflavones, 351
    isoflavonoids, 349
    isoformononetin, 351
    phaseol, 351, 359
    pterocarpans, 351
    sojagol, 351
Phytophthora rots, 351
Pine, 11, 25, 247, 250, 296, 297
    loblolly, 250, 252, 256, 264
    lodgepole, 25, 53, 247, 250, 252, 254, 255,
        258, 264, 265
Pinosylvin, 141
    methyl ether, 141
Plumbing principle, 30
Polyamine pathway, 64
Poplar, 14, 32, 197, 184, 185
    balsam, 13, 74, 140–142, 146
    quaking aspen, 13, 72, 73, 76, 77, 79–81,
        137, 140
Potato, 223, 226, 228, 230, 232, 404
Predation, 126, 127, 408, 415
Preference index, 360
Prickles, 325, 327, 330, 336, 338, 340, 341
Protease, *see* Proteinase
Proteinase:
    aspartic, 225, 232
    assay, 226–228
    cysteine, 225, 226, 230, 232
    insect digestive, 223–229, 232, 238
    metallo, 225
    serine, 224, 225, 232
Proteinase inhibitor inducing factor (PIIF),
    232, 233, 255, 256

Proteinase inhibitors, 53, 63, 116, 168, 216, 223, 224, 226, 229, 232, 238, 348, 404
  aspartic, 228
  bioassays with, 228–231, 233–235, 237, 238
  cysteine, 226, 228, 230–232
  induced, 223, 224, 228, 232–239
  serine, 224, 226, 228, 229, 231, 232
Protein precipitation, 108, 110–114, 119
Putrescine *N*-methyltransferase, 55

Rapid inducible resistance, *see* Induction, phytochemical, in plant defense, short term
Resins, 146, 184, 339
  acids, 255
  canals, 248, 255, 257
Resistance to herbivores:
  acquired, 407
  constitutive, 12, 35, 47, 48, 58, 85, 86, 90–92, 95, 161, 172, 224, 247, 248, 261–263, 293, 313, 348, 359, 365, 370, 408
  evolution of, 294, 312
  induced, *see* Induction, physical
  morphological, 304–306
  from mutualistic microorganisms, 380–395
  phenological, 304–306
  qualitative, 48, 250, 252–255
  quantitative, 48, 250–255
Resource:
  acquisition, 5, 7–11, 29, 32, 35, 36
  allocation, 5, 7–11, 30–36, 85–101
  suitability, 10, 11, 35
Respiration, 9
Rhizobacteria, 380

Salicin, 73–77, 79, 185, 197
Salicortin, 73–77, 79–81, 197
Salicylaldehyde, 141, 197
Saliva, insect, 116, 118, 128
Saponins, 168, 348
Sawflies, 106, 285, 287, 296, 297
Secondary plant metabolites, 58, 73, 81, 86, 87, 89, 93, 97, 99, 100, 142, 144, 146, 187, 294, 303, 314, 379, 380, 388, 389, 394, 395
Sheep, 146, 327, 329
Silica, 185, 331, 332
Somatic mutations, 94, 95, 97
Soybean, 210, 226, 347–373, 408
Soybean chemistry, 348
  benzoic acid, 350, 362
  cinnamic acid, 350, 362
  gentisic acid, 363
  pinitol, 348

reducing sugars, 384
  salicylic acid, 363
  saponins, 348
Spatial scale, 405, 406, 410, 411
Spider mites, *see* Mites
Spines, 325–333, 337, 338, 340
Spotted cucumber beetle, 160, 171, 172
  feeding behavior, 172
  reaction to cucurbitacins, 171–173
Squash beetle, 24
  foraging behavior, 164–168, 171
  mandibular specializations, 161, 163, 168
  trenching behavior, 155, 156, 161–171, 177
Starch, 256
Stress, 414
  abiotic, 4, 5, 7, 13, 33–36
  abiotic X biotic, 10
  biotic, 4, 5, 7, 13, 34, 35
Sugars, 142

Tannins, *see* Phenolics
Tent caterpillars, 20, 21, 137, 297, 299, 405, 408
Terpenes, 387, 391, 392, 393. *See also* Monoterpenes
Thorns, 325–331, 334, 335, 337–340
Tobacco, 47–53, 55, 56, 58, 63, 159, 392
  tree, 55
Tobacco hornworm, 48, 53, 56, 57, 60, 225, 228, 233–238
Tomato, 12, 22, 62, 168, 216, 217, 223, 226, 228, 232–236, 238, 404, 414
Translocation, 30, 36
Tremulacin, 73–77, 79–81, 185
Tremuloiden, 73–77, 79, 185
Trenching behavior, 155, 156, 161–167, 177, 257
Trichomes, 326
Triticale, 226
Turgor pressure, 158, 170, 171
Tyrosine ammonia lyase (TAL), 350

Ungulates, 329, 331

Vaccination, 414, 415
Vascular development, 30
Verticillium wilt, 296, 382

Willow, 9, 20, 21, 33, 93, 94, 140, 141, 297
  bebb, 140, 142, 180
  felt leaf, 20, 140–142, 145, 146, 184
  sitka, 184
  weeping, 21, 184–187, 189, 192, 193, 196, 197

Yield, 408, 413